Biology, Brains, and Behavior

**Publication of the Advanced Seminar Series
is made possible by generous support from
The Brown Foundation, Inc., of Houston, Texas.**

**School of American Research
Advanced Seminar Series**

Douglas W. Schwartz
General Editor

Biology, Brains, and Behavior

Contributors

C. G. Anderson
Department of Ecology and Evolutionary Biology
University of Tennessee, Knoxville

Elizabeth Bates
Center for Research in Cognitive Development
University of California, San Diego

Christopher Boehm
Department of Anthropology, Jane Goodall Research Center
University of Southern California

S. E. Cates
Department of Ecology and Evolutionary Biology
University of Tennessee, Knoxville

Terrence W. Deacon
Department of Anthropology, Boston University

Jeffrey Elman
Department of Cognitive Sciences
University of California, San Diego

Lynn A. Fairbanks
Department of Psychiatry and Biobehavioral Sciences
Neuropsychiatric Institute
University of California, Los Angeles

John Gittleman
Department of Biology, University of Virginia, Charlottesville

Patricia Greenfield
Department of Psychology, University of California, Los Angeles

Jonas Langer
Department of Psychology, University of California, Berkeley

H.–K. Luh
Department of Ecology and Evolutionary Biology
University of Tennessee, Knoxville

Ashley Maynard
Department of Psychology, University of California, Los Angeles

Michael L. McKinney
Department of Geological Sciences
University of Tennessee, Knoxville

Sue Taylor Parker
Department of Anthropology, Sonoma State University

Emily Yut Schmidtling
Los Angeles, California

Brian Shea
Department of Cell and Molecular Biology
Northwestern University

Biology, Brains, and Behavior

The Evolution of Human Development

Edited by Sue Taylor Parker, Jonas Langer,
and Michael L. McKinney

School of American Research Press
Santa Fe

James Currey
Oxford

School of American Research Press

Post Office Box 2188
Santa Fe, New Mexico 87504-2188

James Currey Ltd

73 Botley Road
Oxford OX2 0BS

Director of Publications: Joan K. O'Donnell
Editor: Jane Kepp
Designer: Context, Inc.
Index: Wyman Indexing
Typographer: Cynthia Welch
Printer: The Maple-Vail Book Manufacturing Group

Library of Congress Cataloging-in-Publication Data:
Biology, brains, and behavior : the evolution of human development /
edited by Sue Taylor Parker, Jonas Langer, and Michael L. McKinney.
p. cm. — (School of American Research advanced seminar series)
Based on a seminar held in Aug. 1995 in Santa Fe, N.M.
Includes bibliographical references (p.) and index.
ISBN 0-933452-63-2 (cloth) — ISBN 0-933452-64-0 (paper)
1. Genetic psychology. 2. Developmental psychology. I. Parker, Sue Taylor.
II. Langer, Jonas. III. McKinney, Michael L. IV. Series

BF701 .B56 2000
155.7—dc21   00-030123
                                                                 CIP

British Library Cataloguing in Publication Data:
Biology, brains, and behavior : the evolution of human development
1. Behavior evolution 2. Developmental psychology 3. Physical anthropology
I. Parker, Sue Taylor II. Langer, Jonas III. McKinney, Michael L.
155.7

ISBN 0-85255-907-0 (James Currey cloth)
ISBN 0-85255-908-9 (James Currey paper)

1   2   3   4   5 04 03 02 01 00

Contents

Illustrations

Tables

Acknowledgments

The editors of this volume would like to thank the School of American Research (SAR) for hosting the advanced seminar on which the book is based. We especially thank Douglas Schwartz and his staff for their gracious hospitality in a lovely setting. We also thank SAR Press director Joan O'Donnell for her helpful editorial guidance, and two anonymous reviewers for their insightful comments. In addition, we thank our respective institutions, which have provided the setting for our scholarly work over the years: Sonoma State University, University of California, Berkeley, and University of Tennessee, Knoxville. Finally, we thank the contributors for their collaborative efforts.

Biology, Brains, and Behavior

1

Comparative Developmental Evolutionary Biology, Anthropology, and Psychology

Convergences in the Study of Human Behavioral Ontogeny

Sue Taylor Parker

This book arose out of a multidisciplinary seminar entitled "The Evolution of Behavioral Ontogeny" hosted by the School of American Research in Santa Fe, New Mexico, in August 1995. The seminar proposal posed the following questions: What kinds of heterochronic mechanisms have been involved in the evolution of hominoid behavioral ontogeny? Are humans underdeveloped or overdeveloped apes? How do various developmental systems interrelate in the evolution of ontogeny? What is the role of the brain in life history? How did brain development change during hominid evolution?

During the week-long seminar, the participants—evolutionary biologists, biological anthropologists, and comparative and developmental psychologists—agreed that comparative studies of brain development in human and nonhuman primates held the key to understanding the evolution of primates' behavioral ontogeny. This focus led to methodological questions about relationships among the various theoretical models—heterochrony, allometry, life history strategy theory, and cladistics—that provide the tool kit for comparative developmental evolutionary studies. These questions, in turn, led to others about relations among the various levels of organization addressed by the different models—genetic, morphological, behavioral, and life cyclical.

The subject of the seminar reflected a resurgence of interest in ontogeny and phylogeny and the emergence of the new evolutionary fields of comparative developmental psychology and biology. Comparative developmental psychology—the use of psychological stage models to compare cognitive development in monkeys, apes, and humans—has emerged from a union of ethology and developmental psychology (Parker 1990). Developmental evolutionary biology—the comparative study of the genetics of embryological development—has emerged out of the integration of developmental genetics and evolutionary biology (Pennisi and Roush 1997; Purugganan 1996; Raff 1996). Both of these disciplinary realignments have been reflected in biological anthropology.

THE NEW BIOLOGICAL ANTHROPOLOGY

In 1951, Sherwood L. Washburn described the way in which the comparative study of human anatomy was being transformed from a static, measurement-oriented field of physical anthropology into a new, dynamic, process-oriented approach to the study of human evolution. He noted that the new physical anthropology was based on evolutionary biology and particularly on population genetics. In the fifties and sixties, his diagnosis, along with his projection of the impact of the modern synthesis on physical anthropology, was fully borne out in the growth of the fields of functional anatomy and primate behavior, as well as human genetics and modern paleoanthropology. Washburn's description of the collaborative nature of the new physical anthropology was also vindicated by widespread collaboration among geneticists, anatomists, paleontologists, and geologists.

The new physical anthropology continued to develop in the seventies and eighties as cladistic taxonomy became the norm in hominid paleontology and primate molecular biology (Kimbel and Martin 1993; Skelton, McHenry, and Drawhorn 1986). Concomitantly, allometry became a major topic in primate morphology (Shea 1983a, 1984, 1992b), and heterochrony became a major research topic in comparative studies of primate morphology (Keene 1991). In the 1990s, life history became a popular topic in primate studies (Morbeck, Galloway, and Zihlman 1997; Pereira and Fairbanks 1994).

Now, as the twentieth century gives way to the twenty-first, these various approaches are converging and coalescing into the study of the evolution of hominoid life history, particularly the evolution of the brain and cognitive development. This emergent focus is the outcome of a logical progression of study from the evolution of functional anatomy to the evolution of behavior, from behavioral evolution to brain evolution, and from brain evolution to the evolution of brain development. A parallel progression has been occurring, from the study of anatomical and behavioral evolution to the study of life history evolution. The strands meet in the comparative study of the evolution of primate life histories and ontogenies, especially neuro-ontogenies. Like the earlier transformation, these transitions are based on currents in evolutionary biology.

Ironically, this emerging focus is bringing biological anthropologists back to some classic questions about the evolution of the brain, language, and intelligence that occupied their predecessors at the turn of the last century. Consequently, this new transformation entails new cross-disciplinary collaborations, in this case with comparative and developmental psychologists as well as neuroscientists, developmental biologists, paleontologists, and developmental geneticists. In sum, a broad new developmental evolutionary synthesis is emerging that integrates behavioral ecology (including life history theory), developmental genetics, paleontology (including heterochrony theory), neurosciences, primatology, and developmental psychology.

Still following in the footsteps of evolutionary biology, the new biological anthropology has turned its attention to the long-neglected study of the evolution of development. In light of its focus on the comparative study of embryos and especially embryonic and neonatal brains, this new biological anthropology might be described as evolutionary epigenetics. Because of its developmental focus, evolutionary epigenetics turns investigators back toward developmental psychology, with its emphasis on the construction of increasingly complex and integrated behavioral schemes during ontogeny. If we are back to the past, however, we bring new theories and methodologies to bear on the old problem of the evolution of ontogenies.

THE NEW COMPARATIVE DEVELOPMENTAL PSYCHOLOGY

Oddly, during the twenties and thirties, when the modern synthesis of biology was emerging, the study of development was superseded in psychology as well as biology. Behaviorism and learning theory supplanted the speculation-laden theories of development and evolution that had dominated the turn of the last century. The "science" of behaviorism washed away the taint of Granville Stanley Hall's naive psychological recapitulationism, in which a focus on universal laws of learning gave little impetus to the study of development.

Like developmental biology, developmental psychology was overshadowed by nondevelopmental approaches. Just as developmental biology was marginalized by the rise of molecular genetics, developmental psychology was marginalized first by the rise of learning theory and then by cognitive psychology, with its focus on adult cognition. Even the new subfield of "evolutionary psychology" focuses primarily on the achievements of adults, postulating that innate modules underlie linguistic and cognitive abilities in our species (Cosmides and Toobey 1991; Pinker 1994a).

The return to development in psychology has come from two sources: studies of brain development and studies of language and cognitive development. In recent years, three new approaches have revolutionized the neurosciences. The first of these is developmental genetic studies of the embryonic nervous system (Deacon, this volume). The second is the use of imaging techniques to record brain function in living animals (Elman et al. 1996), and the third is computer modeling of neural networks using parallel processing designs (Elman et al. 1996). This last approach has restored the lost luster of connectionism. These three approaches combine to reveal a great deal about the brain's role in cognitive and language development.

These new studies have stimulated interest in constructionist models of cognitive development that see psychogenesis as a continuation of embryogenesis (Butler and Hodos 1996; Trevarthan 1973). According to these Piagetian and neo-Piagetian models, both prenatal development and postnatal development are epigenetic and constructive (Parker 1996b; Parker and Gibson 1990). The major difference lies in the enlarged scope of feedback in postnatal life.

Like embryogenesis, psychogenesis proceeds through a sequence of developmental stages marked by increasing complexity of organization. According to Piaget (Butler and Hodos 1996; Piaget 1966, 1970; Piaget and Inhelder 1969; Trevarthan 1973), development occurs through integration of feedback from self-generated activities and their effects on the world. Cognitive development begins at birth with simple reflexive and spontaneous schemes and gradually constructs more differentiated and coordinated schemes through the reciprocal processes of assimilating objects and events and concurrently accommodating to those objects and events.

These developmental models have stimulated a new wave of comparative studies of primate cognition aimed at reconstructing the evolution of hominoid intelligence using new techniques of cladistic analysis and concepts of heterochrony.

THE NEW EVOLUTIONARY DEVELOPMENTAL BIOLOGY

Evolutionary developmental biology is distinguished from other subfields of biology by its focus on comparative studies of closely related species and their near relatives—for example, all the great apes and Old World monkeys. This phylogenetic approach contrasts with those of molecular geneticists and comparative psychologists, who have focused on model organisms—a few species that are ideal for laboratory research because of their short life spans and rapid and reliable development (Raff 1996). In molecular genetics these include the bacteria *E. coli*, fruit flies, and mice. In developmental genetics they include nematodes and zebra fish, and in comparative psychology they include pigeons, white rats, and rhesus monkeys (Beach 1965).

The Evolution of Ontogeny

As McKinney and McNamara (1991) emphasized in their book *Heterochrony*, not only the bodies of adults but also ontogenies evolve. Focusing on the evolution of ontogenies is the key to evolutionary developmental biology.

Animals go through three major developmental stages between fertilization and birth: the maternal stage preceding embryo formation, the philotypic stage of early embryo formation, and the divergence stage of later embryo and fetal development. Each living phylum is characterized by a highly conservative body plan that emerges during the so-called phylotypic stage of early embryo development.

Embryo development occurs through a series of cascades of genetic effects. Comparative developmental research has revealed that in contrast to the highly stable phylotypic stage, during which the body plan emerges, both earlier and later stages of embryogenesis in related species within a phylum or even a family are much more variable. The early development of various vertebrates, for example, involves very different patterns of egg size and yolk content, cleavage, gastrulation, and cellular movement, as is expressed in the differences between amniote egg and placental membrane development in reptiles and birds versus placental mammals. Likewise, the later stages of development in various vertebrates involve divergent patterns of head and limb development, as is revealed in the differences in locomotion among fish, reptiles, and mammals (Raff 1996).

The opportunism of the egg stage may arise from its relative simplicity and autonomy. Raff (1996) cited work suggesting that egg development occurs under the influence of the maternal genome, which encodes instructions for cleavage and cellular movements. Mutations in these patterns are unlikely to disrupt these early stages. This may explain the evolutionary flexibility in early development that is seen across phyla. The opportunism of the later developing embryonic stages also arises from their relative autonomy. In this case autonomy follows from the increasingly modular organization of body development after formation of the basic body plan.

In contrast, the conservatism of the intervening phylotypic stage follows from its highly interactive, interdependent mode of development: "It is at this developmental midpoint that ultimately widely separated and distinct modules interact with one another in complex and pervasive ways" (Raff 1996:205). Such interactions at this developmental stage may constitute developmental constraints on evolutionary change.

The phylotypic stage was the primary focus of Haeckel, Garstang, von Baer, and other pioneers in the concept of *heterochrony*, or the evolution of differences in the timing of development. It was the universality of this stage that led Haeckel (1874) to propose the idea that ontogeny recapitulates phylogeny and that led von Baer to argue that the earliest stages of development were the most stable and conservative. Since earlier stages of embryo development rarely fossilize, the phylotypic stage has also been the beginning point for paleontological studies of the evolution of development. The neglect of earlier stages of egg development has been reflected in the terminology of heterochrony (Raff 1996).

Comparative developmental research focuses on cascading pathways from genes to gene products to morphology in developing embryos of closely related species. Following is an outline of the organizational levels (from lowest to highest) involved in the development of metazoan organisms:

Genes in cells:
  Control genes (e.g., Hox genes)
  Producer genes (e.g., insulin genes and keratin genes)
Gene products:
  Growth factors (e.g., growth hormone, insulin growth factors)
  Other factors (e.g., structural proteins such as keratin)
Cells and tissues:
  Epigenetic interactions among gene products within cells
  (e.g., biosynthetic pathways to break down testosterone into
  dihydrotestosterone)
  Epigenetic interactions among gene products between cells
  (e.g., testosterone-induced differentiation of internal genitalia)
Morphology:
  Organogenesis (e.g., prenatal development of hands and fingers)
Behavior:
  Central nervous system motor programs (e.g., thumb sucking,
  scratching)
Life history patterns:
  Growth and development (e.g., prenatal differentiation of the
  reproductive organs; pubertal activation of sex hormones)

Comparative developmental studies are beginning to reveal how genetic differences result in developmental differences. Genetic differences encompass differences in the rate, sequence, and timing of gene activity as well as differences in alleles and genes themselves. Chief among the organs being studied from this perspective is the vertebrate brain (Butler and Hodos 1996).

Evolutionary Approaches to Ontogeny

Evolutionary biologists have taken three primary approaches to the study of ontogeny. The first of these, heterochrony, has been defined more or less narrowly (see Shea, this volume) and more or less broadly (see McKinney, this volume). According to the broader definition, heterochrony refers to changes in the timing or the rate of developmental events in comparison with the same events in the ancestor (McKinney and McNamara 1991). According to the narrower definition, heterochrony refers to "changes in the relative time of appearance and rate of development for characters already present in ancestors" (Gould 1977:2, quoted by Shea, this volume). The latter definition disallows characters that have undergone fundamental transformations—that is, new characters (Shea discusses the question of how much change is required under the definition of "new"). Therefore, it might disallow some cases of terminal addition or extension, depending upon their magnitude. The appropriate narrowness of scope of heterochrony is a crucial but unresolved issue.

The evolutionary developmental approach to neontology complements paleontological models of heterochrony constructed through comparative studies of fossil lineages. Heterochronic models classify kinds of changes in developmental timing by their outcomes. The two major classes of change from ancestral to descendant forms are *paedomorphosis*, or juvenilization, and *peramorphosis*, or adultification. These involve changes in onset and/or offset of growth and in rate of growth. Juvenilization results from earlier offset of growth (progenesis), delayed onset (postdisplacement), or reduced rate of development (neoteny). Adultification results from later onset of growth (predisplacement), delayed offset (hypermorphosis), or increased rate of development (acceleration) (McKinney, this volume; McKinney and McNamara 1991; McNamara 1997).

Heterochrony can occur globally, changing the whole descendant organism relative to its ancestor, or locally, changing some part of the descendant organism relative to that part of its ancestor (mosaic evolution). These two modes are known as global and dissociated heterochrony, respectively. Paleontologists have long studied heterochronic changes in fossil lineages. More recently, comparative anatomists and evolutionary developmental biologists (Raff 1996) have studied such changes by comparing the development of living forms.

The second main approach that evolutionary biologists have taken to the study of ontogeny is that of allometry, or study of the growth of a part of an organism relative to the whole or to some standard. At least since the time of Galileo, biologists have known that changes in proportions of certain body parts—that is, allometric changes—may occur when organisms become larger through evolution or through individual development during ontogeny. Sex differences in body size produce allometric differences between males and females of the same species. These changes have been explained as the product of physical phenomena that accompany changes in scale.

According to the classic explanation, as both living and nonliving objects grow larger, their mass increases by the cube of either their radius or their cubic length, while their surface area increases by the square of their radius. Consequently, as objects enlarge, their surface area increases less rapidly than their mass by a factor of two-thirds (the equation is $y = ax^b$, where y = variable organ size, a = a proportionality coefficient, x = body size, and b = exponent). This law affects the evolution as well as the architecture and engineering of three-dimensional objects (Calder 1984; Schmidt-Nielsen 1984).

According to a recent, more powerful explanation, the surface area of an organism increases less rapidly than its mass by a factor of three-quarters ($b = 3/4$) because of laws relating to the transport of essential materials in three-dimensional animals and plants through branching linear networks such as blood vessels, air passages, and xylem. Specifically, these laws are, first, that the network must fill the entire volume of the organism in a fractal branching pattern in order to supply the organism; second, that the final (smallest) branch is scale-equivalent to the largest unit; and third, that the energy required for distributing the resources must be minimized (Cates and Gittleman 1997; West, Bronw, and Enquist 1997). This new explanation purports to explain a wide range of allometric life history phenomena, including scaling of the rate and duration of embryonic and postembryonic growth and development, age at first reproduction, and life span.

Whereas some organs and life history patterns scale proportionally (isometrically) with increased body size, others scale nonproportionally (allometrically). Organs such as the lungs and the intestines, as well as the cross sections of long bones and the surfaces of molars that function on their surface area, must increase more than proportionally (positive allometry) in order to keep up with the greater volume of a larger body. Other organs, such as the brain, whose functions are not directly related to body volume, may increase less than proportionally (negative allometry). Recognizing which species differences are due to allometry is a prerequisite for distinguishing variations due to scale from those due to new adaptations (Pilbeam and Gould 1974; Shea, this volume).

Differences in size and proportion are related to differences in rates and timing of development (Calder 1984). Both are related to life history features—that is, to gestation period, age of first reproduction, number of offspring per reproductive effort, birth interval, parental investment, and life span. These and other features reflect selection for optimal allocation of energy across the life span into maintenance, growth, defense, and reproduction, constrained by phylogenetic and developmental inertia. Life history theory and associated concepts arose out of demography—study of the age and reproductive structure of populations—and behavioral ecology—study of the ecological and reproductive consequences of behavioral strategies (Cole 1954).

The third approach, life history strategies, was popularized by MacArthur and Wilson (1967), who characterized two contrasting life history strategies that are widespread among plants and animals. The r strategy of rapid maturation and high reproductive output results in explosive population growth under favorable circumstances. The K strategy of slow maturation and low reproductive output results in nearly steady-state population levels under stable circumstances. In contrast to r strategists, K strategists are large organisms that mature slowly and live for extended periods. Among the primates, the gorilla exemplifies a K strategy, and the small, rapidly developing mouse lemur exemplifies an r strategy (Harvey 1990).

Although K and r strategies are useful for conceptualizing life histories, they do not exhaust the range of life history patterns. Nor do the associated concepts of r and K selection adequately explain the evolution of these patterns. Life histories are shaped by a variety of extrinsic (selective) and intrinsic (phylogenetic and developmental) forces, which are incompletely understood (Stearns 1992).

Reconstructing the evolution of genes, characters, life histories, and allometric and heterochronic changes or any other traits requires the mapping of trait variants onto a branching diagram of species evolution—a cladogram—that has been constructed from other data (Gittleman, this volume; Parker, chapter 10, this volume). These branching diagrams are constructed on the basis of homologous characteristics that are unique to a group of related species (shared derived characteristics), as opposed to homologous characteristics that are not unique to that group. Uniqueness is determined by comparison with more distantly related groups (outgroups) (Hennig 1979; Wiley 1981). This cladistic technique of classification dominates systematics today. Typically, cladograms and the derived phylogenies (which show ancestors) are based on comparative analysis of DNA (Harvey et al. 1996).

These and other biological theories and methodologies provided the background for the discussions that took place during our advanced seminar. The other dominant theories and methodologies are represented by Piagetian and neo-Piagetian models for human cognitive and linguistic development and by comparative neurological and psychological approaches to primate development.

CONVERGING CONTEXTS FOR EXPLORING THE EVOLUTION OF HUMAN BEHAVIORAL ONTOGENY

Perhaps the most surprising and gratifying aspect of the seminar was the rapt attention participants accorded to one another's ideas. All participants came equipped with sufficient knowledge of the basic concepts of the other disciplines to be able to communicate across academic boundaries. The three developmental psychologists (Liz Bates, Patricia Greenfield, and Jonas Langer), with backgrounds in Piagetian theory and psycholinguistics, were able to talk with the evolutionary biologists (John Gittleman and Mike McKinney) and the biological anthropologists (Terry Deacon, Sue Parker, and Brian Shea), whose backgrounds lay in comparative methodologies for reconstructing morphological and behavioral evolution.

Reciprocally, the evolutionary biologists talked developmental psychology. Everyone was able to follow discussions by the cross-cultural psychologist (Patricia Greenfield) of cultural learning and cultural change in humans and chimpanzees. And everyone, including a comparative psychologist (Lynn Fairbanks), was able to follow the implications of revolutionary new findings by the cognitive neuroscientists (Liz Bates and Terry Deacon) about the genetic and cellular bases of brain development.

The Evolution of Brain Development

Appropriately, given its intermediating role between genes and behavior, brain development turned out to be a major focus of the seminar. Much of the discussion revolved around evolutionary and physical constraints on brain development and the implications these have for understanding processes of behavioral and cognitive development. Six contributors to this book—Bates, Deacon, Fairbanks, Gittleman, McKinney, and Shea—address brain development and evolution.

Terrence Deacon's chapter focuses on developmental constraints on brain evolution that flow from the scale of embryonic brain development in relation to gene action. He notes that all animal embryos, from the smallest to the largest species, progress through the steps that determine their cell fates while they are less than a centimeter long. This is necessarily so because the molecular mechanisms of embryogenesis involve chemical diffusion and cellular contact, which set a limit on the size of cell populations. These constraints result in evolutionary conservatism in body plans across taxa of vertebrates of vastly different sizes. They also result in allometric variations among these taxa.

Deacon cites evidence that the volumes of adult brain structures are determined by the timing and duration of neuronal generation. The later the onset of neurogenesis, and the longer its duration, the larger the structure. The correlation of size with the later onset of neurogenesis follows from the fact that the longer period before neuron production allows more mitotic cycles for production of stem cells. In other words, the cell division clock and the cell differentiation clock can be set independently. In general, cells seem to have an intrinsic, autonomous timing mechanism that is genetically determined.

Deacon notes that among mammals, brain and body growth patterns all follow the same course during the prenatal period. Postnatal growth curves differ in extent but have similar shapes. He suggests that differences in encephalization follow from modification of postnatal somatic growth curves rather than of brain growth curves. Primate embryos begin with less total body mass and the same brain mass as other mammals. The increase in human brain size results from prolongation of prenatal growth rates into the postnatal period. Finally, Deacon argues that this effect could result from a heterochronic shift in the time course of the expression of segmental genes.

Whereas Deacon focuses on prenatal genetic and developmental constraints on brain evolution, Elizabeth Bates and coauthor Jeffrey Elman focus on postnatal developmental constraints on brain evolution and their implications for the validity of competing developmental models. Their analysis of the neural bases of language acquisition integrates data from neuro-imaging of the brains of children with language deficits in different languages and from computer simulation of language learning by neural networks. Brain imaging reveals that language processing occurs in many different brain regions that change with shifting task demands. It also reveals that grammatical systems are neurally encoded differently in different languages.

Bates and Elman's research suggests that cortical specialization occurs through competition among regions whose slightly different architectures suit them for particular kinds of information processing. In response to predictable environmental inputs and feedback patterns, these biases lead to specialized representations. They call these factors *computational biases*. Their research also suggests that particular brain regions are recruited for particular tasks because they are ready at the appropriate time. Bates and Elman call these factors *chronotopic constraints*. Finally, they suggest that developmental changes in these computational and chronotopic constraints converge in similar outcomes, given similar inputs at critical periods.

According to their model, innateness in language acquisition arises from species-specific architectural and temporal brain design rather than from the innately programmed language modules proposed by Pinker (1994a). Their simulations show that these networks can learn key grammatical and semantic characteristics of words simply from repeated exposure to sentences. Their most tantalizing discovery is that the capacity to construct these grammatical categories obtains only when the network is designed with a narrow window of attention that matures slowly over time. Under these circumstances, it is able to construct hierarchical trees of animate and inanimate nouns and transitive and intransitive verbs. Computer simulations of learning by neural networks support the soft architectural model rather than Pinker's hard architectural model of language development.

In a complementary formulation, Lynn Fairbanks proposes that the development of play in primates is timed to occur during the periods when brain development is most susceptible to influence. She argues, first, that comparative data on primate play are best explained by the hypothesis that the timing of various forms of play—nonsocial activity, social play, and object play—functions to modify synapse formation in an adaptive manner. Second, she notes that the frequencies of various forms of play rise with synaptogenesis and fall with completion of myelination. She argues that these patterns are inconsistent with the alternative model for the development of play, the practice model. Fairbanks's model is significant to our volume because it suggests a specific link between brain development and the timing and duration of the juvenile period in primates. This in turn suggests one way in which life history evolution is tied to the evolution of brain development.

In a complementary approach to this issue, John Gittleman, H.–K. Luh, C. G. Anderson, and S. E. Cates use a multivariate analysis to examine the relationship between brain size and life history traits in mammals, contrasting various alternative statistical approaches to this issue. They conclude that evolution of brain size is most closely associated with evolution of litter size, lactation period, and juvenile and adult mortality, rather than with gestation period. (Sacher reached a different conclusion based on a different methodology: see Sacher 1982; Sacher and Staffeldt 1974.) Although they recognize that their specific conclusion may not hold, they show how their methodology can uncover causal connections among multivariate problems.

Michael McKinney, who uses the broader of the two definitions of heterochrony, addresses the relative merits of two competing models for the role of heterochrony in the evolution of the human brain—the underdevelopment, or juvenilization, model versus the overdevelopment, or adultification, model. Arguing from comparative data on increased brain size and slowed brain development from living and fossil primates, McKinney concludes that brain evolution occurred through overdevelopment rather than underdevelopment. He notes that brain complexity has increased through increased numbers of neurons and neural synapses and that this correlates with increased numbers and complexities of stages of cognitive development.

Brian Shea, in contrast, lays out the case for using Gould's (1977) narrower definition of heterochrony—"changes in the relative time of appearance and rate of development for characters already present in ancestors." This definition excludes evolutionary novelties and reduces the scope of heterochrony to the treatment of relatively minor transformations that have little relevance to understanding the evolution of many derived characteristics in humans, including the vocal apparatus and the brain. This definition also excludes character changes described by single-dimensional plots of growth against time. Shea emphasizes that heterochrony refers to changes in shape, which is inherently multidimensional (described by ratios or regression lines), and therefore excludes proportional giantism and dwarfism.

In an extended analysis of the misunderstandings inherent in recent critiques of "allometric heterochrony" by Godfrey and Sutherland (1994), Shea notes that shape is latent in multivariate plots of dimensionless ratios. Likewise, information on absolute and relative time is only indirectly available in allometric plots and tells us nothing directly about these factors. This, he stresses, is why allometry does not equal heterochrony (though it plays a key role in one central component of heterochronic analysis), a point he says Godfrey and Sutherland miss in their critiques.

Shea also emphasizes Godfrey and Sutherland's failure to understand the inherent relativity of results of allometric and heterochronic analysis; he writes that "heterochronic categories will necessarily vary depending on both whether and how we assess the four primary inputs into the comparative analysis—that is, size, shape, age (absolute), and relative developmental stage." This variable outcome occurs because assessments of change depend upon the developmental standard adopted. Following Gould (1977), Shea argues for the use of developmental stage rather than chronological age or size as the optimal criterion for assessing heterochrony whenever possible (see Deacon,

Fairbanks, Parker, this volume). He also notes that even with a stage criterion, heterochronic classifications of the same species may differ depending upon which stage is chosen. When human brain development is compared in ancestral and descendant species, for example, it is peramorphic shortly after birth and paedomorphic at sexual maturity.

Shea concludes his chapter by saying that although an understanding of the relationship between ontogeny and phylogeny is crucial to unraveling the mysteries of human evolution, "the evidence seems not to implicate any simple heterochronic transformation, nor any coordinated neotenic paedomorphosis in particular, as the dominant change involved."

Comparative Approaches to Behavioral Development

The other major focus of the seminar was on the evolution of behavioral development, particularly cognitive and brain development. Implicit in the discussion was the comparative method. Whether at the level of cross-species comparisons, cross-language comparisons, or cross-cultural comparisons, the comparative method lies at the heart of evolutionary reconstruction. In recent years, phylogenetic systematics—cladistics—has contributed greatly to the rigor and scope of evolutionary reconstruction based on comparative data. First, it has produced more reliable cladograms (branching diagrams of related species) based on shared derived characteristics. Second, it has provided a systematic method for reconstructing behavioral evolution through identification and mapping of shared derived character states onto these cladograms (Brooks and McLennan 1991; Harvey and Pagel 1991).

Most of the chapters in this volume use the comparative method. Bates and Elman, for example, employ a cross-language developmental methodology to assay the mechanisms underlying language acquisition. Their gloss on the constructivist approach to cognitive and linguistic development paraphrases Piaget's comment, "That which is inevitable does not have to be innate." Fairbanks compares patterns of development of physical, object, and social play in vervet monkeys and other Old World monkeys with those of prosimians, great apes, and human children. She demonstrates that these forms of play correspond to periods of brain development in the related modalities and that development in both brain and behavior scale with life history—that is, they are proportionally similar at different developmental scales, a pattern McKinney and McNamara (1991) described as *sequential hypermorphosis.* Deacon's chapter, too, is based upon comparative data on brain development in primate and other mammalian species. He focuses on the early embryonic stages of development and their differences in timing and apportionment of tissue. This allows him to identify a major change in primate brain evolution.

Jonas Langer uses comparative data on cognitive development in monkeys, apes, and human children to identify the heterochronic processes involved in their evolution. He concludes that the rate of cognitive development in humans in the logicomathematical, physical, and social domains is accelerated in comparison with that of nonhuman primates. He also concludes that the rate of development of logicomathematical cognition is more accelerated than the rates of development of physical and social cognition in nonhuman primates. He hypothesizes that temporal realignments among these cognitive domains that occurred during human evolution transformed them into simultaneously developing systems. In other words, heterochronic displacement among domains resulted in multiple cascading possibilities for information flow between logicomathematical and physical domains absent in other primates (see Bates and Elman, this volume).

In an exercise in evolutionary reconstruction, Patricia Greenfield, Ashley Maynard, Christopher Boehm, and Emily Yut Schmidtling compare the cultural transmission of tool use and technology in humans and chimpanzees. Noting species-specific connections among cognition, apprenticeship, and behavioral ontogeny, they identify six mechanisms of apprenticeship shared by chimpanzees and humans and two that are unique to humans. Then they discuss the role of apprenticeship in cultural innovation and culture change. Finally, they speculate on the co-evolution of technology, cognition, and teaching in the evolution of human ontogeny.

In my own chapter, I use cross-species comparisons to reconstruct the evolution of human cognitive development. Specifically, I use cladistic methodologies to reconstruct the stages of evolution of human cognitive development on the basis of comparative data from monkeys, apes, and human children. Similarly, I employ comparative data on dental development to reconstruct the evolution of life history. I extrapolate from great ape and human development to that of *Homo erectus*, concluding that its cognitive abilities and life history lay midway between those of modern humans and ancestral great apes. According to my heterochronic model, new cognitive stages appeared at the end of development in a series of primate ancestors. This pattern of terminal addition of a series of stages, and their subsequent acceleration, is reflected in recapitulation of the stages of cognitive evolution during human development.

REACHING AGREEMENT

In response to the questions posed by the seminar organizers, participants agreed on the following broad areas. First, human cognitive and linguistic development occurs through a process of active construction that begins prenatally and continues through life. Second, the neurological basis of this construction is achieved through epigenetic mechanisms that produce subtle architectural differences leading to biases in information processing that function predictably in consistent contexts. Third, the constructive nature of neurological and cognitive development is necessitated by genetic, physical, and developmental constraints operating during early stages of embryogenesis and neurogenesis. Fourth, these constraints limit the range and direction of evolutionary changes that have occurred in the past and might occur in the future.

Finally, we agreed that cognitive development seems to have evolved through terminal addition of new stages (which arose from differentiation, extension, and hierarchical integration of ancestral stages) in descendants, and that human cognitive development is accelerated relative to that of ancestors (Parker and McKinney 1999).[1] Therefore, human cognition involves overdevelopment rather than underdevelopment, or what is popularly known as neoteny. On the other hand, participants thought that new data on brain development suggested that human brain evolution might be too complex to be fully described in terms of heterochronic mechanisms (Deacon, this volume).

We also agreed on several general conclusions and future agendas. First, we found that the disciplines of the assembled participants had much of relevance to offer one another. Second, we agreed that comprehensive theoretical frameworks from developmental evolutionary biology are vital to progress in understanding the evolution of ontogeny, and we should focus on heuristics and hypothesis generation. Third, we agreed that the comparative cladistic method is necessary for reconstructing the evolution of behavioral ontogeny. Fourth, we concluded that models of life history, allometry, and heterochrony relevant to reconstructing behavioral ontogeny need to be dissociated and refined within the larger framework of developmental evolutionary biology in the interest of selecting appropriate tools for analyzing various aspects of the evolution of ontogeny. Shea's chapter on the relationship between heterochrony and allometry is an important step in this direction, as are the chapters by Fairbanks and by Gittleman and colleagues on the relationship between life history and brain development.

Fifth, we agreed that analyses of hierarchically nested phenomena, from the genetic to the cellular to the morphological to the behavioral and life history level, are crucial to understanding the evolution of behavioral ontogeny. Sixth, we agreed that the developmental genetics and histology of the brain (as the organ of behavioral organization) are crucial to model building. Finally, we agreed that although the processes involved might be different, it would be valuable to develop a taxonomy of processes of evolutionary change within each level. A taxonomy of heterochronic phenomena at each hierarchical level of development would help identify the limits of heterochronic concepts of the evolution of development. Table 1.1, a revised version of a table circulated for discussion at the seminar, reflects an initial attempt to sort out heterochronic effects at various levels of organization.

It is important to note in this context that similar morphological and behavioral systems can develop in different species through different mechanisms (Raff 1996). There is no a priori reason to expect identity between the mechanisms operating at various levels—the genetic, the morphological, and the behavioral—even when they are involved in the sequential construction of an emergent outcome. Terminal addition of new stages of cognitive development, for example, need not imply terminal addition of new genes or tissues (Deacon, this volume).

Recent work in evolutionary developmental biology has begun to pinpoint the role of genes in embryo development, and specifically in embryonic brain development. These studies reveal that heterochonic

TABLE 1.1

Suggested Mechanisms Underlying Heterochronic Effects at Several Levels of
Organization in Animals (in Comparative Frame between Species or between Societies)

	Level of Organization	
Heterochronic Changes in:	*Genetic (Control & Producer Genes)*	*Cellular (Tissues)*
Onset or offset time	Variation in timing of on and off switches on genes	Variation in timing of cell differentiation & division
Rate	Variation in transcription rates	Variation in rate of cell differentiation & division
Terminal stage	Gene duplication and and deletion; switching on by sequential position	Addition or subtraction of cell type
Sequence change, or sequential hypermorphosis	Variation in sequence of gene transcription	Variation in sequence of cell differentiation
Organization (from all the previous factors)	Variation in genetic response of target tissue	Developmental variation among target tissues

Note: Phenomena do not necessarily correspond across columns. Heterochronic changes can be either global, involving consistent change in an entire descendant organism relative to its ancestor, or local, involving dissociated changes among different systems in the same organism relative to those of its ancestor (the last row refers to dissociated changes).

Level of Organization		
Organ (Somatic, Reproductive, Brain)	*Life History and/or Cognitive Stages*	*Social-Cultural*
Variation in duration of growth	Variation in timing of stages	Variation in historical time of innovations
Variation in rate of growth	Variation in developmental rate	Variations in rate of innovation
Addition or subtraction of segment or target-tissue connection	Addition or deletion of terminal stage	Addition or deletion of recent innovations
Variation in sequence of target-tissue differentiation	Variation in developmental sequence	Variation in sequence of activities
Developmental variation among organs	Developmental variation among domains	Variable change among domains of social organization

changes occur in evolution at genetic as well as morphological levels (McKinney and McNamara 1991; Raff 1996). Raff (1996), for example, discusses heterochronic mutations in one species of nematode, *C. elegans*. The lin-14 gene causes displacement of cell lineage differentiation in the larvae. Failure to turn off the lin-14 gene (which encodes a nuclear protein) in the somatic cells of the first larval instar results in reiteration of the larval pattern (neoteny). Raff notes, however, that heterochrony is only one factor in the evolution of development at the genetic level (also see Deacon, Shea, this volume).

Most of the data on heterochronic changes in evolution have come from comparative studies of the anatomy of fossil lineages by paleontologists and of related species of living organisms by neontologists. Living organisms have the advantage of providing direct data on developmental stages, which are rarely available in the study of fossil lineages (Raff 1996).

Although heterochronic changes in postnatal morphological development are well documented, they constitute but one class of evolutionary change. Other changes fall under the category of allometry. The relationship between these two factors has been a matter of some interest (Shea, this volume). The relationship between heterochrony, allometry, and life history changes in evolution is also a matter of considerable interest (see Fairbanks, Gittleman, this volume). Life history changes resulting from selection for earlier or later maturation or higher reproductive rate, for example, are often associated with heterochronic differences in development. Table 1.2 offers a preliminary approach to sorting out these relationships.

Heterochronic changes in the evolution of behavior are less studied than those of morphology. Psychologists have been suspicious of the idea of peramorphosis and attendant recapitulation since G. Stanley Hall's (1897, 1904) ill-founded claims for such phenomena in child development. Recently, heterochronic explanations for evolutionary changes in cognitive development have been proposed on the basis of comparative developmental studies of primate cognition (Langer, Parker, this volume). These studies suggest that the sequence of cognitive development in humans roughly parallels the sequence of its evolution in ancestral forms (Parker, this volume).

Finally, participants discussed whether the concept of heterochrony

TABLE 1.2

Suggested Effects of Various Evolutionary Forces on Various Kinds of Developmental Change in Descendant Forms

	Extrinsic Factors Effect of Selection		Intrinsic Factors Constraints on Selection	
Type of Change	Direct Effect	Indirect Effect	Phylogenetic Inertia	Developmental Constraints
Small or large body size	Life history shift	Allometry, heterochrony	Limitation due to body plan	Limitations due to body plan
Early or late sexual maturity	Life history shift	Heterochrony	Limitation due to mode of reproduction	Limitation due to energy constraints
Change in reproductive rate	Life history shift	?	Limitation due to mode of reproduction	Limitation due to energy constraints
Specialized organ size change	Dissociated heterochrony	Life history shift	Limitation in evolutionary potential due to body plan	Limitation in developmental potential due to body plan
Change in life span	Change in life history	Heterochrony	?	?
Changes in proportion of life stages	Changes in life history	Sequential hypermorphosis	Limitation in evolutionary potential due to body plan	Limitation in developmental potential due to body plan

23

is useful for comparing behavior at the level of culture change in different societies (Greenfield et al., this volume). Certainly anthropologists and psychologists have long agreed that humans' advanced cognitive abilities underlie their capacity for culture (Goodenough 1981; White 1959). Lumsden and Wilson (1981) argued that epigenesis of mind is the key connection between genes and culture. Most behavioral scientists also agree that the step from individual behavior to cultural phenomena depends crucially on symbolically mediated social learning through imitation and teaching (Donald 1991; McGrew 1992; Tomasello, Kruger, and Ratner 1993). In recent years they have emphasized the notion of co-construction and apprenticeship inspired by Vygotsky (Rogoff 1990). Although it is possible to contrast the nature of change in two societies using heterochronic categories, whether this has heuristic value remains to be seen.

Overall, participants agreed that heterochronic concepts are useful for describing, classifying, and comparing some evolutionary changes in development at several levels of organization, from genes to behavior, though we differed on the scope of its utility. We also agreed that heterochrony is only one of several concepts—including allometry, cladistics, life history, and developmental genetics—that are useful for understanding the evolution of development. Finally, we agreed that developmental evolutionary perspectives offer a rich and stimulating agenda for research in biology, anthropology, and psychology in the next century.

Note

1. Children traverse four major periods of cognitive development: the sensorimotor period, from birth to 2 years; the preoperations period, from 2 to 6 years; the concrete operations period, from 6 to 12; and the formal operations period, from 12 onward. Each succeeding period is characterized by more flexible and powerful instruments for assimilating and accommodating to the world, culminating in the capacity for hypothetical-deductive reasoning and logical necessity. Linguistic development partially parallels cognitive development. The first representational communications develop at the transition between the sensorimotor and preoperations periods, coincident with tool use, object seriation, and symbolic play (Elman et al. 1996; Greenfield and Smith 1976).

2

Evolving Behavioral Complexity by Extending Development

Michael L. McKinney

> Indeed, since human babies have extremely large brains, one might say exactly the opposite of what Bolk has said and state that they have become adultified.
>
> —*Ernst Mayr,* Animal Species and Evolution

In this chapter I discuss recent work by paleontologists, anthropologists, developmental biologists, neurobiologists, and psychologists showing that the evolution of hominid brain ontogeny has been produced largely by the extension of brain growth and development. In addition to evidence discussed elsewhere (Langer, this volume; McKinney 1998; McNamara 1997; Parker 1996b; Parker and McKinney 1999; Shea 1989, this volume), I incorporate exciting new evidence on the ultimate cause of our extended growth in brain and cognition: *predisplacement*. Predisplacement, or "starting with more," is a common developmental change in evolution (McKinney and McNamara 1991; McNamara 1997). In this case, it occurs from a change in homeotic gene expression that alters the proportions of embryonic neuron stem cells to produce a larger, slower-growing brain (Deacon 1997, this volume). This initial change in embryonic neuron proportions has many cascading effects on later human brain and cognitive development, nearly all of which can be characterized as *sequential hypermorphosis*, referring to extended (prolonged) duration of stages in brain development (McKinney and McNamara 1991; McNamara 1997).

Because of recent comments correctly noting that many developmental changes in evolution do not reflect such time-related, or heterochronic, causes (Raff 1996; Rice 1997; Zelditch and Fink 1996), I hasten to add that brain evolution is one of the most likely places where one might find heterochronic causes. This is because heterochronic processes are most clearly defined as measurable changes in covariant growth patterns (Rice 1997), and mammalian brain development shows one of the most highly conserved (that is, covariant) growth patterns of any organ (Finlay and Darlington 1995). Such a developmental system is highly prone to modification by tiny perturbations that can have huge cascading impacts on morphology and, in this case, cognition and behavior. I also hasten to add, however, that as one would expect for such a complex organ, other brain developmental changes have also occurred as specific localized adaptations superimposed on the general pattern of extended growth of the whole organ.

WHAT ARE JUVENILIZATION AND OVERDEVELOPMENT?

A main source of confusion in discussions of developmental evolution has been their excessive jargon and a basic ignorance of the mechanisms that produce evolutionary changes in development (Gould 1977; Raff 1996). *Juvenilization* is just one type of heterochrony, which is often defined as evolution via change in the rate or timing of development (McKinney and McNamara 1991; McNamara 1997). Juvenilization (also called paedomorphosis, underdevelopment, and terminal truncation) can occur through three basic rate or timing changes: slower growth (neoteny), early termination of growth (progenesis), and late initiation of growth (postdisplacement).

The opposite process, *overdevelopment* (also called peramorphosis or terminal extension), can occur through the opposite changes: faster growth (acceleration), late termination of growth (hypermorphosis), and early initiation of growth (predisplacement). These terms and their history are reviewed by McKinney and McNamara (1991). *Terminal extension* is probably a better term than the traditional *terminal addition* because development involves many complex cell interactions, and so the result of extending a developmental process is rarely additive. I will say more about this later.

In addition, terminal extension need not be limited to only one stage of development. Hominid behavioral and cognitive development is but one example of the relatively common evolutionary process of sequential hypermorphosis, whereby each developmental stage is terminally extended (McKinney and McNamara, 1991; McNamara 1997). Rice (1997) provides a brief mathematical description of sequential hypermorphosis.

Juvenilization, overdevelopment, or any other single heterochronic change rarely affects the entire organism. Raff (1996) and Zelditch and Fink (1996), for example, provide ample evidence from many taxa that few cases of developmental changes in evolution consist of alterations that affect the entire organism in the same way. Instead, "dissociated" changes occur in which some traits may show paedomorphosis, others peramorphosis, and still others nonheterochronic changes such as novel tissue arrangements.

Such mosaics of dissociated changes in evolution have long been known (Gould 1977; McKinney and McNamara 1991; Shea 1996). These phenomena show that heterochronic terms are most useful and informative when there is a general pattern of similar heterochronic changes present among many traits. For example, for some descendant taxa, such as domesticated dogs, many behavioral and morphological traits may be explained by a general process of juvenilization, whereas other traits do not fit this pattern (Coppinger and Schneider 1995). Human brain evolution also follows this dissociated pattern, such that the general process of overdevelopment is accompanied by deviations of dissociated changes in some parts of the brain (Deacon 1990a, 1997; Finlay and Darlington 1995).

Relevance to Human Mental Evolution

Human evolution has clearly been characterized by heterochronic dissociations. It has been widely noted by both proponents (e.g., Gould 1977; Montagu 1981) and opponents (e.g., McKinney and McNamara 1991; Shea 1989) of human juvenilization that many traits have evolved relatively independently of others. Most of these researchers, however, have acknowledged such dissociation only in the context of morphological evolution. For example, various aspects of limb, pelvic, dental, and skull morphology all show a variety of distinct developmental changes during human evolution (Shea 1989).

In contrast to this mosaic of dissociated somatic changes, human brain evolution has been relatively simple, mainly because change in the shape of the mammalian brain is constrained in comparison with other organs (Finlay and Darlington 1995). This reduces the amount of independent change in various regions of the brain. The important evolutionary result is that increasing size has produced relatively consistent shape changes throughout the brain. Many of our mental abilities are largely attributable to extension of brain development to produce a proportionately scaled-up version of the ancestral ape brain. Sequential hypermorphosis of behavioral and cognitive development is accompanied by prolonged stages of neurogenesis, dendritogenesis (and dendrite pruning), synaptogenesis, and myelination. These, in turn, are ultimately traceable to our originally larger endowment of embryonic neurons, as I discuss later.

Overdevelopment = Delays without Rate Reduction

A delay in the termination of a developmental event will not, by itself, necessarily cause overdevelopment if there is also a concurrent reduction of growth rate. This potential decoupling of rate of growth in a stage from termination of growth in a stage is another key factor in the confusion over juvenilization. Humans have a sequentially delayed life history pattern in comparison with other living primates. Infancy, puberty, and life span in general are prolonged in humans. There is also evidence, especially from fossil dentition, that human evolution, from australopithecines to modern humans, has involved a progressive delay in major life history events (review in Smith 1992). Proponents of human juvenilization have seen these life history delays as reflecting a reduction in the rate of development (e.g., Gould 1977; Montagu 1981), leading to permanent underdevelopment. Underdevelopment, however, would occur only if humans also experienced a slower growth rate during each prolonged stage. But the rate of growth in humans is relatively unchanged compared with that of close living relatives (e.g., Shea 1989) and hominid ancestors (Smith, Gannon, and Smith 1995). Life history delays thus produce overdevelopment, not underdevelopment, in both somatic and brain morphology. By extending growth in each stage without reducing the rate of growth per stage, humans grow to large body size and have larger brains (McKinney and McNamara 1991). This also causes "overdeveloped" cognitive capacities.

OUR "OVERDEVELOPED" BRAIN: PROLONGING GROWTH

Given the often complex developmental origins of many heterochronies (McKinney and McNamara 1991; Raff 1996), it is perhaps ironic that the evolution of life's most complex organ, the human brain, has a comparatively simple basis. Production of cortical neurons in mammals is limited to early prenatal developmental. Mitotic rates of neuron creation in mammals are relatively constrained, perhaps reflecting limitations on neuron mitotic processes in the prenatal environment (Sacher and Staffeldt 1974). Thus, as discussed by Finlay and Darlington (1995), it is developmentally "easier" to produce a larger brain by extending prenatal brain growth than by altering the rate of growth. In the case of modern humans, this fetal brain growth phase is extended to about 25 days longer than that of living monkeys. This delay in fetal brain growth has had profound cascading consequences that underlie all later major delays in brain and cognitive development and increases in brain complexity. Specifically, this fetal growth delay causes the following key traits that characterize brain "overdevelopment" in humans.

High Brain/Body Ratio

As noted by Shea (1989), the brain grows faster than the body during the extended fetal phase, producing a high brain/body ratio in humans. After this phase, especially postnatally, brain growth is much slower than body growth. If body growth were accelerated, as it is in the relatively fast-growing gorilla (McKinney and McNamara 1991), a lower brain/body ratio could result. But in humans, postnatal body growth is relatively slow, so our bodies stay relatively small. Thus, modern humans have a brain size that would be predicted for a much larger primate (Deacon 1990a, 1997).

This high brain/body ratio has sometimes been used as evidence for juvenilization in human evolution—that is, as evidence that humans have a juvenilized "shape," as defined by the brain/body ratio (Gould 1977). I suggest, however, that it results from a brain that is "overdeveloped" in both absolute and relative (brain/body) terms. Indeed, many workers have noted that shape is a misleading way to identify juvenilization, or any other heterochrony, because the same shape can be attained by many different combinations of rate or timing changes (Raff 1996; Shea 1989, this volume). As Deacon (1990a:270) noted: "Paedomorphism of human brain/body proportions...is an artificial correlate of brain size evolution....The fact that humans exhibit paedomorphic brain/stature proportions with respect to other apes... in no way implies any corresponding arrest of brain differentiation."

Furthermore, one could not even argue that human postnatal body size is juvenilized, because humans have generally been increasing in body size over the last five million years as a result of progressively delayed puberty (Smith 1992). In other words, both brain size and body size have increased, but brain size has increased relatively faster, giving us, as adults, a higher brain/body ratio than our adult ancestors had.

Perhaps most important, brain/body size ratios are often poor predictors of cognitive abilities. As discussed by Deacon (1997), small domesticated animals, for example, have relatively large brains but are not necessarily more intelligent. Indeed, as I discuss next, it is developmental changes at smaller scales, such as number of neural interconnections, that underlie much of humans' cognitive "overdevelopment."

More Neural Complexity

A relatively large brain is obviously a main component of human cognition, but brain size alone is a coarse way to define "overdevelopment." A finer definition would also note that the large numbers of neurons produced during prolonged fetal growth also undergo a prolonged period of postnatal growth and maturation. Delay in offset of the infancy and juvenile stages of development leads to a more complex brain because dendritic growth in the cortex is extended to 20 years or more in humans (Gibson 1991). Similarly, glial cell growth and synaptogenesis are prolonged and thus extend beyond ancestral developmental patterns (Gibson 1990, 1991). This extension of dendritic and synaptic growth explains why neurons have many more dendritic and synaptic interconnections in the human brain than they do in other primates (Purves 1988). Stanley (1996) noted that such interconnections are essential for "intelligence."

Brain myelination is also prolonged in humans. Because it promotes more effective nerve transmission, myelination plays a large role in improvements in memory, intelligence, and language skills during development (Case 1992; Gibson 1991). Myelination is highly canalized; that is, it follows the same specific sequence among primates. So regional delays are readily seen in humans. In both monkeys and humans, myelination begins in the brain stem and proceeds through the subcortical areas. The neocortex and especially the prefrontal cortex are among the last to myelinate (Gibson 1991). But rhesus monkeys show myelination up to only about 3.5 years, whereas in humans, myelination continues until well over 12 years of age (Gibson 1991). Even peak synaptic complexity is delayed in humans. Dendritic complexity in the human brain peaks at about two years of age and begins to decline thereafter as "pruning" begins (Edelman 1987). In contrast, in apes and monkeys, dendritic pruning begins well before the age of two (Purves 1988).

More Neocortex and Prefrontal Cortex

Finlay and Darlington (1995) show that increase in brain size, from shrews to primates, accounts for an enormous amount (over 96 percent) of the variation in size of individual brain regions. Of special relevance to humans is that the order of events in neurogenesis is phylogenetically highly conserved. As a result, evolutionary brain size increase occurs by disproportionately more growth in areas of the brain such as the neocortex that are generated late in development (Finlay and Darlington 1995). In terms of behavior and cognition, the neocortex (or "isocortex") is a "general-purpose integrator," and increasing neocortical size produces an increasing capacity to process information of all kinds (Finlay and Darlington 1995). Tool use, language, and social behavior may all increase in complexity as the neocortex increases in size because, as Gibson (1993) reviewed, they all share common neocortical substrates that promote mental constructional skills.

It has long been thought that an area of the neocortex that is crucial for cognitive function is the prefrontal cortex (Stanley 1996). In particular, the prefrontal cortex is the center for short-term memory (the "central executive system"), in which information is temporarily stored and manipulated (Case 1992). This function has been confirmed by magnetic resonance imaging of prefrontal brain activity during various cognitive tasks (D'Esposito et al. 1995). How has prolonged fetal growth affected this key area of the neocortex? Deacon (1990a) calculated that the prefrontal area in modern humans is 202 percent the size of the prefrontal area in an anthropoid ape of our body size. This percentage increase is much greater than that in any other area of the brain. Some areas have even decreased relative to their proportion in the apes, especially areas associated with the sensory and motor systems: olfaction (32 percent the size of its counterpart in a same-sized ape), vision (60 percent), and motor (35 percent).

On the basis of such evidence, Gibson (1991) concluded that "on neurological grounds...there is nothing paedomorphic about the adult human brain." In direct contrast to the brains of infant primates, adult human brains are "large, highly fissurated and myelinated with complex synaptic morphology."

OUR "OVERDEVELOPED" COGNITIVE SKILLS

Human mental development follows the same general pattern as the brain: a prolonged learning stage with no reduction in the rate of development (e.g., learning), which ultimately produces an "overdeveloped" adult. Indeed, a period of accelerated cognitive development is prolonged in humans (Langer, this volume).

A main obstacle to documenting this cognitive acceleration has been that of objectively measuring cognitive skills. One way is to use Piagetian stages of human development to compare development in humans and in living primates. There are a number of reasons for the use of Piaget's approach, as discussed by Parker (1996b). In particular, Piaget's approach has been extremely influential, and a huge body of human developmental data has been collected using his framework. Criteria that define Piaget's stages in humans can be observed in the development of other species, so that rate and timing of stage development can be measured.

Comparison of cognitive stages between humans and other primates indicates that monkeys and apes follow the same general sequence of cognitive development as humans (Langer 1996; Parker

1996b). There is a clear developmental difference, however, between monkeys, apes, and humans in terms of both the rate and the ultimate level of terminal cognitive development. Monkeys have the lowest rate and ultimate level of development, whereas humans have the highest rate and ultimate level (Parker 1996b). This indicates a correlation between rate and level of cognitive attainment (just the opposite of what is predicted by juvenilization).

Monkeys (macaques) do not complete the Piagetian sensorimotor period. This is the first Piagetian period, characterized by such cognitive traits as rudimentary imitation and perceiving logical relationships among objects. The highest cognitive level achieved by monkeys in the sensorimotor period is about equal to that of two-year-old modern human children (Parker 1996b). Apes do complete the sensorimotor period, at about three years of age, and proceed into the next Piagetian stage, the preoperations period. The preoperations period is characterized by such cognitive traits as pretend play and ability to discern immediate causes and reactions (as used in basic tool making). Apes do not, however, complete the preoperations period, and they terminate cognitive development at a level approximating that of four-year-old children (Parker 1996b). Modern human cognitive development thus goes "beyond" that of other primates. It is also accelerated, because the same stages are attained at younger ages in humans.

It is highly relevant that mathematical models of sequential hypermorphosis, as discussed in McKinney and Gittleman (1995) and especially Rice (1997), can produce the same pattern seen in human evolution. In such models, each developmental stage of the ancestor is extended to produce a descendant growth-trajectory curve that not only eventually exceeds the ancestral trait but also shows acceleration (higher slope) of preterminal growth stages. This occurs because each succeeding stage "starts off with more" of whatever trait is being measured (in this case, cognitive abilities traceable to more neural circuitry and related attributes).

Evolution of Cognition

Such comparative data on cognitive development in humans and living primates can be used to reconstruct the evolution of human cognition. Cladistic mapping of Piagetian stages indicates that the early

Piagetian preoperational period is a shared derived character state that was present in the common ancestor of all great apes (Parker 1996b). The adult ancestor of all hominids apparently attained this stage, which is characteristic of cognition in two- to four-year-old children. Thereafter, human ancestors, from australopithecines to *H. erectus*, apparently evolved progressively through a sequence of cognitive developmental stages lying somewhere between those of apes and modern humans (Parker 1996b). This progression implies prolongation and acceleration of cognitive development during human evolution. There was apparently an intermediate cognitive development for *H. erectus*, judging from both cultural and biological evidence (Parker 1996b, this volume; also see review in Mithen 1996).

Prolonged Brain Maturation as a Cause of Prolonged Learning

Humans' accelerated cognitive abilities are ultimately attributable to our having a larger endowment of neurons, dendrites, and synapses, which allows us to store and manipulate information at a faster rate, especially in the prefrontal area. This initially large endowment (pre-displacement) produces prolongation of an accelerated learning sequence because each stage of neural maturation is delayed. Thus, peak rates of neurogenesis, dentritogenesis (and pruning), synaptogenesis, and myelination occur later in modern humans, and each process is itself extended (Gibson 1990, 1991). Myelination, for example, continues well into adolescence in modern humans.

Langer (1996) noted that a key aspect of human cognition is the potential for information flow among cognitive domains. This potential arises because many cognitive domains, such as physical ("cause-effect") and logicomathematical cognition, mature early in humans. This leads to cascading effects such as the ability to perform logical reasoning and experiments on physical events. In contrast, different domains of cognition mature at different times in other primates, so that information flow among them is not possible until later in development, if at all (Langer 1996).

MECHANISMS OF BRAIN "OVERDEVELOPMENT"

The mechanisms that result in developmental evolution can be described at many levels, ranging from genetic to behavioral (Raff

1996). At a coarse level, we might describe the evolution of our cognitive "overdevelopment" as being "caused" by the combination of accelerated learning plus the prolongation of childhood and our life history in general. But this accelerated and prolonged learning ability can be traced back one step farther, to the prolongation of fetal growth, which extends the duration of neuron mitosis (Finlay and Darlington 1995). This extended fetal growth not only produces the large number of cortical neurons characteristic of the human brain but also seems to have cascading effects on neuronal complexity by prolonging the development of individual neurons, allowing more complex dendritic and synaptic outgrowths and connections.

Furthermore, many aspects of brain complexity seem to originate as cascading effects that ultimately derive from prolongation of fetal brain growth. As Deacon (1990a:270) noted, "In all the major indicators of mammalian brain development and maturation, including the level of differentiation of brain structures, the morphological maturation of neurons, the level of myelination of axons, the rates and specificity of neurotransmitter synthesis,…as well as many other measures, the adult human brain has at least achieved the level of…mature mammalian brains and typically has carried these trends much further."

That many aspects of brain size and complexity are "overdeveloped" at so many scales, from the cellular to the gross morphological, is indicative of a highly integrated system. This high integration agrees with the highly covariant, constrained patterns found by Finlay and Darlington (1995) in mammalian brain evolution. Or, as Deacon (1990a:277) noted in his overview of brain ontogeny and phylogeny, "brain development parameters are remarkably similar in different species and across a phenomenal range of sizes." In such highly integrated systems, we might expect, for example, the pattern seen in human evolution, in which prolonged duration of axon growth causes not only longer axons but increased myelination of those axons as needed for effective nerve transmission.

Can the sequence of causation be taken back one step farther to specify what causes the prolongation of fetal brain growth? Deacon (1997, this volume) has suggested that prolonged fetal brain growth occurs because of changes in segmentation in very early embryonic development. Changes in homeotic gene expression may alter the initial proportion of late-maturing embryonic stem (parent) neurons. If so, then the ultimate heterochronic event underlying human brain evolution would be traceable to mutations in homeotic genes that produce "predisplacement," the heterochronic term referring to "starting off with more" (McKinney and McNamara 1991; McNamara 1997).

It is unclear whether the prolonged fetal growth of humans is also responsible for our prolonged childhood and life span. For example, a variety of mechanisms is involved in aging, and many of those mechanisms are poorly understood (Wolpert 1991). Is our long life span traceable all the way back to a resetting of the "mitotic clock" during fetal growth? Or has selection acted upon a different set of mechanisms (e.g., endocrine timing) that are activated postnatally?

It is important to note that the very early developmental mechanisms just mentioned for producing our overdeveloped brain and cognition have little or nothing to do with the known mechanisms that produced the juvenilized morphology and behaviors of domesticated dogs. Wayne (1986) showed that the juvenilized morphology of dogs is produced by growth rate changes occurring mainly during late fetal development. Furthermore, the juvenilized social behaviors of domesticated dogs, in comparison with those of wolves (Goodwin, Bradshaw, and Wickens 1997), are apparently unrelated to any change in cognitive abilities.

Although developmental extension explains much about our cognitive evolution, there are clearly deviations from this pattern. I reemphasize that such deviations (dissociations) are extremely common in evolution (McKinney and McNamara 1991; McNamara 1997; Raff 1996). In the case of the brain, Deacon (1997) described a number of disproportionate elaborations (and reductions) of neural circuitry (e.g., Broca's area) beyond those expected from the extrapolation of ancestral brain growth. These features appear to be related to natural selection for language and other forms of internal symbolic manipulations. Similar localized neural adaptations are well documented for many other species—for example, changes in brain circuit development in owls as adaptations to hearing (Nishikawa 1997).

DOES COGNITIVE RECAPITULATION OCCUR?

The idea that human mental evolution is recapitulated during ontogeny has long been popular. That the development of children may repeat the mental development of early humans was an idea favored by Darwin himself and by many early psychologists such as James Baldwin and G. Stanley Hall (Morss 1990). It is a theme in the writings of Freud and Piaget and continues to be proposed today (e.g., Ekstig 1994). Some anthropologists have even begun to interpret human cultural evolution in terms of mental ontogeny (review in Mithen 1996).

Although it may be tempting to interpret evolutionary brain overdevelopment as a process of developmental terminal addition that produces evolutionary recapitulation in modern human ontogeny, at least two key caveats accompany this interpretation. First, brain over-development is not a simple process of terminal addition. Second, although general aspects of brain and cognitive evolution are likely to be recapitulated, it is unlikely that all the fine details of brain and especially cognitive evolution are repeated. We must therefore specify exactly what is being recapitulated.

Terminal Extension, Not Terminal Addition

Human cognitive "overdevelopment" has been produced largely by prolonged fetal brain growth. This prolongation produces not only the addition of more neurons at the end of brain growth but also "overdevelopment" in many aspects of brain (and cognitive) development, such as neuron, dendritic, and synaptic complexity and connectedness. For this reason, the expression *terminal extension* is more accurate than *terminal addition:* it better describes the process whereby many aspects of the developmental trajectory, from cellular to brain morphology to behavior, are extended.

General Brain and Cognitive Recapitulation

MacLean's (1990) "triune brain" hypothesis is the most popularized expression of the "overdeveloped" brain concept (e.g., Damasio 1994a; Sagan 1977). The triune brain refers to three grades of brain evolution, represented by the protoreptilian (R-complex), the paleomammalian (limbic system), and the neomammalian (neocortex) structures (MacLean 1990).

The details of the triune brain hypothesis have been strongly criticized (Deacon 1990a). But the evidence that brain development is highly covariant and that brain evolution has been highly constrained (Deacon 1990a, 1997; Finlay and Darlington 1995) implies that one might expect brain ontogeny to recapitulate general evolutionary patterns. Raff (1996) has described such patterns as the result of "phylotypic" embedding, in which ontogenetic traits become entrenched among higher taxa. Traits that develop late in ontogeny are "modularized" and can be modified independently by selection. Traits that develop earlier, during organogenesis, are less liable to modification because modularization has not yet occurred. In this light, the "triune brain" appears to describe phylotypic embedding representing reptilian, early mammalian, and late mammalian brain evolution, respectively.

At a finer level, I have reviewed evidence that the origination and maturation (e.g., synaptogenesis, myelination) of various areas of the brain correspond roughly to their order of evolutionary appearance (see especially Gibson 1990, 1991). Konner (1991) discussed how such brain recapitulation might translate into behavioral recapitulation. Ritualized motor displays found in reptiles are among the first displays to appear in human infants due to maturation of the basal ganglia. In reptiles and higher vertebrates, these are followed by bonding, attachment, and other, more complex emotional behaviors evoked by maturation of the limbic system (Konner 1991). Finally, maturation of neocortical areas initiates cognition, a maturation process that lasts for a substantial part of the individual's life span, especially in humans (Gibson 1991).

This scenario of recapitulation is very general indeed. It focuses mainly on the fact that humans are recent mammals and thus developmentally repeat the mammalian sequence of mental evolution, with neocortical areas being the last brain areas to originate (Deacon 1990a), enlarge (Finlay and Darlington 1995), and mature during development (Gibson 1990, 1991). Do the finer details of development of areas within the human neocortex exactly recapitulate our evolution? Unfortunately, we cannot be sure. I noted earlier that some localized neural deviations from simple size extrapolation do occur, especially as related to language and symbol manipulation (Deacon 1997). In addition, although human cognition rests largely in the prefrontal area, which has increased in size about 202 percent relative to the prefrontal area in an average living anthropoid (Deacon 1990a), the exact evolutionary sequence that produced that increase is unknown. Fossil crania do not preserve the details of neocortical structure. We can speculate, however, that the steadily increasing brain size of human evolution largely translated into a steadily increasing prefrontal area (Deacon 1997). Fossil crania, for instance, show a disproportionate increase of the prefrontal cavity during human evolution (Stanley 1996).

So far I have focused on general recapitulation in brain structure. What about brain function? Cognitive abilities are related to neocortical size and complexity, which permit increasingly complex mental constructions (Gibson 1990, 1991). Such mental constructions include increasingly complex combinations of objects, words, and, ultimately, ideas that are created and manipulated (Gibson 1993). It may be that as our complex brain matures (e.g., synaptogenesis, myelination), especially the prefrontal cortex (Case 1992), our general ability to "think" by manipulating ideas (and objects and words) recapitulates that ability of our ancestors (Ekstig 1994). At a finer level, this is reflected by our increasing ability to perform Piagetian tasks, to produce cognitive recapitulation. Physical evidence for this includes similarities in tool use and stone artifacts made by early humans, living apes, and modern children (reviews in Mithen 1996; Parker 1996b).

HOMINID BRAIN EVOLUTION IN THE HISTORY OF LIFE

A profound implication of this "overdevelopment" scenario of human brain evolution is the surprising ease by which complexity can evolve. As life's most complex organ (Katz 1987), the human brain has always been at the center of heated debates about evolution and progress (Gould 1977; Morss 1990). In addition to promoting the "juvenilization" scenario of human evolution (Gould 1996a), Stephen Jay Gould has spent many years attempting to refute the concept of evolutionary progress by arguing that evolution is dominated by "luck" (e.g., Gould 1996b). The evidence discussed here not only invalidates the juvenilization scenario of human origins but also seems to contradict many of Gould's (1996b) assertions about the overwhelming role of chance in evolution and the improbability of complexity and intelligence as evolutionary products. The relative developmental ease by which our complex brain can be produced seems to argue that the common process of "overdevelopment" (McKinney and McNamara 1991; McNamara 1997) can readily evolve (or re-evolve, given a future mass extinction) morphological complexity and intelligence where environmental selection favors them.

It is also interesting that the evolution of life's most complex organ occurred by developmental changes that are typical of many morphological traits. In his overview of evolutionary developmental biology, Raff (1996) described three basic rules that characterize the developmental evolution of life: modularity, cooption, and dissociation. Modularity—the production of redundant parts—is seen in the increased multiplication of neurons in human brains. Cooption—the acquisition of new functions by these neurons—is seen in the new cognitive abilities acquired by our enlarged brain, especially its prefrontal area. Indeed, a main problem in explaining the selective forces producing hominid brain evolution is that because brain growth is very conservative (covariant), selection for enlargement in one area of the brain may have incidentally produced enlargements in other areas that subsequently became coopted for new functions that were not originally selected for (Nishikawa 1997). Dissociation is seen in the deviated growth or reduction of brain structures that do not follow the same "overdeveloped" trajectory of the bulk of the brain.

Another typical developmental-evolutionary pattern that is followed by hominid brain evolution is that of a *peramorphocline*. As described by McNamara (1997; see also McKinney and McNamara 1991), a peramorphocline is an evolutionary trend in a group that shows a progressive increase in "overdevelopment." The pattern of brain increase from australopithecines to modern humans is a classic example of this. Such trends can of course be driven by a variety of selective forces, often from geographic gradients (McKinney and McNamara 1991). In the case of human brain evolution, it has been suggested that the selective forces involved the advantages of culture, technology, social skills, or symbol manipulation (Deacon 1997), to name but a few suggestions. The point here is that these forces have produced a consistent pattern of progressive extension of brain development in a manner typical of peramorphoclines throughout the history of life.

3

Heterochrony in Brain Evolution
Cellular versus Morphological Analyses

Terrence W. Deacon

One of the great mysteries of animal development is how the cells in mouse embryos and elephant embryos know how many times to divide before committing to fixed cell fates in each tissue in order to produce proportioned adult bodies of vastly differing sizes. Given the incredible difference in scale from zygote to adult organism, it is remarkable that many comparative morphological relationships exhibit highly predictable scaling patterns in animals of different sizes. Such regularities imply that the control of differentiation and growth is highly conserved and tightly regulated even when extrapolated in time and space over many orders of magnitude. Not surprisingly, elephants take longer to grow than do mice, and they also take longer to reach developmental stages comparable to those of mice. Surprisingly, the two animals reach these different end points from highly similar beginnings and by using almost identical mechanisms extrapolated to different degrees during development. Thus, the timing mechanisms underlying development play critical roles in animal design, and their variations are widely cited as likely major factors in morphological evolution. Time, growth, and developmental differentiation of tissues are linked variables, but the mechanisms linking them, the extent to which each is dissociable from the others, and the role these relationships play in the evolution of animal forms remain poorly understood.

When Ernst Haeckel coined the term *heterochrony*, he applied it to either the accelerated or the retarded maturation or growth of an organ with respect to its embryonic developmental timing in an ancestor. He was particularly interested in cases that disrupted true recapitulation as he conceived of it. Although critical of the assumptions behind Haeckel's "biogenic law," which served as the reference frame for the definition of heterochrony, many subsequent writers (such as de Beer, Gould, and McKinney) have argued for the value of retaining the term heterochrony irrespective of his larger theoretical scheme. They have advocated using it more generally, however, to refer to any shift in the timing of developmental events relative to one another as compared with their developmental relationships in some ancestral condition. These more contemporary heterochrony theories echo Haeckel's view that mosaic changes in developmental timing can explain many important trends in morphological evolution, and their proponents have collected diverse examples of organisms that exhibit apparently paedomorphic (developmentally regressive) or peramorphic (developmentally extrapolative) traits in adulthood (for more precise definitions, see the chapters by McKinney and Shea, this volume).

Yet the precise identification of these developmental modifications in different phylogenetic comparisons has not been uncontroversial. Humans, in particular, have been variously described as neotenous, paedomorphic, or peramorphic by different researchers over the past century (and also in different chapters in this volume), often with specific reference to the growth and developmental timing of the brain and brain functions. Many of these apparently incompatible conclusions reflect differences in the ways these terms have been defined over the many generations in which they have been used. Efforts to systematize the use of these concepts in recent years have helped to resolve some of the confusion. Some of the variety of interpretation, however, may also reflect deeper theoretical disagreements about how these concepts should apply to different levels of cellular, morphological, and functional development.

Stephen Jay Gould introduced a widely influential way of conceiving of morphological heterochrony with his notion of linked developmental clocks for size, shape, and age (e.g., time to sexual maturation). This scheme for identifying the quasi-independent variables of morphological change during development has been the starting point for most subsequent analyses. An account of morphological evolution, however, is framed in terms of structures, organs, and developmental stages. It is thus defined in terms of indices of epigenetic results, not in terms of any specific epigenetic mechanisms, and this may be a source of some difficulty. Though framed in appropriate terms for paleontological comparisons, where the data are morphological states (mostly of adults), it may be less useful for investigating the molecular and cellular developmental mechanisms that determine these states. Though it is possible that there are simple correspondences to be drawn between these levels of analysis, it is also possible that what appears heterochronic at the morphological or functional level of analysis does not result from any underlying modification of rates or durations. Alternatively, it is possible that heterochronic reorganization at the level of molecular and cellular development could express itself in whole organism morphological differences that do not obviously admit to a heterochronic description. Unless there are corresponding quasi-independent, clocklike processes at work at all levels, there is no a priori reason to expect that what expresses itself as temporal displacement at one level should exhibit a parallel temporal logic at lower levels.

Brain development offers a particularly useful system in which to investigate this between-levels problem, because research in neuroembryology spans the entire range from molecular to behavioral development. The brain's component cell types and subregions develop through complex sequences of phenotypic states to achieve highly differentiated organization in adulthood. In addition, brain function typically unfolds over time, especially in mammals, for which learning and memory superimpose an additional cumulative process on top of simple cellular and morphological development. To investigate whether heterochronic interpretations of morphological and behavioral development can be reconciled with heterochronic cellular and molecular developmental parameters, I first consider developmental timing processes in a number of central nervous system (CNS) developmental contexts in vertebrates. I then return to the question of whether human brain evolution can be understood as some sort of heterochronic modification of more typical great ape brain development patterns at this submorphological level.

GROWING FROM SMALL TO LARGE

Development and size are complexly interrelated. In order to map heterochronic morphological accounts onto cellular-molecular accounts of development, it is first necessary to disentangle these two aspects of development and to identify major factors influencing their interrelationships.

Though there are aspects of organism design and function that are fractal in nature—that is, are expressed in a self-similar or isometric way in both small and large organisms[1]—there are also many nonisometric influences imposed by differences in scale. Many features of organism design scale allometrically with respect to one another. The changing structural requirements of physical support affected by geometric scaling of lengths, surfaces, and volumes are responsible for some of the better-known allometric relationships observed in animal morphologies. These constraints on physical support mechanisms, cellular physiology, metabolic resource distribution, and information processing are highly predictable correlates of size, irrespective of other adaptational variations. This is true even at the microstructural level. For example, within vertebrates, the sizes of genomes, molecules, and cells are remarkably similar, irrespective of vast differences in overall organism size. One obvious consequence of this is that smaller organisms have organs that are inevitably "grainier" at a cellular level and are likely to be more undifferentiated as a result. This produces a kind of allometry of differentiation. Embryogenesis involves both an increase of size, with the addition of cell numbers, and a correlated increase in the degree of cellular and tissue differentiation. Small adult species and small embryos of larger species may thus share resemblances due both to the paired allometries of structural-functional geometry and to the degree of cellular graininess and differentiation.

But in addition to this, simple differences in size place animals in very different adaptational contexts. The texture, or "graininess," of ecosystem variations and major environmental fluctuations differs radically for animals at different extremes of size and life span. For example, the viscosity of water or air is very different for small and large species and affects their locomotor apparatuses and streamlining differently.[2] Larger animals are much more resistant to variations of temperature and food availability. And long-lived animals are potentially exposed to far more seasonal and spatial variations, though each phase occupies a much smaller fraction of their total life span. In terms of CNS functions, then, there are corresponding differences in functional demand with respect to overall size. Larger animals typically take longer to mature, are subject to lower infant mortality, have a greater range of experiences, and have a greater opportunity to profit from trial-and-error learning than smaller species.

Finally, growth to large size is a more complicated matter than just growing bigger. All animals begin life as a single cell, and so large animals inevitably have to pass through the size ranges of smaller species on their way to a mature state. Each individual that grows to a large size must temporarily pass through physiological and environmental conditions that correspond to those confronted by other, smaller species in adulthood. They must therefore adapt at least partially to each of these diverse contexts, and they must do so in a specified order from one stage to another in the life cycle. Many of the variations on the theme of assuming distinctive morphologies at different phases during development, punctuated by metamorphosis, reflect these scaled selective forces.

In addition to the fact that large species must develop through size ranges occupied by smaller species, all large species ultimately evolved from smaller ones. The last half-billion years of evolution have been characterized by an increase in the size range and the upper size of most metazoan animal forms, particularly vertebrates. Because of this evolutionary bias with respect to size, all species had ancestors with smaller adult sizes, and few had ancestors with larger bodies. This means that during the last half-billion years, selection on features relevant to small body size was essentially ubiquitous, whereas selection on features relevant to large body size was both less common and less extensive. It also means that different species' adaptations to the demands of large body size are far more likely to diverge and also to be homoplastic (for example, the result of convergent or parallel evolution in different lineages) as opposed to synapomorphic (the result of shared primitive traits). So across species, adaptations to scale-correlated constraints should be ontogenetically entrenched at the smaller scale and more diverse and flexible at the larger scale.

These are all powerful sources of embryological parallelism that will generate apparently recapitulatory patterns irrespective of any intrinsic developmental constraints. It is important, then, not to jump to the conclusion that what appear to be highly conserved embryogenetic mechanisms reflect developmental processes that are in some way intrinsically constrained. This is an important caveat that needs to be considered when mapping developmental mechanisms to phylogenetic patterns, because ignoring it increases the probability that we will mistake homoplasy for synapomorphy with respect to early embryogenetic events.

Conversely, this caveat has a bearing on the probability that linkages between different aspects of developmental timing could have evolved. Adaptive radiations might be aided by the availability of internalized ontogenetic information about size-correlated functional constraints, but only if it were possible to dissociate developmental processes to express these adaptations outside of their typical developmental context. Theoretically, these developmental trends could provide an important reservoir of preadaptations. Many examples of morphological variation among living and fossil species appear to exhibit dissociations of developmental events. The question in each case is whether this reflects modification of the respective developmental processes or an independent adaptation whose phenotype is merely convergent on the morphology of a particular developmental stage for extrinsic reasons. Mutations that modify the relationships between developmental mechanisms are likely to produce numerous other linked morphological and physiological consequences as well, whereas merely convergent adaptations should produce more superficial and mosaic juxtapositions of traits. So although reutilization of existing developmental mechanisms in novel ways may be hit upon more easily by random mutation, there may be costs imposed by correlated traits that cannot be dissociated. The evolutionary trade-offs between these two pathways to adaptation depend on the extent to which the timings of different developmental events are linked and on the flexibility of the developmental clock or clocks that regulate them. The weaker the linkage and the more flexible the timing mechanisms, the more opportunities there are for development to offer a palette of options for adaptation.

GROWTH VERSUS DIFFERENTIATION

Though it may be possible to determine the extent of this flexibility of timing mechanisms simply by comparing development in different species, comparing superficially similar morphological traits may obscure deeper dissimilarities. To determine whether there are specific developmental mechanisms that correspond to the clocklike processes that define heterochrony, we need to be able to translate the classic morphological analysis of shape, growth, and time-to-maturity into accounts phrased in terms of molecular and cellular biology—that is, into temporal and regional patterns of gene expression, molecular communication, cellular differentiation, cell migration, and metabolic and mitotic rates. I suggest that there is no clear correspondence between molecular-cellular descriptions of development and morphological descriptions of evolution, and yet there may be a partial translation of the concept of a developmental clock into the molecular-cellular domain that preserves some aspects of its explanatory power for evolution.

We now are entering a new phase in our understanding of development at the cellular and molecular level. Many previously mysterious developmental processes can now be reduced to the level of molecular events. Unfortunately, the historical development of the heterochrony concept long preceded any of these insights, and so heterochrony was primarily conceived in terms of organism-level morphological descriptions. The major consequence of this is that the dimensions used to define heterochronic relationships bear only a superficial resemblance to any of the cellular-molecular developmental mechanisms that must underlie them, even though all are expressed across the same developmental time course.

One of the principal methodological tools used to delineate and classify forms of heterochronic species differences is the allometric comparison of relative growth trends of different body structures (e.g., Alberch et al. 1979; Godfrey and Sutherland 1996; Gould 1977; McKinney and McNamara 1991; Shea 1988). This methodology assesses growth in terms of the measurement of some weight, length, or volume of a given structure (e.g., brain weight, long-bone length). It assesses shape in terms of some changing morphological index (e.g., relative proportions), and time in terms of real clock time or else some surrogate for reaching a given stage of development (e.g., eruption of teeth, sexual maturity). What is the relationship of these terms of heterochronic description to the underlying mechanisms that produce these patterns? It turns out not to be obvious.

Translating descriptions of morphological change into descriptions of molecular and cellular processes was not, of course, a possibility until very recently. With the advent of numerous methods for tracing cell lineages and for mapping the time and place of the expression of distinct molecules within a developing embryo, these processes can now be approached in at least a limited way. Methodological and theoretical differences, however, become obvious as soon as we attempt to compare the basic processes of embryogenesis with their morphological consequences that are used to analyze heterochronic relationships between species. Comparative morphological descriptions and cellular-molecular descriptions are not concordant. They analyze development with respect to very different kinds of dimensions, and it is not clear that there can be a descriptive translation that preserves the categorical distinctions implicit in each analysis. Though this is not a reason to abandon any of the useful features of heterochronic analysis, it may mean that there will be strict limits on its predictions of processes occurring at the lower level.

In describing embryogenesis at the cellular level it is useful to distinguish between *differentiation* processes, which determine regional differences and subdivide the whole of some developing tissue into collections of cells with divergent developmental fates, and *growth* processes, which consist of mitotic cell multiplication and synthesis of nuclear and cytoplasmic constituents and which produce quantitative increases in cell numbers, cell sizes, and body structure sizes. Of the three dimensions of developmental change (age, growth, and shape) modeled by Gould's clock metaphor of heterochrony (variants of which are now widely used as the basis for most heterochrony accounts), only growth has a clear correspondence with embryonic cellular development, and even that relationship is ambiguous to the extent that cell differentiation processes can alter cell growth and mitotic cycling. So our first step must be to consider how each of these macromorphological variables might map onto lower-order epigenetic processes.

Total body growth is largely determined by the number of mitotic cycles that characterizes the average cell lineage of the body from zygote to adult, though cell enlargement can also play a minor role. For many cell types, there is a modest positive correlation of average cell size with total body size in vertebrates, but there is considerable variation in this trend in different tissues and cell types. For example, in the brain, increasing body size is associated both with increased size of the largest neurons and with increase in the range of cell sizes, because many of the smallest cells in large and small brains are comparable in size. The molecular machinery of cell metabolism undoubtedly sets an upper limit on cell size that is approached asymptotically. One obvious source of such a limit is the surface-to-volume relationship. To the extent that the ratio of cell surface to cell volume is critical to both the function and structural integrity of a cell, the exponentially diminishing surface-to-volume ratio with increasing size imposes costs on large cell size that probably sets an asymptotic upper limit. In addition, there is a size-correlated reduction in the rate and therefore the efficiency of intracellular molecular communication, transcription, and synthesis processes that are dependent on molecular diffusion.

Increase in the number of mitotic cycles from fertilization to the point at which cells reach mature phenotypes is thus the predominant determinant of adult size. The rates of mitosis and cell growth, however, are linked in complex ways. Metabolic support plays a direct role in regulating the rate of both cell growth and mitotic cycling in some species. In animals that face considerable fluctuations of metabolic rate during development (e.g., in response to external fluctuations of temperature, oxygen, and nutrients), mitotic growth may vary considerably as well. In contrast, in mammals the protected environment of uterine development mostly insulates early development from metabolic variance (except in unusual circumstances), and so both cytoplasmic growth and mitotic activity can be largely internally regulated. Nevertheless, intrinsic constraints on mitotic rates appear to be correlated with size. Within the cerebral cortex of mammals, for example, the average length of the mitotic cell cycle appears to be longer in larger mammals and slows later in development (Finlay and Darlington 1995). This may be a reflection of the negative allometry of basal metabolism with increasing size and the corresponding reduction of energy available to each cell, but the functional linkage remains unclear.

The rate of cell division, the rate of synthesis of new cytoplasm, and the amount of time that elapses from mitotic to nonmitotic stage in a cell lineage all play roles in determining differences in relative sizes of adult body structures. The existence of similarly organized body plans in animal species of very different sizes demonstrates that the timing of cell differentiation across species is not strictly tied to numbers of mitotic divisions since conception. The corresponding intermediate and terminal differentiation states that are reached by cells after relatively few divisions in small animals are reached after many more divisions in larger ones. There does, however, appear to be some rubber-sheet-like regularity in the *relative* spacing of cell fate transitions with respect to mitotic cycles, so that the spacing in time of these events is greater in larger species. In the case of cell lineages that produce nonmitotic phenotypes, differentiation controls mitosis because cell differentiation halts mitotic activity. But conversely, differentiation mechanisms that depend on concentrations of intercellular signaling molecules may also depend on cell numbers, since concentration gradients across intercellular distances may determine threshold effects, and so reaching some concentration threshold may be a function of the number of prior cell divisions.

One critical question for completing the analysis of morphological growth in terms of cellular multiplication is whether growth and differentiation are merely the sum of many independently regulated processes in different tissues throughout the body, are globally regulated by hormonal or growth-factor effects that are diffused throughout the body, or are controlled by a sort of mitotic "clock" synchronized in all cells at a point shortly after conception and maintaining this preestablished synchrony independent of specific context within the many different cell lineages of the developing body. To the extent that the number of cell divisions prior to cessation of mitotic growth is an independent function in each cell lineage, growth processes will play a major role in determining body proportions. To the extent that it is more globally determined (or predetermined), parcellation processes will be more important. There are now some intriguing hints that the latter may be predominant in mammals and possibly in most vertebrate lineages. If so, there will be significant implications for heterochronic explanations of developmental differences.

In contrast to growth, age and shape are far more difficult to translate into cellular and molecular terms. In morphological accounts of development, age is typically defined as time to some corresponding point in the life span (e.g., to cessation of growth, to sexual maturation, or to maximum life span), to the extent that such developmental homologies can be independently determined. In cellular terms, age can be represented by differentiation status—for example, time to expression of some cellular marker of differentiation, such as final cell division in a cell lineage destined for a nonmitotic fate, such as a neuron, or the onset of expression of some receptor molecule that signals mature cellular functions, such as a neurotransmitter synthetic enzyme. Life history events such as sexual maturation can also be described in terms of cellular differentiation processes such as the hormonal activation of mature cell phenotypes correlated with the onset of fertility or the appearance of secondary sex characteristics. But these may be special cases that should be treated as separate from background epigenetic processes to the extent that they show a degree of independence from other growth and differentiation processes due to responsiveness to external signals.

Hormonally regulated differentiation may be important for controlling the differentiation of classes of cells that need to be sensitive to environmental conditions to some degree. The onset of reproductive function, for example, may need to be independently fine-tuned to match food resources or factors such as temperature, humidity, or light period. It may even need to be sensitive to the presence of a potential mate. Hormonally regulated differentiation processes are typically terminal cell-fate decisions that are held in check until the body receives the necessary external "all clear" signal to synchronously activate receptor-bearing cells in widely different parts of the body in order to produce the new global phenotype. As such they are the special exceptions to the general ontogenetic plan, in which the control of mitotic activity and the timing of differentiation events during development proceeds independently of external environmental factors, except as limiting resources. Hormonally regulated epigenetic transitions effectively define the facultative "joints" between distinct, functional life history phases. So it is not surprising that hormonally triggered epigenetic transformations, such as those that control metamorphic transitions in many insects or amphibians, have been the basis for many classic examples of heterochronic species differences. The twin requirements of environmental tracking and partial independence from background growth and differentiation processes make hormonally regulated epigenetic transitions the most susceptible loci for heterochronic developmental differences to evolve. Because of this, however, they may be misleading as exemplars of developmental timing mechanisms in general.

Shape, too, is a complex developmental result that cannot be mapped to any single differentiation or growth mechanism. In general terms, shape is a function of the relative positions and sizes of body segments and organs, and sectors of these, with respect to one another. Regional differentiation is what ultimately determines the boundaries between body structures and organs, but differentiation is analytically orthogonal to growth. It results in *parcellation* of a larger structure into smaller structures. Thus the relative size of any given structure is not just a function of growth but rather a function of the interaction of growth, differentiation, and parcellation. How much each of these mechanisms contributes is seldom obvious, except by detailed microscopic analysis. Moreover, parcellation processes that are critical to development have no straightforward counterpart in the variables used to investigate heterochrony. This complication is further exacerbated by the fact that some parcellation processes are mediated by regressive events—that is, reductions. Since classic categories of heterochronic species differences focus on rates of growth and timing of differentiation processes, species differences in parcellation processes must be taken either as unanalyzed givens (that is, assuming unchanging body segmentation, such as limbs versus trunk) or as surrogates for "shape" change. Like growth, parcellation processes occur over time, which further complicates the process of distinguishing them operationally. For the most part, parcellation tends to be a remarkably conserved process that is completed at relatively early stages of embryogenesis, and so it may be taken as a shared initial condition. Ignoring the possibility of early parcellation differences can, however, be a source of serious confusion.

The relative positions of organs and body segments are almost entirely a consequence of parcellation of cell lineages that subsequently track different differentiation trajectories. But it is not only position that is determined. The ultimate sizes of different organs are significantly influenced by the ways in which this initial parcellation interacts with subsequent mitotic and cell growth processes within each of these cell lineages. Though early parcellation processes are highly conserved, even slight differences in the initial segmental populations can become extrapolated to appear as if these cell populations are dividing at different rates, as opposed to merely starting from different initial population sizes.

To complicate matters, eliminative or "regressive" processes such as "programmed" cell death can also play a significant role in these early parcellation processes by differentially "sculpting" away certain local cell populations with respect to others. Such regressive developmental processes exhibit features of both growth and parcellation. Cell loss reverses the effects of mitotic growth and may suggest an apparent modulation of growth rates within a given region, when in fact two distinct processes may be involved. Regressive processes can also be critical for parcellation. For example, by eliminating intervening regions (as in finger digit formation), "programmed" cell death can provide natural boundaries that separate cell lineages and promote lineage-specific differentiation. Regressive processes are, in effect, secondary parcellation processes that provide a means of fine-tuning both growth and parcellation.

Incorporating the role played by secondary parcellation processes into the analysis of development makes the comparison of morphological heterochrony to cellular-molecular heterochrony even less clear-cut. Recognition of the importance of normal regressive processes is relatively recent, but it turns out to be of considerable significance. Developmental cell death mechanisms are "programmed" to the extent that they are not pathological but rather essential elements of normal development, without which a functioning adult could not be produced. In many ways, however, they are not the result of cell-intrinsic genetic "programs" but rather are the expression of cellular-level adaptive mechanisms. Regressive processes of this sort offer a substantial "savings" in the amount of genetic information necessary to control the development of complex structures from initially undifferentiated beginnings. Regressive fine-tuning is regulated by a sort of post hoc Darwinian selection that recruits "free" contextual information to contribute to the determination of patterning. Not surprisingly, this contribution is greatest in the development of systems that require a very high level of structure, such as the central nervous system, where, for example, cell death is used to fine-tune the relationships between functionally linked neural cell populations in different separated regions. Overproducing and then culling cells on the basis of their interactions produces matching cell populations from initial populations that were independently extrapolated by mitotic growth processes in very different systems.

This is quite dramatically demonstrated in the case of the development of connections between motor neurons and their target muscle fibers in vertebrates. Initial production of spinal-cord motor neurons occurs in excess of the number of target muscle fibers available in the trunk and limbs. During development, muscle fibers initially may receive axonal inputs from many motor neurons. As the muscle fiber matures, it progressively eliminates all but those connections originating from one single neuron. When, inevitably, some motor neuron loses all links with targets in this way, it appears to initiate a spontaneous self-destruction process called apoptosis. If, however, an additional limb bud is grafted onto a developing embryo before this competition begins, the extra targets provided by its muscles enable many more neurons to survive and mature.

The importance of programmed cell death for influencing apparent growth has been particularly dramatized by the development of a transgenic mouse that does not express a protein, CPP32, that is critical to the cascade of cellular events leading to apoptosis. These CPP32-deficient mice develop huge brains that cause the fetal skull to bulge. Numerous structural relationships within their overgrown brains are abnormal, with large, disorganized, and sometimes even doubled structures, leading to disastrous consequences.

Thus, the actions of programmed cell death can mimic either differentiation or growth processes in the results they produce. Cell death is also more context sensitive than either growth or parcellation, because of its post hoc implementation and the roles played by surrounding conditions. This makes such regressive processes relatively more likely mechanisms to get recruited to produce species differences in development.

In summary, shape change in development is a crude index of the interactions of growth, parcellation, and regressive processes, and as such it is a proxy index that confounds all three. To the extent that the underlying relationships among mechanisms remain unanalyzed in the assessment of shape change, there is a danger of incorrectly assuming that shape is locally regulated and independent of these more global processes. In any quantitative morphological analysis of shape change during development, there is no a priori way to discern which process is of primary importance and which processes are of secondary importance. With respect to the relative timing of morphological change during development, then, merely tracking the divergence in time of some index of shape (e.g., segmental proportions) is insufficient to resolve true heterochronic transformations, such as paedomorphosis or peramorphosis, from shifts in the differential contributions of these three developmental mechanisms extrapolated over the growth period.

These classes of cellular-molecular processes do, however, exhibit a degree of temporal asymmetry across ontogeny and so at least have the potential for exhibiting heterochronic variations. Segmental parcellation of major body divisions and organ systems by cell-lineage differentiation tends to occur relatively early in embryogenesis, while the embryo is still very small, whereas mitotic growth is ubiquitous, continuing throughout the lifetime in many species. And because mitotic growth is a multiplicative process, its largest effects tend to get expressed comparatively later in growth. Cell death and other forms of secondary parcellation and reduction of mass also tend to be more significant in early phases of development. As a result, major differences in comparative adult body proportions may become evident only at late developmental stages and yet be the result of mechanisms activated much earlier in development. Though determined early in embryogenesis while the embryo is very small, parcellation differences can become vastly extrapolated over many orders of magnitude by mitotic processes.

The development of the central nervous system exhibits both forms of parcellation, though the role of regressive parcellation processes appears somewhat exaggerated with respect to other regions of the body. The CNS is initially partitioned into distinctly differentiated segments arranged along the early neural tube, like segments of a worm, with characteristic restricted cell fates in each segment. This is followed by considerable cell mitosis and subsequent subdifferentiation into mature cell types (neurons and glia). But there are significant features of brain architecture that result from late-occurring cell death and structural elimination processes including apoptosis, axon invasion, competition for synapses and trophic support, and connectional pruning. Untangling these many developmental threads in order to test the claim that a general change in one or two rate variables could account for species differences such as those between humans and other great apes is no simple matter.

EMBRYOLOGICAL PARCELLATION OF THE BRAIN FROM THE REST OF THE BODY

One of the most highly conserved features of epigenesis is the parcellation of the major body segments during very early stages of development. Regional specialization of the body and organogenesis can be traced to these early epigenetic events. Biologists have long been aware of morphological evidence suggesting that mammals share a highly conserved embryogenetic plan with many other animal phyla—even beyond vertebrates. Over the course of the last decade, powerful new cellular and molecular evidence for this commonality has been discovered. The major mechanisms that determine body axes, segmental subdivision, and within-segment patterns of cell fate determination depend on complex molecular-genetic circuits in which DNA-binding gene products turn other genes on and off, en masse, like orchestras of genes. These early-expressed genes have been dubbed *homeotic* genes, after William Bateson's (1894) description of animals in which whole body segments had been displaced or duplicated as a result of what he called homeotic mutations. Interactions among these genes set up complex gradients, thresholds, and semi-repeating patterns of molecular expression throughout the embryonic body as a consequence of their competitive and cooperative activation and inactivation of one another, as well as of other genes. These regional expression patterns establish segmental territories in which patterns of cellular differentiation are orchestrated in parallel so that each segmental region contains similar structures (e.g., vertebrae). These interactions take place via cell contact and diffusion of regulatory signaling molecules (including homeotic gene products themselves). The patterns that result appear to be a function of the relative concentrations and timing of production of these molecules in different parts of the embryo.

The discovery of homeotic genes and their functions has opened a new window onto the study of development and has revealed some totally unexpected phenomena. One of the most remarkable features of this class of genes and their expression patterns during development is their similarity in species from distant phyla. There is cross-phylum similarity in the structure of these genes that is so close that fruit fly and roundworm homeotic genes have provided the major probes for discovering homologous vertebrate homeotic genes. In addition, their temporal and positional expression within embryonic bodies in different taxa is also highly similar. For example, in both fly and mammal embryos the homologous HOM/Hox genes are expressed in the same serial order along the body axis (see review by Holland, Ingham, and Krauss 1992). Their role in tissue determination is demonstrated by the production of abnormal segmentation of the developing body when these genes are blocked or expressed ectopically (e.g., Dolle et al. 1993). Transgenic experiments that insert genes from one species into another at an early embryonic stage have further proved that both structure and function are conserved. Indeed, insertion of human homeotic genes into cells of mutant fly embryos where the corresponding genes are damaged can restore lost function and partially "repair" the embryos' structural defects (McGinnes and Kuziora 1994).

The conservatism of these genes and their functions during development probably accounts for the extensive similarities between developing embryos, especially at early stages, that a century ago suggested the theory of recapitulation. Subtle differences in gene structure or expression patterns may also explain some of the structural differences that distinguish the major animal groups. At the present time, however, the comparative analysis of these genes' expression patterns is very limited, and differences in homeotic gene expression that might explain differences in ontogenetic development in the major animal phyla have been only minimally explored. Nevertheless, major changes in the segmental topology of vertebrate brains, such as those that distinguish the major vertebrate classes (e.g., teleosts, birds, mammals), may correlate with differences in homeotic gene expression.

Among vertebrates, the major segmentation phase that determines axial patterning begins just after gastrulation, with the formation of the neural tube. In a relatively short period with respect to the remainder of gestation, embryos complete the differentiation of all the major divisions of the body and nervous system. Because homeotic genes act via cell contact and diffusion of molecules, there is almost certainly an upper limit to the size of the cell population that can be controlled as a whole by such molecular mechanisms. I suspect that this upper limit is approached by the largest vertebrate embryos. Though different species' embryos vary in size at comparable developmental stages, both the smallest and the largest species of animals progress through this major body segmentation phase while their embryos are still quite small. New cell classes and organs are seldom added in vertebrate evolution subsequent to this initial parcellation of the embryonic body, and all further cell differentiation occurs within the constraints laid down at this early stage.

Adult body proportions and the relative sizes and positions of major organs reflect the ways in which subsequent growth has extrapolated this initial partitioning of the undifferentiated body. Because the underlying molecular partitioning mechanisms must operate while the embryo is still quite small, subtle discrepancies in parcellation of different structures may become increasingly exaggerated in animals of progressively greater adult size. Consequently, all other things being equal, smaller organisms within a given order should more closely resemble one another in segmental proportions than larger ones.

Paleontological evidence from the Cambrian period suggests that most of the extant metazoan phyla (along with some that are now extinct) emerged during or just prior to the first flowering of complex body plans. One of the prominent features of this evolutionary watershed was the utilization of a segmental strategy for building large, complex bodies. By reusing the same genetic information repeatedly in slightly different combinatorial patterns to build something approaching interchangeable parts, a remarkable diversity of body organizations was possible with relatively minimal genetic reorganization. This initial diversification of novel animal body plans may reflect the "discovery" of this homeotic segmentation trick and the founder effects produced by the success of a few variations on it.

But why have no new animal body plans been added since this first evolutionary experiment with the homeotic trick? One possibility is that modifications of homeotic gene expression patterns are likely to be lethal because they tend to disturb early-stage ontogenetic events and leave an organism without some critical functional component. This echoes a classic speculation that because late developmental events depend on earlier developmental events, the earlier ones are under more intense selection and will thus be more conservative in evolution (e.g., de Beer 1930). It could also reflect a kind of metastability of these mechanisms that is supported by highly distributed homeostatic relationships. And it might also reflect simply the remarkable flexibility of this strategy, which allowed a few founders to enjoy enormous advantage over any later alternatives.

This body construction strategy is the source of a slight additive tendency in evolution as well. Segmental addition or deletion of redundant body segments or supernumery organs (both of which are typical consequences of homeotic mutations, e.g., changing numbers of limbs, digits, or vertebrae) may be less debilitating because of this redundancy. So although the metabolic expense of redundancy ultimately favors selection to eliminate simple duplication, this leaves open the possibility of additive changes as a source of evolutionary innovation. Though these are additions in structures rather than "terminal" additions in time, they may provide some basis for rescuing an aspect of the classic, if fallacious, accretive view of evolution.

GROWTH AND PARCELLATION PROCESSES IN CNS DEVELOPMENT

Numerous examples of a correlation between size and developmental features of the CNS are known in which small species' brains resemble the immature brains of larger species. An interesting example is provided by comparisons of the brains of crayfish and lobster species (Helluy et al. 1993). Many aspects of crayfish adult morphology are paralleled by lobster immature morphology, including, for example, the extent of limb differentiation. The smaller brains of adult crayfish also resemble the brains of immature lobsters in a number of respects. For example, the stages of development of the major paired olfactory analysis organs (deuterocerebrum) in the brain of the lobster closely parallels the interspecies allometry of this brain structure in the wider clade. These ganglia grow larger and become more complexly structured with respect to the rest of the brain during development in large species, but they also exhibit these trends in comparisons of smaller to larger species. The question is whether this size-correlated pattern in both ontogeny and phylogeny can best be understood as a truncation or extension of a series of distinct developmental stages in these related species or rather as the expression of a common embryonic parcellation processes merely extrapolated to different points via mitosis.

Though macro-allometric trends in global organism design, such as those exhibited by metabolism, brain size, support structures, and so on, have been widely investigated, micro-allometries have generally been overlooked until quite recently. D'Arcy Thompson (1917), one of the first great pioneers in the study of allometric effects, recognized numerous microscopic allometric correlates of size. For example, very small animals do not need circulatory systems or nervous systems to guarantee distribution of nutrients and information. Microscopic organisms can also get by without special physical support structures such as endo- or exoskeletons. And the type of structure that most effectively provides support and mobility differs radically with scale. Even external factors such as the relative viscosity of media such as air, water, and soil are much different for organisms at different size ranges and demand different locomotor strategies. But despite the fact that these multidimensional effects of scale have been recognized for nearly a century, there is still little detailed study of the relationships between these different levels of allometry.

The organization of brains reflects many levels of allometric effects that cannot be easily partitioned. Unlike the physics and geometry of support, heat dissipation, or force production, which inform much of the study of gross structural allometries, factors affecting the cellular architecture of brains are poorly understood. Allometries associated with molecule size, chemical diffusion, limits on cell size, neuronal metabolic demand, and even information processing itself in vastly different-size neural networks likely introduce independent correlates of scale that influence the differences in design of small versus large brains (see fig. 3.1). For example, mammalian brains differ by roughly three orders of magnitude in size from the smallest to the largest brains and yet probably differ by less than a single order of magnitude in neuronal cell body volumes. This results in an increasing disparity between the volume of a given brain structure and the volumes of the cells that comprise it. Small brains thus exhibit a greater cellular "graininess," and individual neurons inevitably must contribute a larger fraction of each neural computational process. Larger brains will consequently have structures that are more clearly laminated, more highly nucleated, more subdivided, and so forth, simply as an artifact of this cell-size versus brain-size allometry.

Another microscopic scaling trend that changes with size is the increase in axonal length required to span larger distances in larger brains. Signal transmission over greater distances tends to degrade and to take more time as well as more energy. This requires that neuronal cell bodies must be larger to support the greater metabolic demand of long axon trees and that the axons must be encased in myelin of greater thickness to increase conduction velocity and reliability. The secondary consequences of these size-correlated demands on axonal conduction ramify throughout the design architecture of larger nervous systems. Among these secondary effects are decreasing neural/glial ratios within gray matter (because proportionately more glia are required to support larger neurons) and decreasing gray/white matter ratios in the whole brain (due to increasing distances and increasing myelination).

These scale-correlated architectonic differences also take longer to be expressed in development in large brains, perhaps giving the misleading impression that they reflect late-expressed design information as opposed to merely extrapolated allometric processes.

In an effort to relate developmental constraints to large-scale quantitative morphology of brain structure, Finlay and Darlington (1995) recently compared the long-recognized predictability of major brain structure proportions across mammal orders to data on the timing of neuronal "birth dates" in species that differ considerably in size.

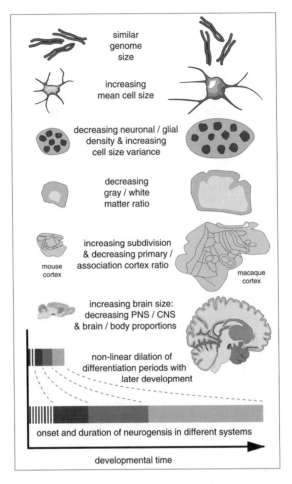

FIGURE 3.1.

Cartoon depictions of some major trends of cytoarchitectonic allometry in mammal brains. The top pair of images represents the lack of correlation between brain size and amount of genetic information available to build a brain. Below these are size-correlated trends in cellular and gross brain structure. For example, the depictions for increasing architectonic subdivision show flattened maps of the mouse (left) and macaque (right) cortex (not to scale), with lines designating discrete cortical areas. All of these trends likely reflect the interrelationships between cellular metabolic allometries, information-processing allometries, and the unequal effects of extending the period of preneuronal mitotic growth on early and late differentiating cell classes. It is argued that all may be pleiotropic consequences of changes in the developmental clock to extend total mitosis. (Summarized from diverse sources listed in Deacon 1988, 1990a, 1990b, 1997; see also Blinkov and Glezer 1968.)

(Note that because neurons are terminally differentiated cells, we can track their production in terms of the last cell division of precursor cells.) These authors demonstrated that, except for a few regions, the timing and duration of neuronal generation predict the relative volumes of adult brain structures. In summary, the later that neurogenesis commences in a region *or in a larger brain*, the longer the duration of its continuation and the larger the resulting structure (and vice versa for early neurogenesis and small brains). The duration-size correlation is not surprising, because producing more cells (assuming the same mitotic rate and cell size) will produce a larger final population of cells. But the correlation with time of onset of neurogenesis is more interesting.

Neurogenesis is the end of the line of a dividing stem cell lineage. Assuming similar rates of stem cell mitosis at any given time, the longer a dividing cell population takes to reach this period of penultimate differentiation, the more precursors will have been produced, and consequently, the larger the resulting terminally differentiated neural population (i.e., brain structure). Starting with similar initial populations, two classes of stem cell lineages with different times to terminal differentiation will produce different numbers of neurons proportional to the differences in these times. These cellular developmental allometries are, however, more complex than this simple model suggests, because later neurogenesis and neurogenesis in large species both tend to be slowed. So large-brained species have more mitotic cycles prior to final neuronal production, have more prolonged production of neurons and glia, and have generally slower mitotic cycling than do smaller species. This little-understood allometry of cellular timekeeping in the development of the brain probably holds the keys to the logic of brain design and underlies much of the allometric and developmental regularity that is observed in brains.

Another correlate of late development in vertebrate brains of all sizes is that the latest neurons to be generated tend also to be relatively smaller neurons—typically interneurons with local circuit connections. The very last mammalian neurons to be generated are the tiny "granule" cells of the hippocampus dentate gyrus and the cerebellar cortex, which continue to be produced long after birth. (Despite having the same name, these are entirely different cell types that share the features of small cell bodies and late development.) One possible explanation for this later-smaller correlation could be that a neuron produced relatively early will add new cytoplasm for a longer period, while other stem cells are continuing to divide, and a neuron produced late will thus begin with less volume and grow in size for a shorter period before time runs out.

There is also another relevant correlate of late development in vertebrate brains that has direct functional consequences. In individual adult brains, there is an apparent correlation between a neuron's soma volume and the size of its axon tree. For example, thalamic neurons that have dense, highly topographic projections to a limited area (e.g., principal nuclei projecting to primary sensory cortices) tend to have small, closely packed somata (comprising parvocellular nuclei), whereas those that tend to project widely over a large expanse of cortex (e.g., interlaminar nuclei and some nuclei projecting widely to association cortex) tend to have larger, more loosely packed somata (comprising magnocellular nuclei). Similarly, within the cerebral cortex, the largest neurons are the cortical pyramidal cells that project to the spinal cord, and the smallest neurons are the local circuit neurons (e.g., cortical granule cells) that receive specific thalamic inputs and redistribute them to very localized regions. This soma size–axon extent relationship almost certainly reflects the immense metabolic demand of supporting a structure (the axon) that is essentially all surface area and that requires energy-hungry molecular transport mechanisms and continuously active ion pumps. It may not, then, be merely coincidental that long-projection neurons tend to be produced relatively early and local-circuit neurons tend to be produced relatively late in ontogeny. Larger neurons should be able to grow and support more extensive axons. (This hypothesis is further supported by xenografting experiments discussed later.)

The linkage between neuronal fate determination and developmental timing is perhaps best exemplified by the determination of neuronal phenotypes in the cerebral cortex of mammals. During the development of this laminated structure the sequence of development appears to be an important determinant of both cell position and cell function. In the earliest stages of neurogenesis there are two layers of neurons comprising the immature cortex (fig. 3.2, left) that appear to get eliminated later in development. These pioneer neurons develop

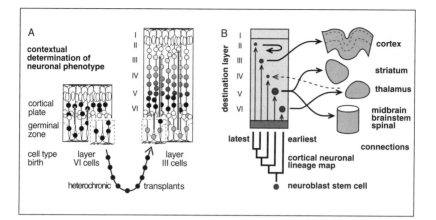

FIGURE 3.2.

How temporal order of cortical neuron production determines laminar phenotype and connec-tions. A: The relative point in development in which neurons are born (i.e., last cell division) determines which layer of cortex they will migrate to, with latest-born neurons migrating to the most superficial layers. Transplanting dissociated cortical neuroblasts into a cortical slice cul-ture at a later stage of development (heterochronic transplant) causes most cells (except those at late stages of final mitosis at time of transplantation) to assume the host stage phenotype and migrate to the appropriate layer. If they have already undergone a last division, they retain their identities but otherwise appear to get their instructions from their surroundings. B: The layer in which a neuron resides correlates with its distinctive connectivity patterns. Thus, time of neuronal cell fate commitment is a critical determinant of axonal affinities as well as migratory tendency. (Data summarized from McConnell 1988.)

functionally active synaptic connections with the first afferents to reach the cortex from the midbrain. The stem cells that will later give rise to most, if not all, of the mature cortical neurons reside in a germinal zone below these pioneering cells. The immature cortex also is spanned vertically by radially oriented filamentous fibers of glial cells on which neurons can migrate out of the germinal zone after they are born. By this route they fill in the space *between* the two pioneering cell layers. As this sandwiched region (called the cortical plate) fills up with cells produced by neurogenesis from below, it does so in a bottom-to-top pattern, so that the neurons that are produced later pass through the layers of neurons produced earlier. Neuronal birth date thus translates into laminar position within the cerebral cortex. This is significant because, as shown on the right of figure 3.2, laminar position is the best predictor of later connection patterns. Thus, layer VI neurons invariably project to thalamus; layer V neurons invariably project to cortical, striatal, and brain stem–spinal sites but not thalamus; and upper-layer cells are mostly intrinsic. Moreover, embryonic cortical neurons exhibit these layer-specific affinities even when co-cultured outside the brain. In contrast, differences associated with tangential position in cortex (which translate into areal differences in function, such as between sensory and motor cortex) do not exhibit these layer-specific affinities. Thus, although we mostly focus on tangential areal differences when comparing species' functional specializations of cortex, areal differences appear not to be strongly prespecified by cell lineage or developmental timing, whereas layer differences determined by neuronal birth date are prespecified this way (Molnár and Blakemore 1991; Yamamoto et al. 1992).

But time is not the sole determinant of laminar position and functional differentiation of cells in each cortical layer. What matters is which cells have been born previously. This was demonstrated in an ingenious experiment by McConnell and colleagues (reviewed in McConnell 1988). They cultured fetal ferret cortical slices and "crossfostered" cells from the germinal zone of a cortical slice at an early stage of development to the germinal zone of a slice of cortex at a relatively late stage of neurogenesis. These "heterochronically" transplanted cells were labeled so that it could be determined what stage of mitosis they had reached at the time of their removal and they could be distinguished from cells in the host tissue slice. What these researchers found was that most of the transplanted cells migrated to the upper layers of the cortex, as was appropriate for the developmental stage of the host tissue but not of the donor cells. But those neurons that had already undergone their last mitosis in the donor cortex retained the characteristics appropriate to that earlier time. One conclusion the researchers drew was that during this process, cell-to-cell contact with earlier-born cortical-plate neurons likely provided important differentiation signals. In this way, temporal sequence information probably enabled both spatial segregation of cell types and regulation of the proportions of these different types with respect to one another.

Though few of the mechanisms behind neurodevelopmental timing regularities are known, there seems little doubt about their conservatism and highly interdependent nature (Satoh 1982). In general, then, a diverse spectrum of size-correlated features cannot easily be dissociated from developmental timing, and vice versa. On the one hand, using these allometries among size-related morphological features of the brain to classify developmental differences between species fails to distinguish the contributions of the very different types of mechanisms involved and often results in a confounding of opposite processes (such as growth and secondary parcellation). On the other hand, these highly conserved developmental correlations provide a background for recognizing exceptions that, when mapped to underlying cellular and molecular mechanisms, may offer valuable hints about the evolution of these processes.

CELLULAR VERSUS ORGANISMIC TIMEKEEPING IN DEVELOPMENT

One global feature of development seems highly variable and capable of wide divergence even in closely related species: mitotic growth. Theoretically, one might expect that a system that creates its basic compartmental divisions early on and then extrapolates the whole structure to one that is hundreds or thousands of times larger by local multiplication of those parts would need to maintain a tight reign on this mitotic growth to ensure predictable outcomes. Growth processes could achieve this by utilizing precisely synchronized cellular clocks that counted cell divisions in order to guarantee that widely separated body parts (for example, paired limbs on either side of the body) grew to equivalent size. This might appear to be necessary because different species must have very different mappings of differentiation events onto numbers of cell divisions from conception. And yet mitotic growth is ultimately subordinate to the differentiation clock, because the mitotic activity of a cell lineage is determined by its phenotypic identity. As cells in a given lineage approach a terminally differentiated state, their mitotic activity will either slow or cease altogether. This provides a useful null hypothesis: mitotic cycling is merely set *on* at conception and is turned *off* only when a terminal phenotypic state of the cell is reached. In other words, mitotic processes might be essentially unregulated, except by metabolic and substrate constraints, but are blocked when a cell is in a particular phenotypic state of differentiation.

The reciprocal relationship is often seen in cell culture. Undifferentiated primary cells and even transformed cell lines (e.g., carcinoma cell lines), which tend to continue dividing in cell culture, can often be induced to assume a specific differentiated state simply by blocking mitotic activity, either by interfering with some molecular process essential to cell division or by altering the growth substrates to limit lineage expansion extrinsically.

Another window onto this relationship between cell metabolic state and differentiation processes is provided by certain dwarfed "paedomorphic" amphibians. There are frog and salamander species that exhibit a combination of slowed maturation, dwarfism, and comparatively simplified brains. Their brains resemble embryonic brains in many respects, notably because of significantly simplified cytoarchitecture (including reduced layering and nucleation). Because of this, it has been suggested that they may be exemplars of ancestral amphibian brains. What makes these animals useful test cases for analyzing the relationship between cellular processes and heterochronic development is that they exhibit an additional molecular correlate that offers a clue to the linkage between the various developmental clocks that may have been modified to produce their peculiar combination of traits.

These frog and salamander species have extremely large genomes (in some cases orders of magnitude larger) in comparison with other vertebrates (Olmo 1983). This is due to the inclusion of much larger amounts of repetitive DNA (fig. 3.3, top). This correlation of genome size with radically slowed growth, dwarfism, and apparent retention of primitive and/or paedomorphic traits (fig. 3.3, middle) can be further correlated with some interesting microscopic and physiological characteristics that offer clues to the developmental linkage between these traits. Their neurons all exhibit comparatively large somas (the cell body center as opposed to neuronal processes) with respect to body and brain size (all cells of the body are likewise large), and this contributes in part to the much simplified brain architecture (Roth, Blanke, and Wake 1994; Roth et al. 1993).

The larger genome produces large chromosomes in large nuclei, and thus large cells. In general, species with brains of the same size but

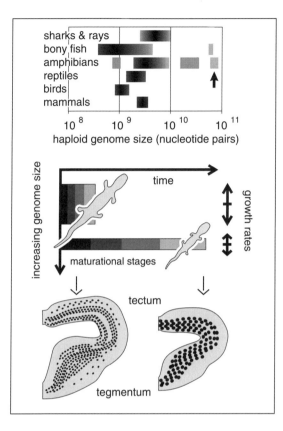

FIGURE 3.3.

Summary of neurological consequences of large genome size on salamander growth, size, and neural architecture. Top: Genome size ranges for vertebrate classes. Gray bars indicate the range of genome sizes; the darkest point indicates the modal range in each. Note the separation of a few groups on the high and low extremes. The arrow points to the extreme high-genome-size amphibian species. The few extremely high-molecular-weight teleost fishes could provide an additional test of the heterochronic development hypothesis about brain structure development. Mammals and birds each appear to represent a rather restricted range of genome sizes at the low and mid-range of genome sizes, respectively, which may be related to their relatively fixed high metabolic rates. Middle: The effect of genome size on metabolism, growth and maturation rates, and adult size in salamanders. Length of the bar indicates duration of development; width indicates rate of growth. Bottom: The comparative midbrain (tectal and tegmental) cytoarchitecture of a low-genome-size (left) and high-genome-size (right) salamander, showing the correlation between relative cell size and cytoarchitectonic complexity. (Data summarized from Roth et al. 1993; Roth, Blanke, and Wake 1994.)

with comparatively smaller cell sizes would exhibit more complex cytoarchitecture. In this special case, since the entire brain has a very small size but very large cells, individual neurons occupy a much larger fraction of the size of any given gross morphological division of the brain—even when corrected allometrically for the small size of such a brain. The resulting cellular "graininess" in these species produces comparatively simpler brain architecture merely because the range of densities, numbers of layers, and diversities of cell sizes are reduced in comparison with the size of any morphological division. For example, these large-genome species have a minimally laminated tectum in comparison with species with small genomes and smaller cells (fig. 3.3, bottom).

This combination of cellular traits results in a dissociation of the typical comparative allometric pattern seen in vertebrates. In most species (with smaller genome sizes), larger body size tends to correlate with prolongation of developmental staging, larger cell sizes, and more complex cytoarchitectonic organization of brain structures. But in these large-genome species, large cell size and slow differentiation are decoupled from large body size and progressive differentiation, so that the inverse relationship holds.

Increased cell sizes also correlate with decreased rates of metabolism and slowed mitotic rates. These differences probably account for the dwarfism and slowed growth in these large-genome amphibians and may also help explain the dissociations between growth, maturation, and brain complexity. Larger animals tend to have slightly larger average neuronal cell sizes. This is not generally produced by the larger species' having larger genomes but rather by some other, as yet undiscovered difference in gene regulation and/or metabolism. Large bodies, produced by prolonged development and by the greater number of mitotic divisions that occur between first cell divisions and final differentiated states, also tend to have slower metabolisms, longer cell cycle lengths in mitotic phase, and larger terminal cell sizes. In these large-genome species, slowed metabolism is linked with slowed mitotic cycling, but there does *not* appear to be a correlated increase in the number of cell divisions to a terminal state. So although the developmental clock may be slowed, this is not the cause of the amphibians' paedomorphic appearance. Their paedomorphism is rather the result of the unusual combination of large cells with comparatively few mitotic cycles to maturity. Their development *is* heterochronic with respect to other species and is secondarily convergent on paedomorphic appearance, but it is not actually arrested at an early stage of development. So these species are not, strictly speaking, neotenous, although they appear superficially immature. This appearance is caused not by the retardation of somatic development but by the inverse combination of large cell size with small body size. They are also not progenic, because they do not exhibit a shift in maturity to a comparatively early stage of development but are slowed overall. They are not, strictly speaking, peramorphic, either, although they exhibit significantly prolonged time to maturation, because they do not correlatively extrapolate growth processes. The term "rate-hypermorphic progenesis" might apply, but it does not seem to offer more than a renaming of the relationship in clumsy terminology that fails to capture all the critical details.

This dissociation of an otherwise widely conserved correlation between adult size, rate of growth, and complexity of cellular architecture of the brain offers some crucial insights concerning putative progressive trends in brain evolution and apparent recapitulatory relationships in brain structure. These large-genome species tend to resemble immature animals in brain structure, as well as other features, as a consequence of comparing them with the majority of species, which exhibit tight correlations between gene replication and decoding rates, genome size, cellular metabolism, cell replication rates, and the timing of terminal cell differentiation. This shows that although the possibilities for heterochronic modulation of vertebrate development are in general constrained by a highly conservative and inflexible interdependence between molecular synthesis rates, metabolism, mitotic growth, and the relative timing of cell differentiation, these can be decoupled.

As exceptions to this general developmental allometry of time, growth, metabolism, and differentiation, these amphibian species provide some important clues to the basis for the more typical linkages among these phenomena. In these frogs and salamanders, the type of heterochronic change that best describes how development has been altered differs at different levels of the analysis. The slowing of cellular molecular processes via the investment of an inordinate fraction of metabolism into them slows both the mitotic clock and the differentiation clock, so all cellular events of development are prolonged with respect to some hypothetical ancestral condition. In contrast, these species' dwarfism results from the fact that the mitotic cycling rate is reduced to a proportionately greater extent than is time to final differentiation, with respect to some hypothetical ancestral condition, and this causes terminal differentiation to occur after significantly fewer cell divisions and less somatic growth. Finally, the immature appearance of the brain results from the unusual combination of inordinately small body and brain size with inordinately large cell size. So this special condition demonstrates that there may be discordant heterochronic assessments at different levels of developmental analysis.

JUXTAPOSING DIFFERENT DEVELOPMENTAL CLOCKS

Another window into the relative autonomy of intrinsic controls of developmental timing in different species is provided by experimental contexts in which it is possible to juxtapose cells from different-size species. This is possible by producing chimeric individuals by mixing blastula cells from different species' strains or by transplanting tissues and cells between species (xenotransplantation). This makes it possible to unmask variables that are only contingently correlated in development and to demonstrate how differences in developmental timing and maturation rates of cells interact.

A remarkable source of evidence for the autonomy of the different developmental clocks is provided by chimeras of Japanese quails and domestic chickens (reviewed in Le Douarin 1993). The cells of these two donor species can be unambiguously identified because of easily distinguishable differences in their nuclei. This makes it possible to trace the fates of each donor's cells and transplanted structures with ease. In experiments carried out by researchers in many laboratories using this paradigm, segments of the neural tube derived from quail embryos shortly after neural tube formation can be placed in a developing chick embryo at the site where the corresponding chick brain region has been removed. Such neural chimeras can reach the hatching stage before immune processes begin destroying the graft. They show that development of a chimeric nervous system can proceed relatively normally and that the grafted region can integrate functionally with adjacent host brain regions—in fact, chicks with quail midbrains have even been shown to produce quail-like vocalizations (Balaban, Teillet, and Le Douarin 1988). But in addition to cellular nuclear structure, there is another important difference between these quails and chicks: the quails mature faster and are smaller at corresponding ages in comparison with chicks. This translates into differences in regional brain development in chimeric animals. The grafted quail brain regions appear to mature at a faster rate than the cells they replace would have matured, and they end up producing a physically smaller brain region with slightly smaller cells, irrespective of the structure replaced. This seems to indicate that at least by the time of neural tube formation, the cells' developmental clocks are imprinted with a preset schedule and stick to that schedule irrespective of the surroundings.

A similar developmental autonomy is exhibited by neurons that have been transplanted into adult host brains. If transplanted at a stage subsequent to cell fate commitment, embryonic neuroblasts stick to that fate and are capable of approximately recapitulating the sequence of maturational stages they would follow in a normally developing brain. The timing of this process, however, differs depending on donor species. For example, when fetal pig (Deacon et al. 1994) or fetal human neuroblasts (Wictorin and Björklund 1992; Wictorin et al. 1990, 1992) are transplanted into adult rat host brains, they mature far more slowly than do cells from corresponding fetal rat brain structures. Even the time course of graft-mediated behavioral recovery from previously lost function is proportionately prolonged in response to fetal human and pig cell grafts in comparison with similar grafts derived from fetal rat and mouse donors (Isacson and Deacon 1996, 1997). Transplanted neurons also grow to a mature state that appears to be mediated mostly by intrinsic information. This is exhibited by the fact that cells reach sizes that are characteristic of the donor species from which they are derived and not of the host brain. Thus, mouse neurons implanted into rat brains do not grow to reach the diameter of corresponding mature rat neurons, whereas pig neurons will grow to a diameter much greater than the diameters of their surrounding rat counterparts. This intrinsic developmental programming has other functional consequences as well.

A major component of this maturational process for neurons is the growth of axons and dendrites. Axonal growth is particularly dependent on internal timekeeping in order to maintain an active growth cone at the axon's growing tip. This complex amoeba-like structure—called a cone because it is shaped like an inverted arrow from which active pseudopodia extend forward, "probing" the intercellular spaces for attractive and repulsive signals, searching for adhesive substrates on which to pull itself forward—is maintained by the immature neuron long enough to reach its destination somewhere else in the brain, sometimes a fair distance from its cell body of origin. Though many of the steps from growth-cone production to establishment of stable synapses with other neurons are not fully understood, it does seem clear that if a growth cone fails to find a final target after a certain period of time, it will spontaneously be shut down. But species with large brains need their cells' axons to grow proportionately greater distances than those in small brains, and correspondingly, this window of growth-cone activity appears to be longer for neurons from larger animals. This is demonstrated by the growth capacities of axons of fetal neurons transplanted into adult brains.

In normal rat development, for example, the time required to grow axons to reach targets in widely separated brain regions is usually just a few days. In the last few days of a rat's 21-day gestation period, most neurons giving rise to "long-projection" axons have matured and begin to grow their axons toward potential targets. The axons will reach their destinations within a few days. In a much larger animal— such as a pig, in which gestation takes 115 days—the corresponding long-projecting axons may begin to grow by 30–40 days of gestation and may continue to grow to distant connections up to or beyond the gestation period. But an additional factor influences axon growth from transplanted neurons. Mature brains resist axon growth. This contributes to the failure of damaged axons to regrow for any significant distance after CNS damage, even though neurons may be spared. Consequently, fetal rat neurons placed in an adult rat brain will generally be unable to grow new axons to "distant" targets because the severely impeded rate of axonal growth causes growth-cone activity to shut down before they get there (fig. 3.4). If pig neuroblasts, however, are transplanted into mature rat brains, their much longer period of growth-cone expression allows them to continue growing for months, and although they are limited to extremely slow progress through a

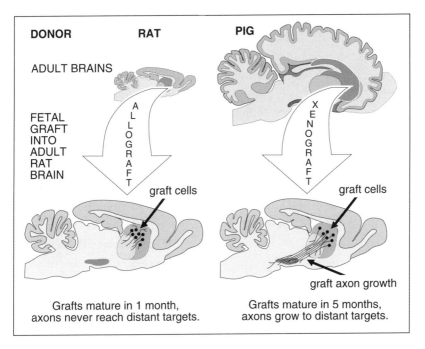

Figure 3.4.

Some results of xenograft studies comparing the development of grafts from comparitively larger and smaller fetal donor species. Transplanted neuroblasts transplanted from rat fetuses to rat adults do not appear to have the capacity to grow some of the long axonal connections (left panel), whereas transplanted neuroblasts from pig fetuses do (right panel) (Deacon et al. 1994; Isacson and Deacon 1996). This occurs because adult brains inhibit and slow axonal growth, and because axon growth processes follow a time course that is regulated by a clock-like mechanism from within the donor cell and is not determined by substrate context. The relatively much slower maturation and larger cell sizes (not shown) of fetal pig neurons, compared to fetal rat neurons, allow them to overcome these growth limitations and unmask target-specific growth tendencies that are conserved across these species' evolution.

nonpermissive substrate, they can grow to reach targets as far apart as those found in immature rat brains. In general, then, it appears that each species' cells march in time to their own developmental clocks, providing a sort of "preestablished harmony" of developmental timing irrespective of immediate context.

A final demonstration of this cellular developmental autonomy is provided by experiments that use cells derived from embryos long before cell fates are determined—that is, cells at the blastula stage (Deacon et al. 1998). At this stage, cells can be dissociated and induced to divide almost indefinitely in culture, thus indicating that they require an embryonic context to set the developmental clock in motion. Despite this requirement for extrinsic signaling to start the clock running, however, it appears that the rate at which it will run is predetermined, because the cells' time to differentiation is species-correlated. Possible evidence for this is provided by cases in which the genotypes of cells in chimeric animals (produced by aggregating blastula cells from different strains or species) can be tagged with molecular probes or other strain-specific markers so that their ultimate placements in the mature body can be compared. In many such chimeric animals, cells tend to coalesce into same-donor clusters. One possible cause of this could be a clonal effect—nearby cells are often daughters of a common stem cell. But timing may play a role as well, because clustering of same-genotype cells within the brain is more pronounced when the donors differ significantly in developmental timing (Goldowitz 1989). This clustering tends to suggest that adhesion affinities of cells are expressed in a time-dependent manner, and so relative synchrony of the cells' autonomous developmental clocks will affect whether cells tend to cohere together or aggregate with nonconcordant neighbor cells.

Xenotransplantation experiments that implant blastula cells or other immortalized cell lines (e.g., carcinoma cell lines or cell lines that have been immortalized with viral vector manipulation) into the brains of adult animals also suggest that the cells mature at rates characteristic of the donor species (Deacon et al. 1998). Thus, mouse-derived cell lines appear to mature in just days or weeks after transplantation, whereas human-derived cell lines may take many months. Because blastula cells are multipotent cells that can differentiate into any cell type in the body, they suggest that the underlying mechanism determining developmental rates must be common to all cells in a developing body. Once set running at some point in the developing blastula, the autonomous synchrony of the clock in each cell guarantees coordination of the numerous interactive processes that will follow. However, the reliance on such a highly conserved mechanism for coordinating all other diverse developmental events is the source of significant constraint on the possible ways in which developmental timekeeping can be modified in evolution.

HETEROCHRONY IN PRIMATE AND HUMAN EVOLUTION?

Although a great deal remains to be learned about the mechanisms underlying the clocks that organize vertebrate development, knowledge of their conservative regularities can provide a useful context in which to reconsider some of the more global differences between species' brain development. In this final section I turn to a classic problem, the nature of human brain evolution, to see to what extent this more reductionist reassessment of developmental timing can offer new hints about the processes involved.

It is a widely shared assumption that anthropoid primate radiations have been defined in part by the evolution of increasing brain size. Relative to the brains of most other groups of mammals, primate brains are nearly twice as large for the same body size. This increased encephalization has naturally led to claims about the evolution of increased intelligence in primates. Various interpretations of this trend have implicated paedomorphosis (maintaining fetal brain/body proportions), peramorphosis (extending brain development by adding new stages of maturation and new structures), and other heterochronic permutations of the nonprimate mammal growth pattern. Analyzing this allometric shift in proportions in terms of cellular developmental processes, however, provides a different interpretation and suggests that a much more complex story underlies the evolution of primate and human brain/body proportions.

Comparisons of brain and body growth of mammals during early development demonstrate a few remarkable commonalities (fig. 3.5A). Because mitotic growth of the brain is entirely concentrated in the prenatal and immediate postnatal stages of development in mammals, the brain/body growth curve is roughly bimodal. Its initial, mostly prenatal phase reflects roughly equal rates of mitotic growth in both the brain and the rest of the body (producing an allometric slope near 1.0). Its later, mostly postnatal phase reflects the continuation of mitotic growth of postcranial regions after mitosis has mostly stopped for neurons (producing an allometric slope much less than 1.0).

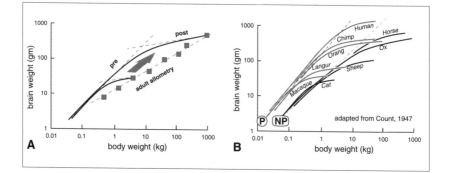

FIGURE 3.5.

Fetal brain growth in a number of mammals. A: Ontogenetic growth of brain and body for two typical mammal species during both prenatal (pre) and postnatal (post) phases of development (depicted by solid curves). The solid squares represent adult brain/body weights for a number of other mammals, showing how they tend to fall along a line in log coordinates. The gray arrow suggests that the larger curve can be derived from the smaller curve by a scalar expansion of mitotic growth. B: Brain/body growth curves for a number of mammals from early embryogenesis to maturity (top right end of each line). These growth curves demonstrate the shift of the entire ontogenetic curve of primates (P) compared with most nonprimates (NP). Primate embryos begin with reduced postcranial mass compared with brain mass (which appears to be similar for similar gestation times in all mammals).

The first remarkable feature to stand out when mammalian brain/body growth curves are compared is that during the prenatal phases there are few distinct growth trajectories—essentially only two, indicated by the dashed lines *P* and *NP* in figure 3.5B. All mammalian brain/body growth seems to follow one or the other trajectory during the fetal phase (Count 1947; Martin and Harvey 1985; Sacher and Staffeldt 1974). Elephants and cetaceans appear to grow along with primates on line *P* (which I will call the "primate" line, for convenience), and essentially all other placental mammals grow along line *NP* (which I will call the "nonprimate" line). Species sharing the same initial fetal growth trajectory (e.g., rat and horse) have the same size fetal brains and fetal bodies at about the same time of embryogenesis, despite the fact that one will continue to develop along this trajectory for longer (to become far larger). In other words, their embryos start out with the same proportions of brain and whole body and grow isometrically from there, throughout the prenatal phase. This suggests that mammal embryogenesis is constrained by a highly conserved initial segmentation of cranial-to-postcranial proportions, with perhaps only two variants.

There is a second, equally remarkable feature of brain/body growth in mammals. Though the curves are of different extents, they appear to be approximately the same shape. The ratio of the prenatal to the postnatal phase (in log-log coordinates) is nearly the same for the majority of species, irrespective of size. This can be demonstrated by treating each curve as two sides of a triangle, with the third side defined by the line described by the adult brain size/body size trend (fig. 3.5A). The brain/body growth curves of different species form similar triangles that differ in scale but not in proportions (Deacon 1990b). The implication is that differences in brain/body growth curves can be described by a single scalar factor, which is an index of relative mitotic growth. Though mitotic growth ceases earlier in the brain than in some other systems of the body, the proportional difference in mitotic cycling in cranial versus postcranial body segments is roughly the same in different species. In other words, it is as if a single mitotic clock is running in all cells of the body, with each tissue type setting its point of last mitosis at the same relative time point with respect to the whole sequence. Following the metaphor used earlier, it is a sort of elastic sheet time line of development that is stretched to different degrees in different species but in which the various stages expand or contract proportionately and apparently the same way, irrespective of prenatal starting proportions (P or NP). This again demonstrates a remarkably conserved mammalian embryology. It appears that the difference between the primate and nonprimate brain/body growth trends is determined by a shift in the initial proportions of brain and body without altering the organization of the developmental time line.

Against the background of these two distinct modal patterns of early brain/body growth, additional minor modifications are superimposed. Within each of the two mammalian groupings distinguished by their embryological proportions (P and NP), divergence of adult brain/body proportions from the mean trend appears to be the result mostly of subtle shifts of the proportions of the prenatal phase relative to the postnatal phase. Evidence from breeding within species (e.g., dogs) suggests that this is an effect that modifies chiefly the postnatal (mostly somatic) growth phase. Dwarfism and giantism produce far more deviation from species-typical values in body weight than they produce in brain weight. It seems relatively easy to select for heritable modifications of overall body size, and it is well known that modifications in the production of growth hormones or in the expression and function of growth hormone receptor molecules in different tissues postnatally are a major cause of such heritable changes in body proportions during later growth stages.

By extrapolation to interspecific variations from the adult brain/body growth trend, I would predict that this sort of secondary modification of postnatal *postcranial* growth is the dominant source of encephalization differences that are observed within groups that share a common embryological growth trajectory. In other words, most encephalization differences across mammalian species (assessed within the primate clade, or among most nonprimate clades, analyzed separately) are probably *not* the result of modifications of brain-growth parameters. This is of critical importance for interpreting claims about the evolution of intelligence based on brain/body proportions. It is not clear that modification of postnatal somatic growth should have a significant impact on parameters of global neurological function. If so, we should expect to find, for example, that small dog breeds are significantly more intelligent than large dog breeds and that human dwarfism is a significant correlate of high IQ. To my knowledge, the evidence does not support these predictions.

But what about the shift in brain/body proportions that distinguishes species on the primate and nonprimate growth trajectories? The difference unambiguously involves the prenatal phase. Can we determine whether this difference is primarily a cranial or a postcranial effect? The divergence of primate brain/body proportions from those of other mammal groups has sometimes been attributed to global shifts in developmental timing. Greater encephalization could be the result of more rapid brain growth or of more prolonged brain growth in comparison with most nonprimate mammals. We can rule out these alternatives, however, by noting that a shift in the rate of brain growth would be likely to increase the slope of the prenatal phase significantly, and a relative prolongation of brain growth would modify the proportions of the prenatal to the postnatal phase significantly. Although some differ-

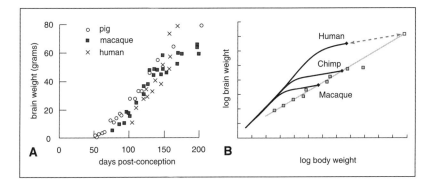

FIGURE 3.6.

A: *A comparison of brain growth in mammals, including primates and nonprimates, indicates that the shift in primate brain/body growth and adult allometry is the result of differences in postcranial growth—specifically, reduced primate somatic growth at all stages of development—and not of more rapid or more extended brain growth in primates.* B: *Human ontogenetic divergence of brain/body growth from the primate trend. Three species' growth curves are compared, showing the common prenatal, parallel postnatal, and aligned adult brain/body allometries (except for humans). The human trend appears truncated postnatally compared to other species' development as would be typical of dwarfism, but this is not the result of dwarfism, but of extended mitotic brain growth with respect to our more chimpanzee-like growth pattern.*

ences in this regard are possible (and have not been systematically explored), it is clear that the major developmental cause of the increase in primate encephalization with respect to most nonprimate mammals is the shift in the initial embryological proportions. The locus of this proportional shift can be determined by comparing growth over time in the brains of fetal primate and nonprimate mammals (fig. 3.6A). Brain growth rates for primates and nonprimates appear to overlap at early fetal stages, when all are still undergoing mitotic growth of the brain (compare, for example, Dickerson and Dobbing 1967; Dobbing and Sands 1973; Holt et al. 1975; Widdowson 1981). The inevitable conclusion is that the locus of primate encephalization is not the brain. The parallel slopes of primate and nonprimate fetal growth further indicate that, for the same body size, primate and nonprimate fetuses grow at roughly the same rates (which may reflect a ceiling effect on growth rates in the near ideal conditions provided by uterine development; see Sacher and Staffeldt 1974).

We must conclude that primate embryos start out with less total body mass though roughly the same brain mass as other mammals, as though at a very early stage in embryogenesis a whole chunk of the postcranially fated tissue mass had been removed. The segmentation of the embryo has been modified, and yet the developmental clock has not been rearranged to produce this primate segmental growth difference. In other words, the primate shift in brain/body proportions appears to involve neither a heterochrony of segmental growth nor a heterochrony of developmental schedules but instead a shift in parcellation of the early embryonic body that simply gets extrapolated by otherwise conserved rates of global mitotic growth.

Does this information provide any clues about probable brain function correlates? On the one hand, it suggests that since the locus of the shift in proportions is postcranial, it may be inappropriate to assume that the shift reflects selection for cognitive factors in primate evolution. We should at least consider the many possible alternative explanations in which somatic reduction could be a significant advantage (e.g., in terms of metabolic investment, advantages for arboreal locomotion, or maternal costs of carrying offspring). On the other hand, the shift in brain/body proportions throughout fetal development is likely to have significant consequences for a number of dynamical processes involved in neural development (e.g., "programmed" cell death and axonal competition). Unlike the case of the secondary reduction in postnatal somatic growth characteristic of dwarfism, we *should* expect the primate postcranial reduction to result in significant differences in brain organization and function (see Deacon 1997 for discussions of the processes involved, the anatomical changes that result, and the possible cognitive implications).

Taken together, the near isometric scaling of prenatal brain/body growth, the conservatism of growth rates for brain and body mass, and the major modification that can be traced to an early embryogenetic change in segmentation all suggest that mammalian embryogenesis is powerfully constrained. The primate exception proves the rule. Though the shift in very early body segmental organization is an exception to the von Baerian rule that very early mutations are unlikely, the general principle is supported by the fact that this sort of modification happened only a few times in mammalian evolution.

This context provides the basis for reconsidering the problem of human brain evolution and the possible role of heterochronic change. It is important to begin this discussion by recognizing that human encephalization is *not* a simple extrapolation of a general primate brain evolution trend, as is so often suggested on the basis of adult encephalization trends alone. Though human adult brain/body proportions have been increased even with respect to what would be predicted from other primates, human prenatal brain/body growth follows the typical primate trajectory. The additional adult human divergence from the encephalization trend is instead a result of a different shift: a shift in the proportional lengths of the prenatal and postnatal phases. The human curve appears truncated (fig. 3.6b), as might be consistent with dwarfism. It is *not* a result of postnatal postcranial growth reduction, however, but rather of brain growth prolongation. This is evident from comparisons of human and chimpanzee brain and body growth patterns, which show that whereas human brain growth extends beyond chimp brain growth (as would be appropriate for a much larger-bodied ape), body growth patterns are similar for humans and chimps. Furthermore, there is no phyletic or fossil support for hominid body-size reduction as a cause of our encephalization. Brains have indeed gotten larger in recent hominids in parallel with a much less drastic increase in stature. Human brain/body growth is analogous to what we might imagine after a whole brain xenotransplant from a huge ape into a chimplike embryo! It appears, on the surface, to be an unprecedented example of a segmental modification of the developmental clock. Is there sufficient information to decide whether or not this is the case?

First we need to determine exactly where this segmental dichotomy occurs. Is the brain as a whole larger, or are some parts enlarged and others not enlarged? As I noted earlier, there is a high correlation between the sizes of major brain structures in primate brains and overall brain size. This exemplifies the global mitotic clock hypothesis of mammalian brain evolution (e.g., Finlay and Darlington 1995). But the human brain as a whole does not share in all these trends, and this indicates that its enlargement is not explained by such a global scaling mechanism any more than by postcranial reduction. A couple of examples demonstrate this point. Figure 3.7 compares the scaling of the diencephalon with two other major forebrain divisions, the neocortex and the corpus striatum. Notice that values for other primate species exhibit little deviation from the general trends. This is also true for the striatum/diencephalon relationship in humans, but not for the neocortex/diencephalon relationship. In the latter case, humans depart significantly from the otherwise tight trend. This mosaicism of correlated and divergent size patterns is found for other comparisons as well, including, notably, the cerebellar cortex, with is also disproportionately large in humans. The otherwise highly conserved patterns in other species of primates suggest that independent size variation of many brain regions is rare (again reinforcing the impression of embryological conservatism in this regard). But it may be possible to delimit the expanded structures in human brains to a single, contiguous domain of homeotic gene expression—that is, a single cerebral "segment" of the embryo (Deacon 1995, 1997; Frantz et al. 1994).

Can we determine whether the cause of this segmental growth change is a change in the clock or a change in initial cell numbers partitioned to this segment? The question might be answered if we could demonstrate a shift in the brain/body growth trajectory. It might also be answered if we could demonstrate that certain cell differentiation benchmarks were reached in disproportionately later stages in expanded than in nonexpanded structures. If not, we would have to conclude that early stem cell production in these brain segments was responsible. The appropriately prolonged development of the human brain for its size suggests the unprecedented: prolongation of the mitotic growth phase in these regions by a localized shift in the differentiation clock.

This hypothesis offers a challenge to the simple model so far developed, in which there is a global setting of both the mitotic and differentiation aspects of the developmental clock. It begs the question of how the segmental differentiation of embryonic cell populations provides cells within a given positional domain with information about when in the sequence they are to initiate terminal differentiation processes. An important clue comes from the fact that segmentation itself is a process that is extended in time. Along the neural tube there

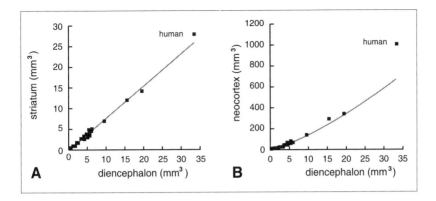

FIGURE 3.7.

Scaling relationships among three functionally interdependent forebrain structures (neocortex, striatum, and diencephalon), showing that there are common allometric trends shared by all primates, including humans, for some structures (A, comparing diencephalon and striatum), but excepting humans in others (B, comparing neocortex and diencephalon). Similar results are obtained comparing neocortex and striatum (not shown). Cerebellar cortex also appears to be a divergently large structure, and olfactory bulbs are divergently reduced. When smaller brain divisions are compared, more diverse patterns are observed, but these are probably determined by second-order processes of axonal competition and cell death superimposed upon initial segmental differences in embryonic neural mitosis (Deacon 1988, 1995). Data are from Stephan, Frahm, and Baron (1981) and are graphed in linear coordinates to show absolute differences. Regression lines were fitted to log transformed data, excluding human data and resulting line and data points, which were included as outgroup comparisons and transformed into linear coordinates. The divergence of the human point for neocortex and diencephalon is statistically significant, even though including this point considerably skews the best fit line because it is such an extreme outlier. (See Deacon 1988, 1990a for discussion of methods.)

are distinct temporal gradients in the expression of many segment-specific homeotic gene families (e.g., Holland, Ingham, and Krauss 1992). The Hox gene families are expressed in a rostral-to-caudal gradient, beginning just behind the region later to become the midbrain, and later the Otx and Emx genes are expressed in a caudal-to-rostral gradient beginning at the caudal midbrain (fig. 3.8; Simeone et al. 1992). In many respects, this parallels the developmental order for most later stages of brain development as well. This brings us back to a

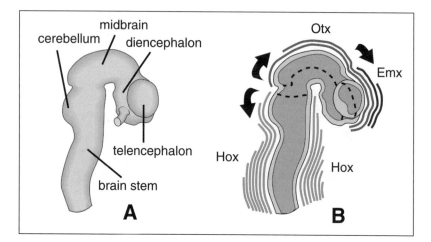

FIGURE 3.8.

Some major families of homeotic genes expressed in the embryonic brain and their correlation with human brain disproportions. A: Embryonic brain showing the major segmental subdivisions. B: Approximate regions in which Hox, Otx, and Emx genes are expressed. Expression domains are indicated by overlapping lines, with earlier expression closest to the embryonic surface. Those that are expressed in a cylindrically symmetric pattern (e.g., Hox genes) are indicated by lines on both sides. The Hox genes are expressed below the midbrain in a nested pattern. The two Otx genes (domains indicated by dark gray lines) and Emx genes (domains indicated by black lines) are expressed from the midbrain to forebrain, also in a nested pattern, but mostly confined to the dorsal surface of the neural tube. Portions of the dorsal forebrain that appear to be relatively larger than predicted in human brains seem to correspond to the same regions demarcated by Otx and Emx expression domains (not including the striatal region that develops from the ventral telencephalon, for example, shown as a bump inside the telencephalon in B). This suggests that there has been a shift in forebrain cell production involving one or more of these genes or in some as yet unidentified factors expressed "downstream" from their effects. Arrows indicate the rough direction of maturation and of gene expression in these gene systems.

developmental correlate noted earlier. As Finlay and Darlington (1995) demonstrated (see also fig. 3.1), later-developing brain structures are positively allometric in their growth with respect to the rest of the brain as the setting of the mitotic clock is extended. Thus, without a resetting of the clock, a shift in the time course of the expression of segmental genes—for example, a delay in rostral expression (an example of heterochronic gene expression) or a shift in expression domains of late homeotic genes to include more of the neural tube—could increase the proportion of the late-maturing component. Either could account for such segmental growth changes without implicating local resetting of the developmental clock.

It is significant, then, that the enlarged regions in the human brain correspond to the latest-differentiating structures in the brain as well as to contiguous homeotic expression domains (Deacon 1995, 1997). At the present time, insufficient information is available to eliminate the possibility that there has been a region-specific change in the developmental clock underlying human encephalization. We can be fairly certain, however, that the enlargement of human brain structures, the shift in segmental proportions of brain and body, and the prolongation of brain growth and development with respect to body growth are not due to changes merely tacked onto the end of development, as suggested in classic recapitulation accounts, nor to some generic heterochronic change of the whole body or even only of the whole brain. The unprecedented pattern exemplified by human brain development and evolution provides another example, like the large-genome amphibians discussed earlier, in which there may be no simple concordance between heterochronic assessments at the morphological and cellular levels. The mechanism underlying this human divergence defies simple classification and remains mysterious.

CONCLUSION

Heterochrony theories of species differences are only as good as the theories of development they are based on. We are currently in the midst of a molecular revolution in developmental biology and are beginning to identify genetic mechanisms that regulate large-scale pattern formation of the body. The problem of translating morphological accounts of the evolution of development into cellular-molecular accounts is not a simple one, and in cases where this can be done in some detail, heterochrony accounts may differ depending on the level of analysis. They may also be confounded by a failure to distinguish the effects of growth, differentiation, and regressive processes. I have reviewed evidence suggesting that primate encephalization likely correlates with one or more postcranial developmental changes and that human brain evolution likely correlates with one or more neurally localized developmental changes in either segmentation or mitosis. These changes have reframed cell fates in these body regions at an early stage in embryogenesis and thus dissociated aspects of allometric growth of body "segments" that are more typically linked. Ultimately, a more careful analysis of such exceptional cases may help unlock the mystery of these developmental clocks.

Notes

1. For example, a recent mathematical analysis of allometric scaling of physiological relationships has suggested that branching of fluid delivery systems may occur according to a fractal geometry in order to maintain equivalent distribution of fluid resources irrespective of scale. This may explain a significant number of allometric trends from metabolic to CNS scaling that tend to fall close to n/4 exponential relationships to size.

2. This was first discussed by D'Arcy Thompson in his seminal book *On Growth and Form* (1917).

4

The Ontogeny and Phylogeny of Language

A Neural Network Perspective

Elizabeth Bates and Jeffrey Elman

How did language evolve? Why are humans the only species on the planet capable of language in its full-blown, fully grammaticized form? This is one of the oldest areas of study in behavioral biology, and it is also the most contentious. Language leaves no fossils (at least not before the development of writing systems), and the neural systems that support language are still so poorly understood that it would do us little good to pore over prehominid skull casts in search of relevant bumps. In the absence of strong paleontological constraints, speculation about the evolution of language has run happily amuck for centuries. The French tried to rectify this situation late in the nineteenth century by outlawing all speculation about language evolution from their academy (Aarsleff 1976; Harnad, Steklis, and Lancaster 1976). This effort fared no better than the Maginot line a few decades later, and speculation about language evolution has continued unabated up to the present time. Candidate explanations range from claims for an innate, detailed, and idiosyncratic "grammar organ" that evolved within a specific part of the brain (e.g., Gopnik and Crago 1991; Lightfoot 1991; Pinker 1994a; Pinker and Bloom 1990) to the claim that language is possible simply because the human brain is very big (for discussion of this claim, see Gibson and Ingold 1993; Jerison 1973).

In the light of all this history, what right do we have to propose a new solution, or anything resembling the outlines of a solution? There are three reasons why we believe the problem of language evolution merits a new look (see also Deacon 1997; Gibson and Ingold 1993; Krasnegor et al. 1991).

First, new technologies for the study of language processing and its neural correlates have provided information that changes our view of language in the adult "steady state" away from theories that presuppose an assembly line of localized and highly specific information types (e.g., separate "boxes" for phonology, grammar, and semantics) and toward theories that make use of broadly distributed and multifunctional neural systems that interact in parallel (Bates and Goodman 1997; Kutas and King 1996).

Second, some dramatic breakthroughs in neurobiology have changed our view of the process by which adult patterns of brain organization emerge across the course of development, moving us away from the notion of a fixed genetic blueprint and toward a more dynamic, plastic, and experience-dependent view (for reviews, see Bates et al. n.d.; Elman et al. 1996; Johnson 1993, 1997).

Third, research simulating human behavior in artificial neural networks has led to new theories of representation and learning that are compatible with the distributed nature of language processing in adults and with the plastic, activity-dependent nature of development (Churchland and Sejnowsky 1992; Elman et al. 1996; Rumelhart and McClelland 1986). As we will see shortly, these neural networks constitute nonlinear dynamical systems. Because of their nonlinear properties, they are capable of producing new and apparently discontinuous behavioral solutions from relatively simple and continuous inputs, providing some insights into the process by which novel and productive linguistic abilities could have emerged from simpler beginnings.

In our view, these three new lines of theory and evidence have profound implications for our ideas about the way language might have evolved in the first place. Specifically, they lay the groundwork for an approach to language evolution that involves relatively small variations in brain architecture and the timing of neural development. As a result of these small changes, multipurpose neural systems in the primate brain arrive at a point at which they are capable of language learning and language use. There is no need to postulate the *ex novo* evolution of a separate system for language; instead, language can be viewed as a new machine made out of old parts (Bates, Thal, and Marchman 1991). To be sure, all we have to offer here is a plausibility argument; we cannot prove that language evolved by this indirect route. If our argument is correct, however, it would solve a host of problems revolving around the apparent discontinuity of language.

In keeping with the central theme explored throughout this volume, we want to underscore that timing is an important part of the story. In the first section, we review evidence on real-time language processing and its neural correlates in normal and brain-injured adults, evidence that (in our view) argues against the fixed, localized, and domain-specific modules presupposed by modern phrenologists, in favor of dynamic, parallel processing across broadly distributed and domain-general neural systems. In the second section, we outline our proposal for a taxonomy of claims about innateness and review evidence against innate knowledge (i.e., cortical representations) and in favor of the idea that cortical development is highly plastic and activity dependent, with species-specific differences produced primarily through variations in timing and regional architecture. With this new definition of the target of language evolution (i.e., distributed neural systems that develop in an experience-dependent fashion), we then present a brief introduction to neural network models of learning and change, with some examples that illustrate the crucial role of timing in the evolution and development of linguistic abilities.

ON THE TIMING AND DISTRIBUTION
OF LANGUAGE PROCESSING IN THE BRAIN

Most of our readers are probably familiar with figure 4.1A, one rendition of the famous Gall-Spurzheim theory of phrenology—that is, the belief that the brain is organized into separate, spatially segregated faculties, each devoted to a specific cognitive function (Goodglass 1993). A variant of this figure appeared more recently on the cover of an influential volume called *The Modularity of Mind* (Fodor 1983), a modern rendition of the doctrine of phrenology in which it is proposed (again) that the brain is organized into domain-specific components or modules that operate independently and obligatorily, in a reflexlike fashion, on specific kinds of content. Fodor's modularity doctrine represents an extension of Noam Chomsky's well-known claims about the autonomy and innateness of language (Chomsky 1957, 1965). Chomsky has unabashedly referred to language as a "mental organ," a metaphor extended and elaborated by Pinker (1994a, 1994b), who also reintroduced the ancient term "instinct" to describe language and other biologically grounded cognitive systems: "It is a certain wiring of the microcircuitry that is essential....If language, the quintessential higher cognitive process, is an instinct, maybe the rest of cognition is a bunch of instincts too—complex circuits designed by natural selection, each dedicated to solving a particular family of computational problems posed by the ways of life we adopted millions of years ago" (Pinker 1994b:93, 97).

In fact, the primary difference between Gall's theory and its modern variant revolves around the content rather than the existence of mental organs. Whereas Gall and Spurzheim included many regions devoted to ethical and aesthetic concerns (friendship, avarice, etc.), the modules of modern phrenology tend to be more cognitive in nature (e.g., language, faces, music, number). The modern variant of phrenology is represented in cartoon form in figure 4.1b, which includes separate regions for separate subcomponents of language, in line with Chomsky's mental-organ proposal. Although this cartoon is of our own making, each entry represents a serious proposal about the localization of specific mental content made by at least one credible source in the last 5 to 10 years. (For proposals about an innate and localized theory of mind, see Leslie 1994a and 1994b; for various proposals about the separate location of words and aspects of grammar, see Ullman n.d. and Ullman et al. 1997; for speculations about the neural bases of free will, see Crick 1994 and Damasio 1994b. A discussion of the neural bases of face perception can be found in Farah 1991. Claims about music, number, and sports are taken from a variety of sources.)

Where do these modules come from? Gall and Spurzheim were largely agnostic with regard to the development and predetermination of mental organs, but today's phrenologists are explicit in their claims

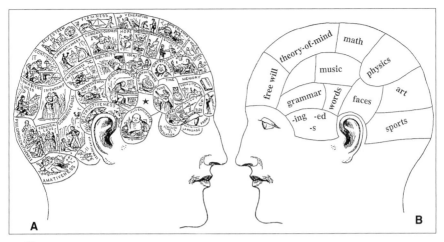

FIGURE 4.1.

A: *The Gall-Spurzheim phrenological model of the brain.* B: *The modern variant of the Gall-Spurzheim model.*

about innateness. Chomsky himself has suggested that the whole idea of learning is an illusion that will one day "go the way of the rising and setting of the sun" (Chomsky 1980). The same theme is reiterated by Piatelli-Palmarini (1989:2): "I, for one, see no advantage in the preservation of the term learning. We agree with those who maintain that we would gain in clarity if the scientific use of the term were simply discontinued."

This emphasis on innate, domain-specific functions leads us to provide a companion to figure 4.1B, the phrenological infant illustrated in figure 4.2.

If language is innate, it must have evolved into its current predetermined form. But how? In a Darwinian framework, new structures have to arise from preexisting forms. And yet it is not obvious how an independent and idiosyncratic capacity for (for example) grammar and semantics could have evolved from preexisting neural material (Bates, Thal, and Marchman 1991; Pinker and Bloom 1990). This problem would be less daunting if we could show that language in general and grammar in particular are neither as discontinuous nor as self-contained as we have been led to believe. In the rest of this section, we provide a brief overview of evidence for and against the modular perspective in three areas: real-time language processing by normal

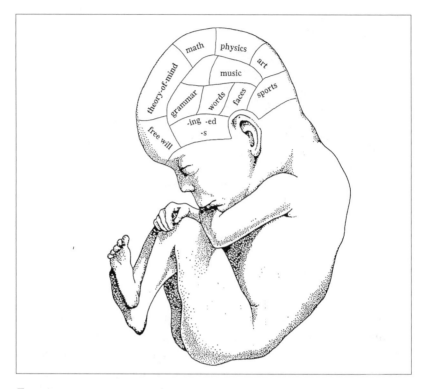

FIGURE 4.2.

The developmental variant of the Gall-Spurzheim model.

adults, language breakdown in aphasia, and neural imaging studies of neural activity during language processing. Our conclusion (not surprisingly) is that the evidence to date does not support a phrenological view. This conclusion prepares the way for a different model of language evolution.

Real-Time Language Processing in Normal Adults

In the 1960s and 1970s, efforts were made to develop processing models of sentence comprehension and production (i.e., performance) that implemented the same modular structure proposed in various formulations of generative grammar (i.e., competence). (For reviews, see Fodor, Bever, and Garrett 1974; Garrett 1980; Slobin 1979.) The comprehension variants had a kind of "assembly line" structure, with linguistic inputs passed in a serial fashion from one module to another (phonetic → phonological → syntactic → semantic). Production models looked very similar, with the sequence reversed (semantic → syntactic → phonological → phonetic).

In the late 1970s and early 1980s, serious concerns were raised against this fixed serial architecture, based on multiple studies showing "top-down" context effects on the early stages of comprehension. These included studies showing that sentence-level information can affect the accuracy and timing of word recognition, thus constituting evidence against a modular distinction between grammar and the lexicon. In response to all these demonstrations of context effects, proponents of the modular view countered with studies demonstrating very early temporal constraints on the use of top-down information during the word recognition process (Kintsch and Mross 1985; Onifer and Swinney 1981). An influential example comes from experiments in which ambiguous words such as *bug* are presented within an auditory sentence context favoring only one of its two meanings (e.g., *bug* in an exterminator context or *bug* in a spy context). Shortly after the ambiguous word is presented, subjects are asked to make a lexical decision (e.g., decide whether the target is a real word or a nonsense word) for word or nonword targets presented in the visual modality. These targets include real words that are related to the contextually inappropriate meaning (e.g., *spy* in an exterminator context), real words that are related only to the contextually appropriate meaning (e.g., *ant* in an exterminator context), and real-word targets that are unrelated to either the word or its context (e.g., *mop*). The term *priming* refers to a reaction-time advantage (i.e., faster response) following either an appropriate or an inappropriate prime (e.g., *spy* or *ant*) compared with a completely unrelated word (e.g., *mop*).

The critical result for present purposes concerns the role of timing. If the prime and target are separated by at least 750 milliseconds, priming is observed only for the contextually appropriate meaning (i.e., selective access). If the prime and target are very close together in time (less than 250 milliseconds), priming is observed for both meanings of the ambiguous word (i.e., exhaustive access). These results have been interpreted as support for a two-stage model of word recognition: a bottom-up stage that is unaffected by context and a second, postlexical stage when contextual constraints can apply. In other words, modularity can be observed empirically only in the first moments of language comprehension, a psycholinguistic analogue to quantum physics and string theory. The finding that both meanings of an ambiguous word are primed regardless of context under some temporal conditions provides much stronger evidence for modularity than do studies showing an absence of context effects, because null effects can be obtained in a variety of uninteresting ways. For this reason, Fodor (1983) used the exhaustive-access finding as one of his primary empirical arguments in favor of the modularity doctrine.

Although the exhaustive-access effect has been replicated many times, it is still controversial (Ratcliff and McKoon 1994). For example, some investigators have shown that exhaustive access fails to appear on the second presentation of an ambiguous word (Simpson and Kang 1994) or in very strong contexts favoring the dominant meaning of the word (Tabossi and Zardon 1993). Van Petten and Kutas (1991) replicated the original finding using event-related brain potentials as well as reaction time; however, they went on to show that the two stages (exhaustive → selective access) are preceded by an even earlier stage at which priming is observed only for the contextually appropriate target (selective → exhaustive → selective). In short, the literature on processing of lexically ambiguous words still provides no clear evidence one way or another for a modular boundary between sentential and lexical priming effects.

Because the evidence for Fodor's favorite form of modularity is mixed, investigators have sought other lines of evidence for early temporal constraints on language processing, using unambiguous words in and out of context. A great deal of evidence has now accrued to suggest that discourse context has effects in the very first stages of word recognition, even before the whole word has been heard (Grosjean 1980; Hernandez and Bates 1994; Hernandez, Fennema-Notestine, and Udell 1995; Marslen-Wilson and Tyler 1987; van Petten et al. 1996).

A number of studies have shown that, in addition to these semantic effects of discourse context, grammatical information can "penetrate" word recognition. These studies have demonstrated the effects of the grammatical gender of a modifier on recognition of nouns in French (Grosjean et al. 1994), on word repetition and gender classification in Italian (Bates et al. 1996), and on picture naming in Spanish

(Reyes 1995) and German (Hillert and Bates 1996; but see Friederici and Schriefers 1994 and van Berkum 1996 for reports indicating that gender priming may be relatively fragile in German and Dutch). Studies in Serbo-Croatian provide evidence for both gender and case priming, with real-word and nonword primes carrying morphological markings that are either congruent or incongruent with the target word (Gurjanov et al. 1985; Lukatela et al. 1983). Unpublished studies from our own laboratory show that repetition of Chinese nouns can be primed by noun classifiers, an abstract grammatical category in Chinese (Lu, Tzeng, and Bates, research in progress).

Returning to English, we have recently investigated the effects of short syntactic frames such as "I like the _____" or "I want to _____" on word repetition (Liu 1996) and picture naming (Federmeier and Bates 1997). Results include significant facilitation of targets that are congruent (object names/pictures in a noun context; action names/pictures in a verb context) and inhibition of targets that are incongruent (object names in a verb context; action names in a noun context), compared in all cases with a neutral prime. Finally, we have uncovered strong sentential priming effects for grammatical function words such as *do* and *which* that are difficult to recognize when they are spliced into a neutral context (Herron and Bates 1997). In all the studies done in our laboratory, the interval between grammatical context and lexical target is very short (set at zero in most cases), and reaction times are in the same range that has been reported for priming in studies that support modularity. Thus it appears that grammatical context can have a significant effect on lexical access within the very small temporal windows that are usually associated with automatic, nonstrategic priming effects. In other words, grammar can penetrate the "lexical module."

This battle for the first milliseconds of language processing often rests on arcane technical details ranging from variations in screen refresh rates and timing devices from one laboratory to another to small variations in the instructions given to subjects or arguments about the stimulus materials used from study to study. Some of this arcana is based upon a more substantive issue. In particular, proponents of modularity have argued that their modules are (by definition) unconscious and reflexlike; such systems coexist, however, with a more general "central message processor" that is capable of conscious reflection on the stimulus to be perceived and/or the problem to be solved. Therefore, any laboratory procedure that allows participants to "catch on" to the purpose of the study could elicit conscious, albeit rapid, strategies that mask the performance of a specialized and encapsulated mental organ. Of course one has to wonder why modules evolved in the first place if the "general cognitive device" is able to fulfill the same functions at lightning speed. Still, if such a thing is possible, then a thousand additional studies will never make the case one way or the other for the existence of independent mental organs. As Altmann and Steedman (1988) have pointed out, the demonstration that context effects operate very early can be used to argue against assembly-line versions of the modularity argument (i.e., "strong modularity"). These demonstrations, however, do not rule out a different form of modularity in which distinct and tightly bounded processors work in parallel on the input stream (i.e., "weak modularity"). In other words, we can never *disprove* the existence of a modular border. Of course one cannot disprove the existence of little green fairies either. The real question is, do we need to postulate a host of separate entities in order to explain how language processing occurs? If we must base evidence for linguistic phrenology on temporal processing alone, the answer is no.

If we cannot use evidence from the time course of language processing to test the modular view, perhaps we can use evidence from the spatial organization that underlies language use. This brings us to the next two lines of evidence, one from studies of patients with brain lesions and the other from modern studies of brain activity in normal adults.

Language Breakdown in Adults

Arguments in favor of the modular account are often based upon findings from adult aphasia—that is, language impairments caused by injuries to the brain. When the basic syndromes of nonfluent and fluent aphasia were first outlined by Broca, Wernicke, and their colleagues, differences among forms of linguistic breakdown were explained along sensorimotor lines rooted in clear, albeit rudimentary, principles of neuroanatomy. For example, the symptoms associated with damage to a region called Broca's area were referred to collectively as *motor aphasia:* slow and effortful speech with a reduction in grammatical complexity, despite the apparent preservation of speech comprehension at a clinical level. A motor definition of this syndrome made sense when we consider that Broca's area lies near the motor strip. Conversely, the symptoms associated with damage to Wernicke's area were defined collectively as a *sensory aphasia:* fluent but empty speech marked by moderate to severe word-finding problems in patients with serious problems in speech comprehension. This characterization also made good neuroanatomical sense, because Wernicke's area lies at the interface between auditory cortex and the various association areas that were presumed to mediate or contain word meaning. Isolated problems with repetition were further ascribed to fibers that link Broca's and Wernicke's areas. Other syndromes involving the selective sparing or impairment of reading or writing were proposed, with speculations about the fibers that connect visual cortex with the classical language areas (for an influential and highly critical historical review, see Head 1926).

In the period between 1960 and 1980, a radical revision of this sensorimotor account was proposed (summarized in Kean 1985). Psychologists and linguists who were strongly influenced by generative grammar sought an account of language breakdown in aphasia that followed the componential analysis of the human language faculty proposed by Chomsky and his colleagues. This effort was fueled by the discovery that Broca's aphasics do indeed suffer from comprehension deficits (Caramazza and Zurif 1976; Heilman and Scholes 1976; Saffran and Schwartz 1988; Zurif and Caramazza 1976). Specifically, these patients display problems in the interpretation of sentences when they are forced to rely entirely on grammatical rather than semantic or pragmatic cues (e.g., they successfully interpret a sentence like "The apple was eaten by the girl," where semantic information is available in the knowledge that girls, but not apples, are capable of eating, but they fail on a sentence like "The boy was pushed by the girl," where either noun can perform the action). Because those aspects of grammar that appear to be impaired in Broca's aphasia are precisely the same aspects that are impaired in the patients' expressive speech, the idea was put forth that Broca's aphasia might represent a selective impairment of grammar (in all modalities) in patients who still have comprehension and production of lexical and propositional semantics (Caramazza and Berndt 1985). Caramazza and colleagues (1981:348) stated this position succinctly: "Although it is possible that Broca patients may suffer from deficits in addition to this syntactic processing deficit, it should be the case that all patients classified as Broca's aphasics will produce evidence of a syntactic impairment in all modalities."

From this point of view, it also seemed possible to reinterpret the problems associated with Wernicke's aphasia as a selective impairment of semantics (resulting in comprehension breakdown and in word-finding deficits in expressive speech) accompanied by a selective sparing of grammar (evidenced by the patients' fluent but empty speech). If grammar and lexical semantics can be doubly dissociated by forms of focal brain injury, then it seems fair to conclude that these two components of language are mediated by separate neural systems. This linguistic approach to aphasia was so successful in its initial stages that it captured the imagination of many neuroscientists, and it has worked its way into basic textbook accounts of language breakdown in aphasia (Newmeyer 1988).

The proposed double dissociation between grammar and lexical semantics has enormous theoretical appeal, not only because of its parsimony (pulling together a range of disparate symptoms) but also because it bears a remarkable resemblance to a modular decomposition of the language faculty proposed on independent grounds in several formal theories of grammar (for a detailed discussion of this point, see Caplan 1987; Garrett 1992; Grodzinsky 1990). The doctrine of central agrammatism, however, has fallen on hard times among students of language breakdown in aphasia—even among those who put that doctrine forward in the first place. For example, in a paper by Zurif and colleagues (1993), the suggestion is made that agrammatic deficits may follow from deficits in information processing that are only indirectly related to grammar: "The brain region implicated in Broca's aphasia is *not* the locus of syntactic representations per se. Rather, we suggest that this region provides processing resources that sustain one or more of the fixed operating characteristics of the lexical processing system—characteristics that are, in turn, necessary for building syntactic representations in real time" (p. 462).

Why has central agrammatism been abandoned? Four different lines of research argue against the idea that grammar can be impaired in isolation and against the related claim that grammatical breakdown is uniquely associated with Broca's aphasia and (by extension) with a particular region in the left anterior cortex.

First, there is the association between grammatical and lexical impairments. Deficits in word production are observed in all forms of aphasia. Indeed, the symptom called "anomia" (word-finding problems) is the one symptom that has been observed in every form of aphasia reported to date. This inescapable fact means, in turn, that grammar is never impaired in isolation, thus greatly weakening claims about a double dissociation between grammar and vocabulary (i.e., the lexicon).

Second, there is the dissociation of expressive and receptive agrammatism. Studies have appeared showing that receptive and expressive agrammatism are fully dissociable. Expressive agrammatism is occasionally observed in individuals who perform within the normal range on tests of receptive grammar (Miceli and Mazzucchi 1990; Miceli et al. 1983), and receptive agrammatism is often observed in patients without grammatical symptoms in production (e.g., Bates, Friederici, and Wulfeck 1987; Lieberman et al. 1992). Indeed, similar patterns of receptive agrammatism have now been demonstrated in normal adults working under perceptual degradation and/or cognitive overload (Blackwell and Bates 1995; Kilborn 1991; Miyake, Carpenter, and Just 1994). Putting these lines of evidence together, we may conclude that expressive and receptive agrammatism are dissociable, which suggests in turn that the co-occurrence of these symptoms in Broca's aphasia may be a coincidence.

Third, there are cross-linguistic studies of patients with expressive agrammatism. A growing body of research comparing grammatical breakdown across natural languages has led to the conclusion that patients with expressive and/or receptive agrammatism still retain a great deal of detailed and language-specific grammatical knowledge (Bates 1991; Menn and Obler 1990; see also Davidoff and De Bleser 1994; De Bleser and Luzzatti 1994; Jarema and Friederici 1994). This conclusion is justified by, for example, dramatic differences in the preservation of language-specific word order patterns in comprehension and production and by detailed cross-linguistic differences in the patterns of omission and substitution that are observed in expressive language.

And fourth, there is preservation of grammaticality judgments. Among the various lines of evidence that have accumulated against the doctrine of central agrammatism, the most devastating comes from studies demonstrating that patients with expressive and/or receptive agrammatism are capable of remarkably good (albeit imperfect) judgments of grammaticality (e.g., Linebarger, Schwartz, and Saffran 1983; Shankweiler et al. 1989; Wulfeck 1988; Wulfeck and Bates 1991; Wulfeck, Bates, and Capasso 1991). Even though grammaticality judgment is a real-time performance domain like any other (i.e., it does not provide a direct assessment of knowledge), linguists working within the framework of generative grammar have always relied heavily on judgments of grammaticality in order to draw inferences about the nature and limits of linguistic knowledge (i.e., competence). If so-called agrammatic patients are able to make judgments of this kind, it is hard to escape the inference that linguistic knowledge or competence has been spared.

Taken together, these lines of evidence have convinced us that grammar and the lexicon do *not* dissociate in adult aphasia. The existence of a separate mental organ for grammar is not supported by research on brain-injured adults, although that literature certainly does suggest that something in the left hemisphere is important for language.

Neural Correlates of Language Processing in Normal Adults

The lesions produced by accidents of nature (e.g., strokes) constitute imperfect experiments. Suppose we had a way to look at the spatial distribution of brain activity during language processing by normal adults. Could we detect evidence for separate language organs that is hard to find in research on adult aphasia? Although research is still in the early stages, techniques that permit us to eavesdrop on the working brain are now available. These include spatially sensitive techniques such as positron emission tomography (PET) and functional magnetic resonance imaging (fMRI). These techniques still have fairly crude temporal resolution because they are based on the detection of relatively slow correlates of neural activity: for PET and fMRI, glucose metabolism and blood flow, respectively. Indeed, it is impossible right now to obtain functional brain images for time slices smaller than 40 seconds—a very long time from the viewpoint of modern psycholinguistics, where (as noted earlier) debates revolve around events that take place 50–100 milliseconds into the boundary of a single word. These spatially sensitive neural imaging techniques can be complemented, however, by the use of event-related scalp potentials, a method that is exquisitely sensitive to neural activity in real time even though the spatial origins of these "brain waves" are still largely unknown. What have we learned so far about the spatial layout of language in the brain using these techniques?

Although it is too soon to draw any strong conclusions, the current state of affairs is discouraging for investigators in search of mental organs. Taking as our text more than 1,000 abstracts from the Second International Conference on Human Brain Imaging (Toga, Frackowiak, and Mazziotta 1996), we can provide the following summary.

First, studies of the brain activity accompanying both linguistic and nonlinguistic processing lead to the conclusion that many different regions of the brain are active simultaneously during any complex mental activity.

Second, in order to isolate specific regions involved in specific tasks, the usual procedure has been to subtract all the distributed activity observed in Task A from the distributed pattern observed in Task B. When such subtractions are conducted, one often does find small, well-localized "points of activation" that distinguish between tasks. The problem is that these points of activation often fail to replicate from one study to another and/or from subject to subject within a single study (Nobre and Plunkett 1997; Poeppel 1996). This is not just a matter of noise in the system. To deal with the vexing problem of "shifting points," a number of investigators have conducted systematic studies in which the same subject is tested repeatedly, using small variations in task instructions (Bub 1994) or stimulus complexity (Just et al. 1996). These small variations often result in a marked shift in brain activity, suggesting that different brain regions are "recruited" as task difficulty goes up and that different regions fire depending not on stimulus content but on the kind of operation that the subject has to perform on that content (e.g., different kinds of memory, attention, rehearsal). The thing-in-a-box metaphor that underlies the phrenological model is hard to defend in the face of data like these. Brain activity seems less like a Henry Ford assembly line and much more like the shifting flow of activity across many muscles that one observes in a dancer or an athlete engaged in a complex physical skill.

What about specific subcomponents of language? Is there any evidence that some "mental muscles" are devoted exclusively to a particular component of language such as word retrieval or grammatical processing? So far, the search for Chomsky's mental organ has not gone well. In a recent review, Poeppel (1996) overlaid the results of multiple PET studies of phonological processing (i.e., processing of linguistically relevant sounds). Of all the claims that have been made about mental organs for language, this ought to be the easiest to verify—that is, the acoustic substrates of language should always engage a common set of processors. To be sure, the primary auditory cortex does seem to be wide awake when sound arrives at the ear, but the same areas are also active for nonlinguistic sounds. Beyond this humble and unsurprising observation, little else can be said. Poeppel found no evidence across studies for a common phonological processor that is uniquely involved in the processing of linguistic sounds, regardless of the task.

Studies of word and sentence processing lead to a similar conclusion. Most studies do show that the left hemisphere is more active than the right during language tasks, although many of these studies provide clear evidence that both sides of the brain are involved, often in homologous patterns (Just et al. 1996; Nobre and Plunkett 1997). There are areas in the left frontal cortex that seem to burn brightly for tasks that involve complex grammar, but these areas have also been implicated in a host of nonlinguistic tasks as well. The left temporal lobe is a frequent "hot spot" in studies that involve processing of individual words, but the same general regions also light up when words are strung together into sentences, with no evidence to date for a single area that wakes up when words turn into grammar. There are also studies pointing to differential centers of activation for different kinds of words, such as names of animals versus names of man-made objects (Damasio et al. 1996; Martin et al. 1996). These studies, however, also differ markedly in detail (i.e., the "tool area" uncovered in one study is not the same as the "tool area" observed in another).

It is possible that we simply have not found the right way to crack the code for language in the brain. The grammar box may yet emerge as techniques are refined. At present, however, it looks as if the burgeoning literature on neural imaging of normal language processing is trying to tell a different kind of story: that language is an activity involving vast stretches of territory in the brain, territory that must be shared with processing of nonlinguistic material. Patterns shift dynamically with task and stimulus complexity, just as we would expect if the phrenologists are wrong and brain activity bears a closer resemblance to the distribution of muscle activity during complex skills than it does to the solemn boxes proposed by Gall and Spurzheim.

To summarize so far, grammatical and lexical structures interact very early in processing in normal individuals, they are impaired or spared together in patients with focal brain injury, and they appear to "light up" the same or similar regions of the brain during processing by normal adults. Even more importantly, the areas that light up during language processing overlap extensively with the areas that light up during nonverbal tasks that make analogous demands on attention, perception, and motor planning (Erhard et al. 1996). We are certainly not trying to argue that grammar does not exist (it does) or that the representations that support grammar are exactly the same as the representations that support simple words and thoughts (they are not). As the literature on neural imaging of language continues to grow, we should not be surprised to find, for example, that syntactic and lexical violations result in somewhat different patterns of cortical activity, including event-related brain potentials (Neville et al. 1991; Osterhout and Holcomb 1993; but see Kluender and Kutas 1993, n.d.) and positron emission tomography (Jaeger et al. 1996; Mazoyer et al. 1993). Grammatical forms, lexical forms, and the nonlinguistic concepts that they represent are not identical. For that reason, deployment of these forms will necessarily involve "nonidentical brain systems" (Mills et al. 1995). We have to configure our arm and hand differently to lift a pin, a book, or a 10-pound box of books. In the same fashion, the brain systems that are deployed during the activation of word endings, root words, whole phrases, and sentences will necessarily differ.

The question we have addressed here is a different one: What evidence do we have to justify the conclusion that grammatical and lexical forms are processed by independent modules, each with its own maturational course, neural representation, and processing profile? We think there is little evidence for a claim of that sort, although there is firm evidence that "different things are processed differently." Language processing involves many different regions of the brain, the patterns of activation that we observe during language processing shift markedly from one task to another, and there is little evidence for the idea that any of these neural systems is unique to language—that is, used for language and language alone.

Having thus redefined the adult steady state, let us turn now to the process by which that steady state develops in the human brain.

BRAIN DEVELOPMENT AND LANGUAGE: LEVELS OF INNATENESS

We have argued against the phrenological picture of the adult brain illustrated in figure 4.1B. In this section, we criticize its developmental analogue, the phrenological infant in figure 4.2. The key to our critique lies not in the rejection of innateness, because something has to be there from the beginning to distinguish between man and mouse. The critical questions have to do with the *content* and the *mechanisms* by which such different outcomes (as man and mouse) are achieved. To answer the questions, we need to analyze claims about innateness into separate levels, permitting us to sort out biologically implausible variants from ideas that could lead to a viable theory of language development and language evolution.

In a recent book called *Rethinking Innateness: Development in a Connectionist Perspective* (Elman et al. 1996), the authors propose three levels at which it would be fair to say that a given brain function is "innate," or at least "innately predisposed": domain-specific representations, brain architecture, and the scheduling or timing of brain development. Let us consider each of these three options in turn, with special reference to claims about the innateness of language.

Innate Representations

Although strong proponents of nativism within linguistics and psycholinguistics are rarely explicit about the level at which innate ideas are implemented in the brain, the usual argument has been that children are born with innate knowledge about basic principles of language in general and grammar in particular (Crain 1991; Lightfoot 1991; Pinker 1991, 1994a, 1994b). To be sure, this knowledge will be shaped by experience to some extent (perhaps in the form of "triggering" or "selecting" among predetermined options [Piatelli-Palmarini 1989]), and some maturation may have to take place before the innate knowledge can be used (Borer and Wexler 1987; Spelke et al. 1992). Most of these investigators, however, have been clear in their belief that children are born with domain-specific representations laid out somewhere in the brain. So to make sense of this claim, we must ask, how might domain-specific representations be instantiated in the brain?

The most likely neural implementation for such innate knowledge would be in the form of fine-grained patterns of synaptic connectivity at the cortical level (i.e., cortical microcircuitry [Pinker 1994b:93]). To the best of our knowledge at the present time, this is how the brain stores its representations, whether they are innate or acquired. But evidence has been mounting against the notion of innate microcircuitry as a viable account of cortical development (i.e., against representational nativism). In a number of recent studies with vertebrate animals, investigators have changed the nature of the input received by a specific area of cortex, demonstrating that these changes result in a very different brain from the one that would have developed under normal circumstances. Here are a few examples:

- Altering the body surface of the fetal animal leads to corresponding changes in the cortical maps that respond to that surface (Friedlander, Martin, and Wassenhove-McCarthy 1991; Killackey et al. 1994).

- When plugs of fetal cortex are transplanted from one area to another (e.g., somatosensory to visual, or vice versa [O'Leary 1993; O'Leary and Stanfield 1989]), the transplanted cortex takes on representations appropriate for its new home and not for its region of origin (i.e., "When in Rome, do as the Romans do").

- When visual inputs are redirected from their intended target to auditory cortex in the infant rodent, auditory cortex takes on retinotopic maps and responds functionally to light (e.g., Frost 1982, 1990; Pallas and Sur 1993; Roe et al. 1990; Sur, Garraghty, and Roe 1988; Sur, Pallas, and Roe 1990; see also Molnár and Blakemore 1991).

- When areas that would normally carry out a specific cognitive function (e.g., the "what is it?" regions of the visual system [Webster, Bachevalier, and Ungerleider 1995]) are lesioned early in life, the same function can be overtaken at a very small behavioral cost by areas with very different processing biases far away from the intended site (e.g., the "where is it?" system).

These experimental results in animal models are compatible with recent findings on the effects of early lesions and/or environmental deprivation in human beings, including the following examples:

- In most cases, children with early injuries to the left or right hemisphere (including hemispherectomy to control seizures) go on to achieve language abilities that are well within the normal range (Bates et al. 1997; Feldman et al. 1992; Vargha-Khadem and Polkey 1992).

- In the early stages of language development, children with focal injuries do display specific relations between lesion site and behavioral outcome; however, these lesion-symptom correlations are markedly different from the ones observed in adults with comparable injuries (e.g., comprehension deficits are actually greater in children with right-hemisphere damage, and lexical and grammatical production are more impaired with left temporal injuries—damage that is usually associated with fluent speech and impaired comprehension in aphasic adults). After five to seven years of age, these patterns are no longer detectable, suggesting that a great deal of reorganization takes place in the first five years of life (Bates et al. n.d.; Reilly, Bates, and Marchman 1998).

- Event-related brain potential studies of congenitally deaf adults reveal response to visual stimuli over areas of auditory cortex that are unresponsive to vision in normal hearing adults, suggesting that unused auditory cortex may have been "recruited" into the visual system (Neville and Lawson 1987).

- Functional brain imagery studies of congenitally blind adults have shown that regions of visual cortex are activated during reading by Braille, suggesting that unused visual cortex has been recruited for this fine-grained somatosensory skill (Elman et al. 1996: chapter 5; Sadato et al 1996; Toga, Frackowiak, and Mazziotta 1996).

These startling findings force us to the conclusion that cortex has far more representational plasticity than previously believed. Indeed, recent studies have shown that cortex retains representational plasticity into adulthood (e.g., radical "remapping" of somatosensory cortex after amputation in humans and in infrahuman primates [Merzenich et al. 1988; Pons et al. 1991; Ramachandran 1993; see also Greenough, Black, and Wallace 1993; Greenough et al. 1986]). Although one cannot entirely rule out the possibility that neurons are born "knowing" what kinds of representations they are destined to take on, the case for innate representations looks poor right now. As Elman and colleagues (1996) note, this means we have to search for other ways in which genes might operate to ensure species-specific forms of brain organization—which brings us to the next level in our proposed taxonomy.

Innate Architectures

Although it now seems unlikely that regions of cortex contain detailed, innate representations, this does not mean that "all cortex is created equal." Regions can vary along a number of structural and functional parameters that have important implications for the kinds of computations they are able to carry out and (by extension) for the kinds of representations they are likely to take on. Elman and colleagues (1996) describe constraints at this level under "architectural innateness." To operationalize architectural constraints in real brains and in neural nets (more on these brain/network comparisons later on), they break things down further into three sublevels:

- Basic computing units. In real brains, this sublevel refers to neuronal types, their firing thresholds, neurotransmitters, excitatory/inhibitory properties, and so forth. In neural networks, it refers to computing elements with their activation function, learning algorithm, temperature, momentum and learning rate, and so forth.

- Local architecture. In real brains, this sublevel refers to regional factors such as the number and thickness of layers, density of different cell types within layers, and type of neural circuitry (e.g., with or without recurrence). In neural networks, it refers to factors such as the number of layers, density of units within layers, presence or absence of recurrent feedback units, and so forth.

- Global architecture. In real brains, this sublevel includes gross architectural facts such as the characteristic sources of input (afferent pathways) and patterns of output (efferent pathways) that connect brain regions to the outside world and to one another. In many neural network models, the size of the system is so small that the distinction between local and global architecture is not useful. However, in so-called modular networks or expert networks, it is often useful to talk about distinct subnets and their interconnectivity.

If we assume that the brain is an enormous and highly differentiated neural network, many parts of which are activated in parallel, then it is reasonable to assume that development is based in part on a process of competition among regions with somewhat different architectures (Changeux, Courrège, and Danchin 1973; Changeux and Danchin 1976; Churchland and Sejnowski 1992; Edelman 1987; Killackey 1990). Through this competitive process, regions of the brain *attract* those inputs that they handle particularly well and are *recruited* for those tasks that require a particular form of computation (not unlike the process by which tall and agile children are recruited to play basketball). In a bidirectional cycle of cause and effect, each region goes on to form representations that are particularly well suited for the tasks that it does best (through additive processes of synaptic growth and strengthening of existing connections and through subtractive processes of synaptic pruning and cell loss; for a review, see Bates et al. n.d.). As a result, the suitability of specific regions for specific tasks increases over time, above and beyond the predispositions that permitted them to "win" in the first place. In this way, an initial architectural bias can result in regional specialization at the representational level.

As Freud (1953 [1891]) and Wernicke (1977 [1874]) argued many years ago, innate constraints on regional architecture and innate variations in input and output can also play a major role in brain organization for language. Specifically, perisylvian cortex may be destined to play a special role in language primarily because of its proximity to the basic input and output systems of speech (i.e., primary auditory cortex; cortical and subcortical speech-motor output systems). If this is the case, then we might expect to find a different pattern of intrahemispheric specialization in visual-manual languages such as American Sign Language—an idea that does have some support (Hickok et al. 1995; Klima, Kritchevsky, and Hickok 1993; Poizner, Klima, and Bellugi 1987). Under this argument, it should also be possible for regions farther away from the privileged perisylvian zones to take over after localized brain injury, if (and only if) they have access to the relevant information. That so many children with perisylvian injuries eventually develop normal or near-normal language provides prima facie support for a process of this kind.

Although the input-output argument can help to explain intra-hemispheric organization for language, it does little to explain why the perisylvian areas of the left hemisphere play a more important role than the perisylvian areas of the right. Presumably, these two areas receive the same kind of information from the speech signal and from the world to which that signal refers. To explain the asymmetry of human language processing, we need to invoke some combination of input-output constraints and innate differences between the left and right hemispheres in local architecture. Together, these initial biases may set in motion a gradual "modularization" process built upon innate predispositions that are only indirectly related to the full form of brain organization for language observed in the adult (see also Karmiloff-Smith 1992). An approach of this kind could explain why it is that the left hemisphere "wins" a primary role in mediation of language in 95 percent of normal adults. At the same time, it is compatible with the fact that children with focal brain injury can develop alternative forms of brain organization for language at surprisingly little cost if the default systems are damaged in some way. However, this still leaves a number of unanswered questions about timing and temporal constraints on cortical plasticity—which brings us to the final point.

Innate Scheduling

Elman and colleagues (1996) underscored the role of timing in all aspects of brain development (cf. Gould 1977), with particular reference to the role that genes play in turning systems on and off at different points in the life span. In addition to the computational biases already described, variations in timing can play a role in the specialization of cortical regions for particular cognitive functions—what Elman and colleagues refer to as "chronotopic constraints" (see also Molnár and Blakemore 1991). For example, regions of cortex may be recruited into a particular task (and develop subsequent specializations for that task) simply because they are ready at the right time. Conversely, other areas of the brain may lose their ability to perform that task because they develop too late (i.e., after the job is filled). Differential rates of maturation have been invoked to explain the left-hemisphere bias for language under default conditions (Annett 1985; Corballis and Morgan 1978; Courchesne, Townsend, and Chase 1995; Kinsbourne and Hiscock 1983; Parmelee and Sigman 1983; Simonds and Scheibel 1989). For example, it has been suggested that the left hemisphere matures more slowly than the right in the first year of life, which could help to explain why the right hemisphere plays a more important role in visual-spatial functions that begin to develop at birth, whereas the left hemisphere takes a greater role in linguistic functions that start to develop many weeks or months after birth. The chronotopic argument may also be related to the subcortical findings reported by Aram and colleagues (1985; see also Eisele and Aram 1995). That is, damage to certain subcortical structures may be more devastating in the early stages of language development (before cortical organization is established) than are homologous subcortical injuries in the adult (after cortical organization for language is complete).

Genetic timing has also been invoked to explain critical-period effects in language learning (Johnson and Newport 1989; Krashen 1973; Lenneberg 1967; Locke 1993). At least two versions of the critical-period hypothesis need to be considered here, one that requires an extrinsic genetic signal and another that does not (Marchman 1993; see also Oyama 1992). According to the "hard" maturational account, plasticity comes to an end because of some explicit and genetically determined change in learning capacity (e.g., a reduction in neurotrophic factors). In this case, the genetically timed stop signal is independent of the state of the system when the critical period comes to an end. In the "soft" maturational account, no extrinsic stop signal is required. Instead, reductions in plasticity are an end product of learning itself, owing to the process of progressive cortical specialization described earlier. In essence, the system "uses up" its learning capacity by dedicating circuits to particular kinds of tasks until it reaches a point at which there are serious limitations on the degree to which the system can respond to insult.

An example of soft maturation comes from Marchman (1993), who simulated aspects of grammatical development in neural networks that were subjected to "lesions" (i.e., random elimination of 2 percent to 44 percent of all connections) at different points across the course of learning. Although there were always decrements in performance immediately following the lesion, networks with small and/or early lesions were able to recover to normal levels. Late lesions (if they were large enough), however, resulted in a permanent impairment of language learning. Furthermore, this impairment was more severe for some aspects of the task than it was for others (e.g., regular verb inflections were more impaired than irregular verbs). Notice that these findings mimic classical critical-period effects described for human language learning (e.g., Johnson and Newport 1989), but without any extrinsic ("hard") changes in the state of the system. Instead, the network responds to the demands of learning through specialization, changing its structure until it reaches a point of no return—that is, a point at which the system can no longer start all over again to relearn the task without prejudice.

As Marchman pointed out, the respective hard and soft accounts of critical-period effects are not mutually exclusive. Both could contribute to the reductions in plasticity that are responsible for differences between children and adults in recovery from unilateral brain injury (see also Oyama 1992). However, if the soft account is at least partially correct, it would help to explain why the end of the critical period for language in humans has proved so difficult to find, with estimates ranging from one year of age to adolescence (e.g., Johnson and Newport 1989; Krashen 1973).

To summarize so far, recent studies of brain development in mammals in general and humans in particular support the idea that cortical specialization is driven by the input that cortex receives. Representations (defined in terms of cortical connectivity) are not innately specified. Current evidence on mammalian brain development is compatible, however, with the idea that there are innate constraints on the architecture and timing of development within specific brain regions (Finlay and Darlington 1995). Thus, arguments about the innateness of language in humans should be rooted in species-specific variations at the architectural and temporal levels. It now seems unlikely that we have evolved a special-purpose stretch of neural tissue dedicated solely to language. But there are good reasons to believe that language has exerted a selective pressure on the timing of development, resulting in some important changes in the information-processing capabilities of various regions in the primate brain (see also Deacon, this volume). From this point of view, a number of facts about the development of language fall into place, including so-called critical-period effects. In the next section, we offer some concrete examples of language learning in neural networks, providing a plausible scenario for the kind of learning device that might have evolved to support language in our species and for the crucial role of timing in the evolution of this device.

CONNECTIONIST MODELS OF LANGUAGE DEVELOPMENT

In the past decade there has been renewed interest in computational models of behavior that have "brainlike" properties. The guiding principle underlying this approach (sometimes called "connectionism," or "parallel distributed processing," or "neural network models") is that many important aspects of human behavior may be explained by understanding the constraints and properties of the hardware (i.e., neural tissue) that supports it. Although it is undoubtedly true that we know far less about brain function than remains to be discovered, there are nonetheless a number of ways in which the brain is clearly different from the standard digital computer. On the one hand, memory access time is dramatically slower in the brain (tens of milliseconds, compared with nanoseconds for the computer); on the other, digital computers execute instructions serially, one at a time, whereas processing in the brain is distributed over millions of units that may be active simultaneously. Programs in the computer consist of lists of instructions; "programs" in the brain are probably stored as modifications in synaptic strengths.

Connectionist models attempt to capture these basic properties by using a large number of simple processing units (each with roughly the computational power of a neuron) in which processing elements may display a graded range of outputs (but with nonlinearities in their output function), with massive interconnections between units. An important feature of this approach is the existence of relatively simple learning algorithms that allow networks of this sort to learn behaviors based on example rather than explicit instruction of rules.

Although one of the early attractions of such systems was their ability to carry out processing in parallel (in a way similar to what is believed true of humans), more recent developments have opened up a new way of dealing with serial processing as well. A variety of approaches have been studied (e.g., Elman 1990; Jordan 1986; Stornetta, Hogg, and Huberman 1987; Tank and Hopfield 1987; Waibel et al. 1989; Watrous and Shastri 1987; Williams and Zipser 1989). The most intriguing are those that treat the processor as a dynamical system (i.e., a system in which the current state depends to some degree on its prior state). These systems are deceptively simple in appearance but have yielded surprising results.

One of the key insights that has emerged from simulations involving these models is that an enormous amount of information can be gleaned merely by attending to serially ordered input and attempting to predict what will come next. In some sense, of course, this is hardly surprising. The order of events in the world (e.g., of words in utterances) is rarely either random or lacking in structure but reflects a rich and complex interaction of constraints. It makes sense that a system that is able to predict which events will follow other events must therefore be sensitive (if only implicitly) to those constraints. What is less clear is how much prior knowledge must be possessed in order to learn these constraints. The simulations we describe suggest that rather less prior knowledge is required than one might have thought.

Learning about Lexical Categories

For instance, consider the problem of lexical semantics: How do children learn the meanings of words? Clearly there are multiple sources of information available that allow children to learn word meanings. But what about the primitive semantic and grammatical categories themselves: animate, human, breakable, noun, transitive verb, and so forth? Are these innately specified? Or can they be induced from the input alone? That languages may differ dramatically in terms of what counts as a basic category (e.g., what is a noun in Turkish versus English) suggests that these categories may be emergent phenomena—reflecting the knowledge that a speaker has acquired—rather than conceptual primitives that are innate.

Elman (1990) attempted to answer this question by presenting a neural network with a simple problem. The network (shown in fig. 4.3) consisted of three layers. The nodes in the input and output layers were used to represent different words (each word was represented by turning on a different node). The intermediate ("hidden") layer was used to allow the network to form intermediate or internal representations of the words it received as input. A context layer was used to recycle hidden-layer activation patterns. This recurrent, or feedback, layer provided the network with internal memory. Note that this internal memory is in no way a verbatim recording of the prior inputs. Rather, it represents a memory for the network's prior internal states (i.e., internal representations). One of the goals of learning is thus to acquire internal representations that not only help produce good outputs (behaviors) but also serve as effective memory.

The network was given a sequence of words as input, one at a time, and asked to produce on its output layer a prediction of which word would follow. There were several rationales for this particular task. As noted earlier, if the order of words reflects some underlying structure (syntactic, semantic, etc.), then presumably the network will need to discover this structure in order to carry out the task. Second, the "teacher" signal (the information required to adjust the connection weights in the network so that it can eventually learn the task) has a plausible source: it is merely the next word, which either confirms or disconfirms the network's prediction. Finally, the task has a psychological plausibility. It is not difficult to believe that young language learners might (covertly) be anticipating the words they hear in an utterance (and we know that adults can do this with a high degree of accuracy; see Marslen-Wilson and Tyler 1980).

The sequence of words that the network "heard" formed simple sentences; these were concatenated so that there were approximately 27,000 words in all (a sample fragment is shown in fig. 4.4).

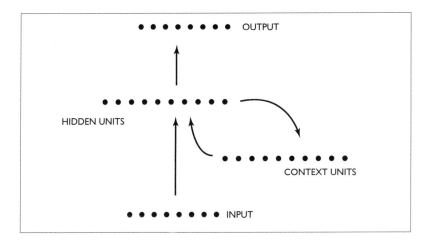

FIGURE 4.3.

Simple recurrent network (SRN). The hidden-unit activation patterns are recycled over time, so that the output of the network depends not only on current input but also the prior internal state.

INPUT	OUTPUT
dog	chase
chase	dog
woman	smash
smash	plate
boy	eat
eat	sandwich
sandwich	dragon
dragon	chase

FIGURE 4.4.

Sample training data for neural network. The input (column 1) was presented to the network, one word at a time. The network's task was to predict the next word in the sequence (column 2). Time goes down the vertical axis.

How were the words represented? Each of the 31 input units was like a neuron that could be turned off or on. Words were therefore represented by assigning to each different vocabulary item a unique

31-bit vector in which 30 bits were off (i.e., had an activation of 0) and a single bit was turned on (i.e., had an activation of 1). Which bit was turned on for a given vocabulary item was decided at random. Thus the representation of words was completely opaque in terms of providing information about a word's grammatical category (e.g., noun, verb) or meaning. The network itself was initially configured with random weights along the connections between layers, so that the predictions at the outset were random (typically all output units would receive some small random activation). As the training proceeded, the network prediction after each new input was compared with the next word that was actually received, and the discrepancy between predicted and actual word was used to adjust the weights in the network (using the back-propagation of error learning algorithm [Rumelhart, Hinton, and Williams 1986]). This procedure allows the network weights to gradually converge on a set of values such that the correct output can be produced in response to any reasonable input.

At the conclusion of training, the network's performance was quite good. Interestingly, the network was unable to predict the exact word that would follow in any given context. Of course, short of memorizing the sentences, this would have been impossible. A sentence beginning "The woman..." might continue in a variety of ways. What the network learned instead was to predict all of the words that might occur in a given context, in proportion to their likelihood of occurrence.

At this point it is worth asking how the network solved the problem. Recall that each word was represented by a vector of 31 ones and zeros in which a single bit—arbitrarily chosen—was turned on. As a result, there was no built-in information about a word's grammatical category or about its meaning. (More precisely, the words were represented by basis vectors, which are orthogonal to each other and lie at the corners of a 31-dimensional hypercube.) Indeed, the very notions of grammatical category and semantic feature were not part of the representations made available to the network.

Grammatical categories and semantic features, however, clearly played a role in the order of words in these sentences. While it is true that many words might occur in any given context, not any word was acceptable. Only words that "fit" by virtue of the grammar or semantics of that sentence could be predicted. So we might expect that in order to solve the prediction problem, the network might have to learn something about grammatical and semantic categories. If so, where would this information be located?

The most likely location would be in the hidden layer. As the network processes a sentence, each new word activates the nodes in the hidden layer. The hidden-layer activations then cause the output nodes to be activated (in order to predict the next word). Although the encoding for words is set in advance, the activation patterns in the hidden units are learned over time by the network, and the network is free to develop internal representations of words that reflect their properties in a way that allows the network to do the prediction task. Thus, it is instructive to look at these internal representations.

This can be done by allowing the trained network to process additional sentences. As each word is presented, we save the hidden-unit activation pattern that it evokes (in the process of producing a prediction). Metaphorically, we might imagine that we are recording event-related potentials (ERPs) from the network as it processes the stimuli, much as we might collect scalp potentials from human subjects as they listen to sentences. Eventually, we end up with hidden-unit activation patterns corresponding to each of the possible input words. These patterns are vectors in 150-dimensional space (because there are 150 hidden units).

Ideally, it would be desirable to visualize the distribution of vectors in this space, since the organization of the space is directly related to the network's notion of similarity. The high dimensionality of the space makes direct inspection infeasible, but we can see how the vectors are distributed in space by looking at the similarity relations between them. A hierarchical clustering tree represents this information by joining vectors that are close in space low in the tree; vectors or groups of vectors that are more distant are joined higher up.

In figure 4.5 we see a hierarchical cluster diagram of a trained network's hidden-unit activations for the words used in this simulation. Remarkably, it appears that the network has discovered a great deal about both the grammatical and semantic characteristics of the words. The hidden-unit activation space is divided into two major regions; verbs (shown on the top of the tree) lie within one region, and nouns (shown on the bottom) lie within the other. Transitive, intransitive, and optionally transitive verbs are grouped separately within the verb space.

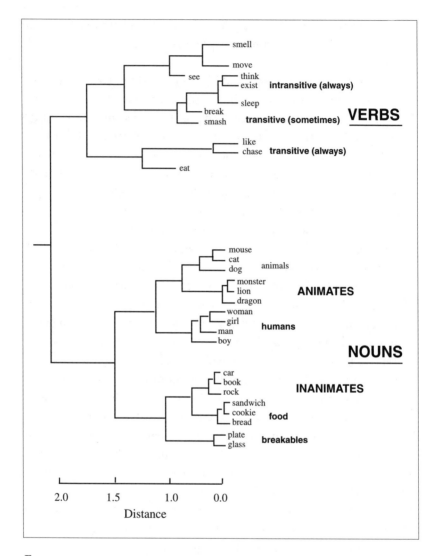

FIGURE 4.5.

Hierarchical clustering of the internal representations (hidden-unit activation patterns) learned by an SRN that learned to predict words in simple sentences. The tree represents the similarity structure of the representations as measured by Euclidean distance between the activation patterns. Although the words do not in any way encode distinctions such as noun, verb, animate, and so forth, the network learns internal representations for the words that reflect their implicit category structure as inferred from the distributional properties of the words.

And nouns are partitioned into regions corresponding to animate and inanimate, with further divisions marking human, nonhuman, breakable, large versus small animal, and so forth.

These groupings are not part of the input in any overt sense, but they do reflect the temporal structure of the sentences. The noun-verb distinction underlies the basic word order; the argument structure of different verbs and the semantic features of different nouns account for other ways in which word order is constrained. So while the network's task did not explicitly involve extracting these distinctions, it is clear that the (apparently simple) task of prediction provides a powerful motivation for inducing much more structure than we might have supposed at the outset.

Rules as Trajectories through State Space

In the last simulation we saw that the network was able to make useful inferences about the lexical-category structure of words on the basis of distributional facts. The representation of the lexicon in this system looks very different from the standard lexicon we are familiar with in psycholinguistics. Rather than being a table of dictionary-like entries, the lexicon here is a region in state space, embodied in the activation patterns on the hidden-unit layer. The hidden units in a network can be thought of as corresponding to dimensions in a high-dimensional space (e.g., with only three hidden units, the state space consists of three dimensions); thus any particular pattern of activations can be interpreted as corresponding to a point in this state space. The network learns to organize this space into regions that represent grammatical and semantic information. Indeed, in this view, words are merely stimuli that induce some change in mental state, and the lexicon is simply the region of space into which the network moves when it "hears" a word. Figure 4.6 attempts to depict how this might look in three dimensions.

What is missing from this picture is a sense of how grammar itself is represented in such a scheme. That is, the lexicon is an important part of language, but we also need a way to represent the constraints on how words may be combined to form interpretable sentences. In traditional schemes, of course, this is accomplished through rules. We might wonder how such constraints might be represented in a system with dynamical properties, such as a recurrent network.

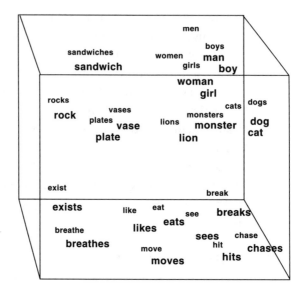

FIGURE 4.6.

A three-dimensional view of the hidden-unit state space developed by an SRN as a way of encoding the category structure of word inputs. The space is structured such that words that are similar (in terms of both grammatical category and semantic features) are close. The precise topology of the space reflects the important distinctions between words.

To explore this question, let us consider a somewhat more complex language than the one used in the first simulation. In the first simulation, sentences consisted of only two or three words. Obviously, natural language typically involves utterances that are not only longer but also more complex. For example, consider a sentence such as "The cat the dogs chase climbs up the fence." In some sense, there are really two sentences here: "The cat climbs up the fence" and "The dogs chase the cat." The sentences are combined here using relative-clause formation (as opposed to simply conjoining them with *and*). As a result, there is a part-whole relationship between the subordinated sentence ("The dogs chase the cat") and the noun it modifies (*cat*). There are not only conceptual consequences to this relationship but also grammatical consequences (e.g., the first noun, *cat*, is in the singular and agrees in number with the last verb, *climbs*, whereas the second noun, *dogs*, is in the plural and agrees with the first verb). An important ques-

Boys who chase dogs see girls.

Girl who boys who feed cats walk.

Cats chase dogs.

Cats who dogs chase chase girls.

Mary feeds John.

Dogs see boys who cats who Mary feeds chase.

FIGURE 4.7.

Sample input used to train a neural network in a prediction task involving complex sentences. Words were presented one at a time; the network's task was to predict the succeeding word.

tion, thus, is how a network such as the one used in the first simulation might represent the structure of such complex sentences. The problem is that the significant structure of the sentence departs from the simple linear order of the words.

This problem was addressed by Elman (1991). A simple recurrent network similar to the one shown in figure 4.3 was trained in a prediction task involving complex sentences. Figure 4.7 illustrates the sorts of sentences that were used.

The prediction task was complicated by the fact that there were many long-distance dependencies (e.g., between main-clause subject nouns and their verbs, often separated by several embedded clauses, or between verbs and their direct objects that preceded rather than followed them). Interestingly, the network was able to learn the grammar only when it began the task with limited resources and a narrow attentional window and then "matured" slowly over time (Elman 1993; this is an example of the chronotopic constraints discussed earlier in the section on innate scheduling). At the conclusion of this training regime, the network not only succeeded in mastering the training data but was able to generalize its performance to novel sentences that contained new lexical items and structures it had not seen before. At least from its performance, it appears that the network learned the complex grammar underlying the data. How was the grammar encoded?

One can answer this question in much the same way as for the first simulation, by looking at the hidden-unit patterns as the network processes the sentences. This time, however, it is instructive to study these patterns as they change over time (and so our network "ERPs" are really more analogous to the time-varying traces recorded with human subjects than to the static traces recorded in the first simulation). This time, to address the problem of visualizing a high-dimensional state space, Elman used principal components analysis (PCA) in order to locate the dimensions along which interesting variation occurred (different principal components encode different information, so depending on what information we are interested in, we may look at different dimensions).

Figure 4.8 shows the trajectory of the network's "mental state" as it processes two sentences: "Boy who boys chase chases boy" (bottom trace) and "Boys who boys chase chase boy" (top trace). The two sentences are almost identical except that the main-clause subject and its verb are singular in the first sentence and plural in the second. The problem for the network is that the intervening relative clause ("who boys chase") is identical in both sentences, and so the network must find some subtle way to "remember" whether the first noun was singular or plural while it processes the relative clause. What figure 4.8 shows is how the network marks the number of the main-clause subject. This is encoded as an upward displacement (for plurals) in the plane of state space.

Figure 4.9 shows state space trajectories for the three sentences "Boy chases boy," "Boy sees boy," and "Boy walks." The verbs in these sentences differ with regard to their argument structure. In this grammar, *chases* is obligatorily transitive and so always requires a direct object; *sees* is optionally transitive (in this example it is followed by a direct object, but it need not be); and *walks* is intransitive. The network learns to encode this difference in argument structure along an axis that runs from upper left (for transitive verbs) to lower right regions of state space (for intransitive verbs).

Finally, in figure 4.10 we see how the network represents levels of embedding. The first part of the trajectory (corresponding to the main clause, "boy chases boy") turns out to be the canonical pattern for all simple transitive sentences (the pattern looks different from that in fig-

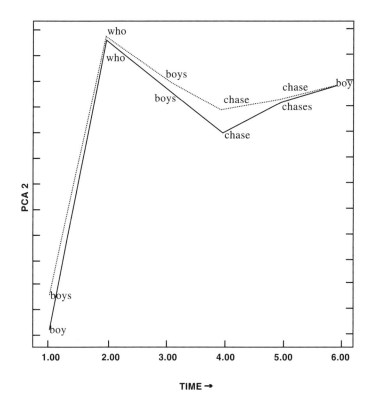

FIGURE 4.8.

The state space trajectories through which a network moves upon processing two different (but similar) sentences: "Boy who boys chase chases boy" and "Boys who boys chase chase boy." The embedded material is identical in the two sentences, and the two sentences contain almost the same words; the network must learn that the main-clause subject in the first sentence is singular, and that in the second, plural. This distinction is marked by a slight displacement in the state space trajectory.

ure 4.9 simply because we have changed our viewing perspective and are looking at a different plane in the high-dimensional state space of the network's "brain"). In this figure, we see that as the network processes additional relative clauses, it replicates the basic canonical pattern, displacing it to the left and downward in state space. This spiraling pattern is typical for embedded clauses and is the way the network represents the hierarchical organization of complex sentences.

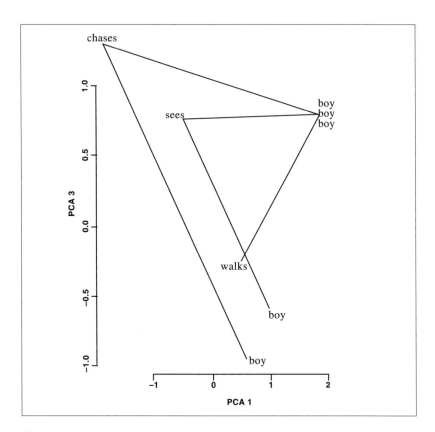

FIGURE 4.9.

The state space trajectories followed by an SRN in response to three sentences: "Boy chases boy," "Boy sees boy," and "Boy walks." The verb chases *is (in the artificial language learned by the SRN) obligatorily transitive and so always has a direct object. The verb* sees, *on the other hand, is optionally transitive and sometimes occurs without a direct object. The verb* walks *is obligatorily intransitive. The network learns to encode these differences in verbs' argument structures along the diagonal axis (running from upper left to lower right).*

Let us now try to summarize what we have learned from these simulations. First, it is clear that there is a great deal of information to be gleaned from the linguistic input. Of course, in some sense we should not be surprised—the information had better be there if communication is to be successful. But what has been unclear in the past is whether this information could itself be used as the basis for a *learning system* in order to develop the knowledge needed to make use of that information.

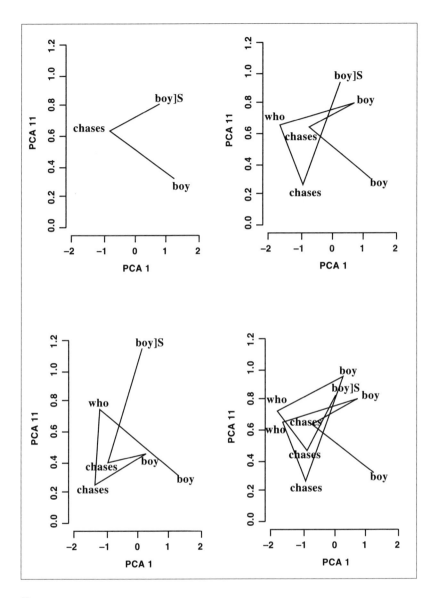

FIGURE 4.10.

The state space trajectories that correspond to processing a simple sentence ("Boy chases boy") alone and when modified with relative clauses in various positions. The network appears to represent embedded sentences by using the same basic pattern for a canonical simple sentence and shifting it over in space; this displacement keeps otherwise identical elements separate while making it possible also to encode the ways in which they are similar.

In other words, how big a bootstrap is required to get things going? Strong proponents of representational innateness have argued that the linguistic input is insufficient for such purposes and that a considerable amount of knowledge must be innately specified.

In these simulations we have seen that networks that start off with minimal a priori information are nonetheless able to discover the lexical-category structure and grammatical structures that are implicit in the input. We do not imagine that children are tabulae rasae in the extreme sense that these networks are; on the other hand, these results suggest caution about exactly how much prior innate knowledge must be attributed to account for language acquisition.

Second, although the prediction task used in these simulations is elementary, and children obviously do more than simply predict, it is also apparent that anticipating how the world will change over time is a powerful driving force for induction and leads to results that are not obviously related to the task itself. We find it plausible to believe that children (and other organisms), particularly at early stages of development, might use prediction as a way to impose basic order on the world and to bootstrap further learning.

Third, these networks give us new ways of thinking about the lexicon and about rules. The view that words are simply sensory stimuli that (through convention) move our brains into particular regions of mental space is appealing and is more consistent with the way we might view other sensory inputs. Sentences are not phrase structure trees but rather trajectories through this mental space. The trajectories are constrained, of course. The dynamics of the system permit certain sequences to be processed; these are the grammatically well-formed sentences. Ungrammatical sentences are sequences of words that move against the dynamics. Thus, rules are simply the sets of attractor and repeller regions that are contained within the system.

Finally—and perhaps most important for our purposes here—the last simulation discussed provides a dramatic illustration of the importance of timing in development and the consequences of timing for behavioral outcomes. One of the particularly important and difficult features of human language is its ability to self-embed sentences (as occurs in relative clauses). Because this property is known to be difficult to learn (in a very precise and technical sense [e.g., Chomsky 1957;

Gold 1967]), some theorists have advanced the strong claim that the ability to self-embed is innate and presumably evolved as a specific cognitive ability.

The simulation involving relative clauses presents a different picture. This simulation demonstrated that in fact embedding *can* be learned—but only if the network undergoes a maturational change in which working memory is initially limited (as it is with infants) and slowly increases over time (as happens with children). In some sense, one might say that this network was innately endowed with the capacity to learn embedded linguistic structures, but note that what is innate (the timing in the increase in working memory) is not specifically linguistic. All that is required from evolution is a small modification in the time course over which working memory reaches adult capacity. Such relatively small changes in developmental timetables are, as we know, common and well-attested mechanisms for evolutionary change and can have far-reaching and dramatic consequences (Finlay and Darlington 1995; Gould 1977; McKinney and McNamara 1991).

CONCLUSION

There is, as we said earlier, no fossil record to support or disprove alternative claims about the evolution of language. All arguments about language evolution are necessarily indirect plausibility arguments that pull together different lines of evidence to support one scenario over another. Our purpose in this chapter has been to establish a plausibility argument for a new view of language evolution, a view in which the language capacity emerges from relatively minor adjustments in the timing of brain development, with implications for the kinds of computations that can be performed by different brain regions. We began by showing that the phrenological view of brain organization for language is simply wrong: Many different regions participate in language processing, in patterns that vary with shifting task demands, and so far as we can tell now, none of the regions involved in language processing is dedicated solely to language itself or to any subcomponent of language (e.g., grammar, phonology, lexical items). We went on to offer an alternative view of the process by which brain organization for language develops in children, an approach based on new ideas about learning and representation in neural networks and on a growing body of research in developmental neurobiology attesting to the plastic and experience-dependent nature of cortical development in mammals. Pulling these disparate strands of evidence together, we conclude that language can plausibly be viewed as a new machine built out of old parts, a capacity that emerged in our species as a result of relatively small changes in the time course of brain development. Although many pieces of the argument are still controversial, all of our claims are testable, providing a viable alternative to the modular (and phrenological) account that has dominated the field for at least two centuries.

Note

Support for this research was provided by the John D. and Catherine T. MacArthur Foundation Research Network on Early Childhood Transitions (R. Emde, director) and by grants NIH/NIDCD 2 R01 DC00216 (Crosslinguistic Studies in Aphasia), NIH/NINDS 050 NS22343 (Center for the Study of the Neurological Basis of Language and Learning), and NIH/NIDCD Program Project P50 DC01289-0351 (Origins of Communication Disorders).

5

The Developmental Timing of Primate Play
A Neural Selection Model

Lynn A. Fairbanks

As a topic, play touches many disciplines. It poses a puzzle to the evolutionary biologist who searches for a biological function for seemingly nonfunctional behavior. It challenges the psychologist to identify developmental consequences and the anthropologist to define its role in the evolution of social and cognitive capacities. In this chapter, I propose that the timing of play is critical to understanding its function and that clarifying the function of play is necessary to enhancing our understanding of the role of play in the evolution of human ontogeny. I present data on the developmental timing of play in one Old World monkey species (*Cercopithecus aethiops sabaeus*), along with results from other nonhuman primate species,. I argue that play evolved to influence neural selection during early brain development and that evolutionary changes in the timing of play reflect changes in the timing of plasticity of relevant neurological and cognitive systems. Finally, I discuss similarities and differences between nonhuman and human play behavior, and I suggest that new forms of play have evolved to promote the acquisition of uniquely human linguistic and cognitive capacities via neural selection.

TIMING AND THE FUNCTION OF PLAY

The function of mammalian play, and nonhuman primate play in particular, has been the subject of a number of excellent books and reviews in the past two or three decades (Bekoff 1974; Burghardt 1984; Caro 1988; Chalmers 1984; Fagen 1981; E. O. Smith 1978; P. K. Smith 1984; Symons 1978). Functions proposed for play have included exercise and physical conditioning; practicing fighting, hunting, sexual behavior, and tool use; learning communication and cooperation skills; developing social relationships; promoting group cohesion; establishing dominance relationships; and learning control of aggressive responses. Discussions of the function of play have typically concentrated on characterizing its forms. Particular play activities are described in detail, and inferences are drawn relating form to function (Symons 1978). Attempts have also been made to estimate the costs and benefits of play in life history models or in natural contexts (Fagen 1981; Martin and Caro 1985). In a few cases, correlational studies have related rates of play to concurrent competence in competitive situations (Caro 1995; Chalmers 1984).

One of the most distinctive features of play is its developmental trajectory. Play is a typically juvenile activity, and rates of play tend to decline steadily throughout the juvenile period (Fagen 1993). Although this feature of play has often been noted, it has rarely been considered in discussions of function. A noteworthy exception is a recent paper by Byers and Walker (1995). They begin with the generally accepted hypothesis that activity play and social play promote motor development. They then use the timing of play to identify the specific mechanisms by which play accomplishes this goal. Behavioral data from rats, mice, and cats indicate that play rises to a peak at about midlactation in these species and then declines rapidly. Byers and Walker argue that play activity should be more effective in achieving the purported benefits at the ages when it occurs than at later ages, and the effects of play should be long-lasting. On the basis of these criteria, the authors examine an extensive list of potential motor training effects of early play. They argue that the frequently cited hypothesis that play is a form of endurance or strength training does not satisfy the foregoing conditions and is inconsistent with the developmental timing of play. Strength and endurance training can be effective at any age, and the effects of strength and endurance training wane rapidly with time.

Byers and Walker (1995) conclude that there are two likely benefits of play that are limited to juvenile ages and that produce permanent effects. These are the effects of play on cerebellar synaptogenesis and on muscle fiber differentiation. In the early development of the cerebellum, more synapses are formed than are retained, and retention is partially dependent on experience (Pysh and Weiss 1979). The cerebellum is important in motor control, regulation of balance, muscle tone, and coordination of fine-skilled movements. Play is timed to occur during the sensitive period of terminal cerebellar development and functions to modify synapse formation and selective retention. At approximately the same time, muscle fibers and the motor neurons that control them are undergoing differentiation (Purves and Lichtman 1985). Neonatal muscle is innervated by several motor neurons. During postnatal development, neuronal endings retract, leaving only one neuron per motor unit. This retraction is in part activity dependent, and once it occurs, the resulting pathways are permanent. Early activity also influences the differentiation of muscle fibers from slow type into a mixture of slow and fast types. The peak rate of play observed in mice, rats, and cats occurs at the end of this period of muscle fiber differentiation. Byers and Walker (1995) conclude that the timing of mammalian activity play has evolved to coincide with these periods of plasticity. They propose that mammalian play occurs at specific ages because those ages bracket sensitive periods during which play activities can permanently influence muscular and neuromotor development.

In this chapter I follow the example of Byers and Walker (1995) in proposing a neural selection model of primate play. Detailed data on play rates of vervet monkeys (*Cercopithecus aethiops sabaeus*) are presented in combination with information from other nonhuman primate species. The results demonstrate that different forms of play each have a distinctive onset, a rate of change, and an offset that is characteristic of the sex and the species. Moreover, the developmental trajectories of different play forms coincide with periods of plasticity in relevant neurological and cognitive systems.

ONTOGENY OF PLAY IN VERVETS

Quantitative data on developmental changes in play behavior were collected during a longitudinal study of social behavior and development of vervets living in captive social groups at the SVAMC/UCLA

Nonhuman Primate Research Facility (Fairbanks 1996; Fairbanks and McGuire 1985, 1988). Vervets are Old World monkeys that have been studied extensively in the field and in captivity (Bramblett 1978; Cheney and Seyfarth 1990; Fedigan 1972; Hauser and Fairbanks 1988; Horrocks and Hunte 1983; Lee 1984). Their behavioral development is similar to that of most other well-studied Old World monkey species. The juvenile period begins at weaning, at approximately six months of age for both males and females, and ends at reproductive maturity. Females reach adult size and deliver their first infant at about four years of age. Males attain adult body weight and secondary sexual characteristics at about five years of age. Tooth eruption patterns, derived from the longitudinal dental records of 104 individuals, indicate that the first permanent molars (M1) emerge at 12 months, the second molars (M2) between 27 and 30 months, and the third molars (M3) between 4.3 and 5.5 years.

The amount of time spent in different activities by individual colony members was determined during instantaneous scan samples taken at six independent time points per week throughout the year. The data set combines longitudinal and cross-sectional samples for all individuals living in four social groups from 1986 through 1989. The sample includes 56 infants (34 male, 22 female), 45 yearlings (19 male, 26 female), 36 two-year-olds (18 male, 18 female), 30 three-year-olds (16 male, 14 female), 33 four- to five-year-olds (9 male, 24 female), and 60 adults six years of age and older (12 male, 48 female). Sampling for infants covered the exact period from birth through 24 weeks. Older animals were categorized by calendar year. The peak of the birth season falls in July, so individuals classified as yearlings were sampled from approximately 7 through 19 months of age.

Four categories of play were coded: activity play, object play, contact social play, and noncontact social play. Activity play includes a variety of locomotor and acrobatic behaviors unrelated to any obvious purposeful activity. Social play, by definition, involves interaction with another individual or individuals and is differentiated into contact and noncontact social play. The object play category includes manipulation and visual, oral, or olfactory exploration of nonfood objects or places. Figures 5.1 to 5.3 show the developmental time courses of these four types of play.

ACTIVITY PLAY

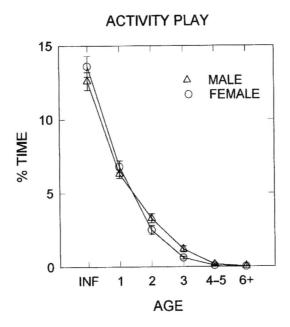

FIGURE 5.1.

Mean percentage of time (± s.e.m.) during which male and female vervets engage in solitary activity play, by age from birth to adulthood.

Nonsocial activity play appears in the repertoire of vervets in early infancy, while they are still spending most of their time in ventral contact with their mothers (fig. 5.1). At first, vervet infants play by bouncing, leaping, and swinging from their mothers' bodies. Later, as infants spend more time away from their mothers, activity play includes a wide variety of acrobatic movements, hanging upside down, and climbing or balancing on unstable surfaces such as the ropes that connect the sides and platforms in the enclosures. Young monkeys use more suspensory postures during activity play than during nonplay activities.

The amount of time spent in activity play rises to a sharp peak in the first six months of life and then declines rapidly after weaning. Further differentiation of play by month of life indicates that individual rates peak between three and five months of age, prior to the emergence of the first permanent molars, when infants are still nursing and are dependent on their mothers for transportation. By one year of age, nonsocial activity play rates have dropped by half. At 2.5 years, the time

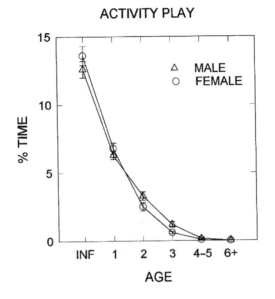

ACTIVITY PLAY

FIGURE 5.2.

Mean percentage of time (± s.e.m.) during which male and female vervets engage in object play, by age from birth to adulthood.

of emergence of the second molars, this form of play occupies about 2 percent of the juvenile's time.

Object play also begins in infancy as young vervets take an interest in handling, exploring, and manipulating nonfood inanimate objects and surfaces. In form, object play includes many movements that will later be used in foraging. Frequent components of object play are rolling or rubbing an object on a hard surface or peeling or prying objects apart with the hands and teeth. Vervets eat an eclectic diet that includes seeds, pods, fruits, and bark, with a small contribution from insects and other animal matter (Struhsaker 1967). Many of their vegetable foods are surrounded by a tough or fibrous exterior coating that must be torn or pried apart to reach the more digestible interior.

Object play peaks at one year of age, when the first permanent molars are erupting, and then declines gradually throughout the juvenile period (fig. 5.2). Object play continues at a measurable rate (2 percent of the time) into adolescence and young adulthood, the time of emergence of the third molars. Male and female vervets engage in

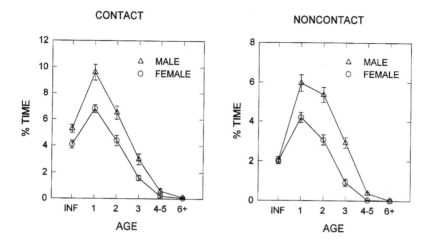

FIGURE 5.3.

Mean percentage of time (± s.e.m.) during which vervets engage in contact social play and noncontact social play, by age and sex.

object play at about the same rate, but males tend to extend object play to a later age than do females.

Social play increases in frequency between 3 and 12 months of age (fig. 5.3), incorporating the vigorous activity and physical exercise of early activity play with the complexities added by competition with social partners. In its earliest form, infants engage in a form of parallel play in which they jump and move next to other infants or juveniles, coordinating the timing of their actions without actually contacting the other animals. Within a few months, they are engaging in true social play. Social play in the vervets consists primarily of rough-and-tumble contact and chases. Young monkeys jump on each other, grapple, wrestle, and manipulate their bodies to get into position to bite their partner's neck without being bitten themselves (Symons 1978). Play movements include defense of the players' own vulnerable body parts at the same time they are trying to gain an advantage over their partners. In play chases, the monkeys run at top speed, often leaping and jumping across chasms or barriers to elude their pursuers.

Interestingly, the chases are usually initiated by the individual being chased. The initiator makes a movement to run away, then looks back to see if he or she is being chased before launching into a full speed escape. Sometimes a chase ends with the lead animal being overtaken, but more often than not the "chasee" stops, turns, and initiates a new bout of contact play.

Both forms of social play peak in frequency at one year of age, coinciding with the emergence of the first molars. Overall, social play declines throughout the juvenile period at a rate that is intermediate between the rapid decline of solitary activity play and the more gradual decline in object play. Within social play, the rate of decline is slower for chasing than for contact wrestling. For infants, the contact forms of social play are twice as frequent as noncontact forms. Thereafter, the relative proportion of social play that involves physical contact declines with age. By three years of age, social play is evenly divided between contact play and chasing.

Unlike activity and object play, there are major differences in rates of social play for males and females. Beginning in infancy for contact play, and at one year of age for noncontact play, juvenile males play more often than juvenile females do. Males spend approximately 50 percent more time than females in social play throughout the juvenile period. Social play rates decline to almost zero for both sexes when the vervets reach their full adult size and completed dentition at four to five years of age.

THE TIMING OF PLAY IN OTHER NONHUMAN PRIMATE SPECIES

The developmental changes in the rate of play just described for vervets are consistent with those reported for other Old World monkey species, including rhesus and Japanese macaques, baboons, patas monkeys, and vervets in other settings. Solitary activity play begins in infancy and is rapidly replaced by social play (Chism 1991; Govindarajulu et al. 1993). Social play peaks at about one year of age and declines throughout the juvenile period, with juvenile males playing more than females (Bramblett 1978; Caine and Mitchell 1979; Chalmers 1984; Fagen 1993; Pereira and Altmann 1985; Rowell and Chism 1986). Similar changes in social play with age have been reported for squirrel monkeys and howlers (Baldwin 1969; Zucker and Clarke 1992). Object play has been studied extensively in capuchins, a genus known for skill in object handling (Visalberghi 1988). In tufted capuchins, the manipulation of food and nonfood objects begins at three to four months, increases in frequency to a peak during the juvenile period, and then declines to adult levels (Byrne and Suomi 1995; Fragaszy and Adam-Curtis 1991), a trajectory comparable to that shown here for vervets.

In relationship to morphological development, activity play in monkeys peaks during midlactation, before the development of the first permanent molars, and then declines rapidly. The peak in social play coincides with the emergence of the first molars, and rates decline to near zero by the appearance of the third molars. Object play has a flatter trajectory. The highest rates occur before and during the appearance of the first molar. Object play declines slowly with age and still occurs at a measurable rate beyond the emergence of the third molars.

Prosimians develop more rapidly than most Old and New World monkeys, typically reaching sexual maturity before two years of age (Harvey and Clutton-Brock 1985). Information on the developmental timing of play for prosimians is scarce, but the few published accounts are generally consistent with the pattern described here for monkeys when scaled to the period of immaturity. Nash (1993) presented developmental data on social play from captive *Galago senegalensis*, a small nocturnal prosimian that reaches adult size and sexual maturity before one year of age. She found that the peak of social play for males occurred between 22 and 38 weeks of age, after weaning and after the emergence of the first permanent molars at about six weeks in this species. Juvenile females had lower rates of social play overall, and play for both sexes declined with age to low levels when adult sexual behaviors appeared at 50 weeks of age. Judging from tooth eruption patterns, the peak of social play for *G. senegalensis* came somewhat later in the juvenile period than that described for monkeys. Further data are needed to determine whether this is a real species difference in the timing of motivation to play or the result of limited partner availability in small family groups.

In a study of free-ranging *Lemur catta*, L. Gould (1990) found that infants peaked in the amount of time spent in environmental exploration (including experimental climbing, jumping, hanging by the feet, and leaping between branches) between 7 and 12 weeks of age, during midlactation. Social play began at 6 weeks of age and increased in frequency until 16 weeks, which is the age of emergence of the first molars in this species (Smith, Crummett, and Brandt 1994). Thus, the peaks of activity play and social play for ringtailed lemurs, relative to molar development, are very similar to those of vervets and other Old World monkeys. The total amount of time devoted to play for prosimian species is limited by the shorter duration of the juvenile period relative to that of monkeys.

The great apes have longer juvenile periods than those of monkeys (Harvey and Clutton-Brock 1985), and they spend more years actively engaged in playful activities. Sample sizes are generally small, and age ranges of subjects are limited, but the available data indicate that rates of social play for mountain gorillas and chimpanzees have their peak in the early juvenile period and decline with age to adulthood. Molar development is similar in these two species, with M1 appearing at approximately 3.5 years, M2 at 6.8 years, and M3 at 11.4 years (Smith, Crummett, and Brandt 1994). Watts and Pusey (1993) reported that social play in free-ranging mountain gorillas declined with age, with the highest rates in the youngest age group studied (3–4 years old). Males in the oldest group (ages 11–13) had the lowest rates but still spent approximately 2 percent of their time in social play. The peak in social play thus coincides with eruption of the first molars in this species. A measurable amount of social play was observed even at the age of appearance of the third molars.

A linear decline in social play between 6 and 10 years of age was also found in a study of free-ranging chimpanzees at Gombe (Pusey 1990). These two ages coincide with the appearance of the second and third molars, respectively. Between ages 10 and 14, the rate of social play in the Gombe chimpanzees was near zero. In a cross-sectional study of play in captive, socially living chimpanzees at the Taronga Zoo, individuals from 1 to 7 years of age were observed (Markus and Croft 1995). At 10 months, rates of social play were low, but they rose rapidly to a peak between 15 and 30 months, before the emergence of the first molar. Social play then declined to age 7, the oldest age period studied, at the time of emergence of the second molars. The frequency of object play also declined with age at the Taronga Zoo throughout the juvenile period, with the exception of one adolescent that spent more time playing with objects than any of the younger juveniles (Markus and Croft 1995). Another study of captive chimpanzees at the Arnhem Zoo found that both social play and object play declined from 3 to 9 years of age (Mendoza-Granados and Sommer 1995). In combination, these studies reflect the general trend for social play to peak at the time of eruption of the first permanent molars, then decline to low levels prior to the emergence of the third molars. Thus, the timing of play in terms of molar development is similar to that seen in monkeys, but the total amount of time spent playing is greater in species with longer periods of immaturity.

This comparison of play in prosimians, monkeys, and apes indicates that the developmental timing of the onset, peak, and offset of play has evolved in concert with the extension of the period of immaturity. The available data are consistent with the view that the onset, peak, and decline in different play types have been preserved across primate evolution, when scaled to maturity as measured by dental development. The comparative data provide no strong evidence for changes in the shape of the play curves or in the relative timing of the different forms of play between prosimians, monkeys, and apes.

NEUROLOGICAL DEVELOPMENT AND THE TIMING OF PLAY

From the perspective of the evolutionary biologist, an activity that consumes a substantial amount of time, is energetically costly, and involves risk of injury or even death must contribute to lifetime fitness, or the process of natural selection would remove it from the repertoire. Combining the data from the four play types in figures 5.1 to 5.3, we find that one-year-old vervets spend approximately 25 percent of their waking hours in some form of play. The widespread occurrence of play across species and the consistency and reliability of its appearance developmentally make it unlikely that play serves no important functions for developing primates. Locomotor ability, fighting skills, and food handling skills are the most likely targets for the functions of activity play, social play, and object play, respectively, judging from the form of each type of play. But how does play accomplish these functions?

As Byers and Walker (1995) point out, physical conditioning and strength training are inconsistent with the developmental timing of play activities. Standard learning models based on accumulation of experience through play fighting would not predict the sharp decline in social play years before fighting skills are needed in adult competition. The developmental trajectory of social play bears no resemblance to a practice schedule that would be used by young athletes training for a competitive sport. A gymnast would not be expected to progressively reduce the time spent training in the years before her first major competition. The timing of play does not coincide with the ages when learning would be most efficient. The arguments proposed by Byers and Walker (1995) for neuromuscular development, combined with new information on development of the neocortex in vervets and rhesus macaques, indicate that the ontogenetic timing of play coincides with periods of maximal plasticity and responsiveness to experience in neurological development. This suggests that the means by which play acts to promote adult competence in physical coordination, fighting, or food handling is through its effects on the developing nervous system.

Solitary Activity Play

Developmental changes in the rate of activity play with age are consistent with the hypothesis that activity play promotes physical coordination by influencing early neuromuscular development. The ontogenetic trajectory shown in figure 5.1 is similar in shape and maturational timing to the age distributions of play in rodents and cats (Byers and Walker 1995), with the peak of activity occurring during midlactation, followed by a sharp decline with age. Activity play includes a variety of actions that would stimulate pathways used for balance, coordination, and agility at the time when exercise influences cerebellar development (Pysh and Weiss 1979). The use of suspensory postures promotes the development of static and dynamic flexibility in developing muscles when muscle fibers are differentiating and peripheral motoneurons are forming their permanent connections (Fontaine 1994; Roy, Baldwin, and Edgerton 1991).

The early peak of nonsocial activity play in monkeys also coincides with developmental changes in the neocortex related to motor activity. Histological data indicate that in rhesus macaques, synaptic density in motor cortex peaks between two and four months postnatally (Rakic et al. 1986; Zecevic, Bourgeois, and Rakic 1989). Experimental deprivation studies have demonstrated that three to six months of age is a sensitive period for the development of spatial and binocular vision in the nonhuman primate visual cortex (Harwerth et al. 1986). Recent studies of brain development of vervets and rhesus macaques using PET technology show a rapid increase in levels of cerebral glucose metabolism between two and four months, with peak activity between three and eight months of age (Jacobs et al. 1995; Moore et al. 1999). A rise in glucose utilization in brain regions has been associated with the emergence of behaviors mediated by those regions in other species (Chugani et al. 1991).

During the period of peak synaptic density and cerebral metabolic activity, vervet infants are still being held and carried by their mothers and other caretakers a substantial percentage of the time (Fairbanks 1990). Activity play dramatically increases the variety, range, and intensity of active movement experienced by the developing nervous system at this age. The stimulation of neuromuscular pathways during this sensitive period can selectively influence which synapses will be retained and which lost, and thereby can have a permanent effect on neuromuscular capacity (Bourgeois and Rakic 1966). Activity play generates the kinds of experiences that will enhance neuromotor development at the time when the underlying structures are most plastic and responsive to environmental influences.

Object Play

Manipulation of objects begins as soon as infant vervets are able to coordinate movement from eye to hand to mouth. Object play occurs at high rates in months three to six, during the time of motor neuron differentiation, rapid synaptogenesis, and peak cerebral metabolic activity in vervets and rhesus macaques (Jacobs et al. 1995; Rakic et al. 1986; Zecevic, Bourgeois, and Rakic 1989). Fine motor control is developing at this time, and it is likely that experience with object manipulation plays a part in the shaping of the neural connections that control precise hand movements. Unlike solitary activity play, however, object play continues to occur at high rates in the second and third years of life.

What is happening during the juvenile period that would lead us to expect the benefits of object play to peak at one year and then decline slowly with age? Neurologically, object play is timed so that it occurs most frequently immediately after the peak in cerebral synaptic density and glucose metabolism described earlier (Jacobs et al. 1995; Moore et al. 1999; Rakic et al. 1986). Decline in the rate of object play coincides with decline in synaptic density in cerebral cortex.

Figure 5.4 shows the density of synapses per neuropil in the motor cortex and primary visual cortex of rhesus macaques (Bourgeois and Rakic 1993; Zecevic, Bourgeois, and Rakic 1989). The number of synaptic connections in developing cortical tissue increases exponentially from midgestation to a peak a few months after birth in these and in other cortical regions (Rakic et al. 1986). Dendritic pruning begins to outpace synaptic formation at five to six months of age and continues throughout the juvenile period. Synaptic density reaches adult levels in rhesus macaques at puberty, about 3.5 years of age.

Experimental manipulations of sensory input to the visual cortex of rhesus macaques have demonstrated that the exponential rise in new synapse formation that occurs before and immediately after birth is not altered by eliminating visual input, but the later decline in synaptic connections is influenced by sensory stimulation (Bourgeois and Rakic 1996). The quality of visual input shapes which synapses will be retained and which will be lost during the period of synaptic pruning. The decline in synaptic density in visual and motor cortex mirrors the curve for object play in figure 5.2. The time spent in object play is directly proportional to the remaining number of extra synaptic connections.

The time course of object play also coincides with the reduction in unmyelinated pathways in the central nervous system (Gibson 1991). Unlike synaptogenesis, which occurs simultaneously over wide regions of the brain, myelination of fiber tracts proceeds in an orderly fashion from peripheral to central pathways and from brain stem to neocortex. Myelination begins before birth and proceeds rapidly in sensory motor tracts that connect the brain stem to primary sensory and motor areas of the cortex. It begins later and proceeds more slowly in tracts that connect within and between cortical regions. Full myelination of all cortical layers is not complete until 3.5 years of age. Myelinated neurons transmit signals more rapidly and more reliably, but unmyelinated neural regions have more potential for change. The

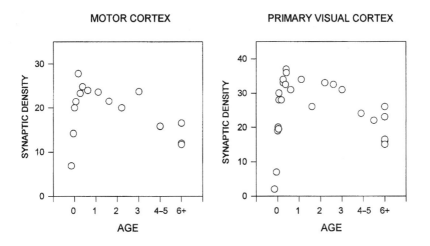

FIGURE 5.4.

Density of synapses in motor cortex and primary visual cortex of rhesus macaques, by age.
Values represent mean density of synaptic contacts per 100 mm² of neuropil per subject. Data
for motor cortex are taken from Zecevic et al. 1989:table 1; the plot for visual cortex is esti-
mated from Bourgeois and Rakic 1993:fig. 5.

rate of object play declines as the number of unmyelinated pathways declines. Thus, object play is timed to occur during the time when the brain is still relatively plastic and when developing neural structures can be modified by experience.

Curiosity, attention to novelty, and the desire to explore are aspects of juvenile psychology that create opportunities to learn about the environment and to take advantage of newly discovered resources (Fairbanks 1993a). A survey of species differences in exploratory approaches to novel objects suggests that there is a relationship between brain size and exploratory tendency: species with large brains relative to body size are more likely to explore novel objects (Glickman and Sroges 1966; Gottlieb 1992). This pattern is consistent with the idea that juvenile curiosity, learning ability, and brain size are linked and that the juvenile period is a time when a heightened motivation to explore enhances input to the developing nervous system.

The exact mechanisms by which social play and object play shape neural pathways are unknown, but the processes are likely to parallel those proposed for cognitive development (Antinucci 1989; Gibson 1990; Parker 1993; Piaget 1952). In Piaget's model of sensorimotor development, individuals construct knowledge and the ability to think and reason through their own actions and through analysis of feedback from those actions. Feedback from repeated concrete object manipulations is internalized, and the effects are permanently incorporated into mental representations or schemes. Piaget's stages 1 and 2 of sensorimotor development involve primary circular actions with the infant's own body. In stage 3, the infant begins to act on external objects and in stage 4 is able to coordinate and transfer actions between objects and across situations. Stage 5 represents the beginning of causal understanding, and by stage 6, the child can solve problems insightfully without having to go through overt trial-and-error processes. Each stage is accompanied by motivation to perform the kind of repetitive physical and object-related activities that provide the feedback to allow progress to the next stage. This same process would also promote selection and reinforcement of synaptic connections and neural pathways to enable the associated cognitive functions.

Parker and colleagues (Antinucci 1989; Parker and Gibson 1977; Parker and Potí 1990) have demonstrated that in capuchins, the process of using an object as a tool to accomplish a goal follows a predictable developmental sequence that parallels Piaget's model of sensorimotor development. When first tested at 5 months of age, infant capuchins were functioning at stage 3 (Parker and Potí 1990). When presented with the task of using a stick to rake in an inaccessible food reward, the infants pulled or slapped at the sticks but made no specific attempts to use them to reach the reward. Between 9 and 14 months, stick manipulation behaviors became more focused on the reward, but the repeated use of wrong movements and ineffective strategies indicated that the young monkeys did not understand how to use the sticks as intermediaries to solve the problem. By 18 months, the capuchins were able to use the sticks consistently and efficiently to rake in the reward. Analysis of the ability to transfer this success to different stick positions and problems indicated that the capuchins were able to achieve an understanding of the stick as a detached intermediary.

Visalberghi and colleagues (Visalberghi, Fragaszy, and Savage-Rumbaugh 1995; Visalberghi and Limonelli 1994) have argued that object manipulation by capuchins leads to success in tool use by an accumulation of trial-and-error experiences. A high motivation to manipulate objects leads to experience with a large number of spontaneous combinations of actions. The capuchin's skill at arriving at new solutions to object problems derives from accumulated experience rather than from the direct influence of object play on emerging cognitive capacities.

The developmental timing of object play is more consistent with the Piagetian view, however, than with the accumulated-experience hypothesis. Object play is most likely to occur when cognitive mechanisms and learning abilities are still immature (Langer, this volume; Parker and Potí 1990; Zimmerman and Torrey 1965). This would be the best time to influence the process of cognitive development, but it would not be the best time to accumulate new information. Learning abilities for complex problems and retention rates are both lower during the period of peak object play than at later ages (Zimmerman and Torrey 1965). Developmental studies of learning abilities of rhesus macaques indicate that infants are able to learn simple visual discrimination tasks as well as adults by the end of the first month of life (Zimmerman and Torrey 1965). The formation of learning sets for greater efficiency in learning new tasks does not begin until four months of age, however, and continues to improve with age. Performance in delayed-response tasks begins at five months and also improves with age. The ability to understand and solve complex problems involving delays and hidden stimuli and the development of problem solving strategies do not reach adult levels until adolescence (Zimmerman and Torrey 1965). The continued development of these more sophisticated learning abilities after 12 months of age suggests that the peak rates of social play and a large proportion of total lifetime play activities are performed when learning abilities are still immature.

After puberty, cognitive capacities are relatively fixed, and the rate of object play declines to less than 1 percent of the time for adult vervets. This developmental trajectory suggests that the benefits of object play are no longer available for adults, a conclusion that is consistent with the Piagetian view that object play contributes to the developmental assimilation of new object constructs and relationships into higher-order cognitive capacities (Antinucci 1989; Parker and Gibson

1977; Parker and Potí 1990). Comparatively, the level of cognitive ability that can be achieved would depend on the species' evolved capacity. The great apes are capable of a higher order of cognitive functioning than monkeys, regardless of the amount of experience the monkeys may have (Parker and Gibson 1977; Spinozzi and Potí 1993; Visalberghi, Fragaszy, and Savage-Rumbaugh 1995). Early object-play activities markedly increase the number and types of object relations that a developing primate experiences at the time when cognitive processes are developing (Fragaszy and Adams-Curtis 1991; Potí 1996) and thus enhance the development of those processes within the range of the species' capacities.

Social Play

During the first year of life, activity play is gradually replaced by social play. For vervets, the amount of time spent in social play peaks at about 12 months of age and then declines steadily throughout the juvenile period to less than 1 percent of the time at puberty. The decline of contact social play follows a slightly steeper slope than noncontact social play, and both reach adult levels sooner than object play, but all three forms of play show a general linear decline from one to four years of age.

The ontogenetic timing of social play coincides with the same changes in development of the nervous system just described for object play. The linear decline in social play rates from 12 months to 4 years of age are paralleled by declines in synaptic density in motor and visual cortex, as shown in figure 5.4 (Bourgeois and Rakic 1993; Zecevic, Bourgeois, and Rakic 1989). Social play continues to stimulate the neuromotor connections for locomotor activity that were activated by nonsocial activity play. It also adds the greater complexity of coordinated physical activity between partners and the strategic assessment involved in play fighting to the juvenile's repertoire during the period when synaptic connections in the prefrontal cortex are being selected (Bourgeois, Goldman-Rakic, and Rakic 1994).

The time course of social play also coincides with the process of myelination of neural pathways in the central nervous system (Gibson 1991). In rhesus macaques, myelination of cerebral cortex is a two-stage process, beginning with a rapid increase up to one year of age, then continuing more gradually to the adult level at 24 months. Social play increases in frequency up to 12 months and peaks at the end of the rapid myelination phase. In the second year, myelination continues in the neocortical association areas in rhesus macaques, and full myelination of all cortical layers is complete at puberty. Social play rates decline as the percentage of remaining unmyelinated fibers declines from one to four years of age.

On the cognitive level, the peak of social play occurs before a young monkey's learning abilities have fully matured and before attainment of adult levels of cognitive functioning. In the first year of life, social play provides feedback to the young monkey about the consequences of engaging in wrestle play with other animals. Social play is comprised of ordered sequences that tend to be enacted over and over again (Stevenson and Poole 1982; Symons 1978). Through repeated experience, monkeys would have the opportunity to develop an understanding of how specific actions influence the reactions of others. Play is the only social context that is as rapid as fighting and requires quick response time and instant social perception and judgment. Effective combat involves precise knowledge of when to escalate and when to retreat. Play sequences provide repeated experiences with the consequences of escalating in different ways and against a variety of different partners. In this way, young monkeys can develop the ability to assess their own and others' relative competitive abilities and to anticipate the outcome of a fight without actually engaging in combat.

Many features of play fighting are consistent with the interpretation that monkeys learn to gauge the consequences of their own and others' actions through play (Bard 1990; Russon 1990). The play face is a communication gesture that allows participants to differentiate play fighting from real fighting (Bekoff 1975). Juvenile monkeys learn to send and receive these signals effectively. They preferentially select play partners that are most closely matched to themselves in fighting ability, suggesting some ability to anticipate and judge the relative attributes of other individuals (Fairbanks 1993b; Owens 1975). When older juveniles play with younger animals, they self-handicap and modulate their actions according to the skill level of their partners. The repetitive experiences derived from social play are likely to play a part not only in the development of the physical skills for fighting but also in the emergence of the mental capacities to know when and where to use those skills to the greatest advantage.

The sex difference in social play is consistent with the relative importance of individual fighting skills in determining male and female adult reproductive success in Old World monkeys. During the time when young males are engaging in high rates of contact play, juvenile females are spending more time in infant handling (play mothering) and in establishing female social and dominance relationships (Fairbanks 1990, 1993b). This differentiation in the behavioral repertoire of juvenile males and females leads to differential selection of neural pathways in favor of those that will be of greatest benefit in promoting adult competence.

The decline of play at puberty is accompanied by increases in sex hormones—estrogens and progesterone for females and testosterone for males—that mediate the maturation of adult sexual characteristics. Experimental studies have demonstrated that these hormones do not initiate the decline in social play with age, but they do play a part in the final reduction of play to adult levels (Loy et al. 1978). The social play of gonadectomized rhesus macaques, like that of normal controls, peaks at one year and declines with age. Gonadectomized males, however, play more often than control subjects at each age, and they continue to play at normal two-year-old levels when they are three and four years of age. Ovarectomy in female rhesus macaques slows the decline of play for three-year-olds but does not prevent play from dropping to adult levels at four years of age.

In summary, the majority of social play happens at a time when monkeys are immature physically, neurologically, and cognitively. Older juveniles and adults are more effective and efficient than infants and yearlings at solving problems and learning new relationships. Social play, therefore, should not have the ontogenetic trajectory that it does if the accumulation of knowledge and information is its primary function. The timing of social play is also not predicted by the physical training hypothesis. An individual training for fighting would be expected to sustain or increase the amount of time spent training as the time of the contest approaches, not to steadily decrease the time devoted to training for years before the first real fight. Other hypothesized functions, such as the development of dominance relationships, are also inconsistent with the developmental trajectory and pattern of social play. Female dominance in vervets and many other Old World monkeys is strongly influenced by matrilineal kinship, and males generally leave the natal group and must establish their adult dominance relationships with unfamiliar animals (Walters 1987).

The developmental timing of social play is consistent with the hypothesis that play functions to influence the development of neural structures and cognitive competencies that promote competitive ability in agonistic conflicts. In the absence of social play, a yearling vervet might never experience intense physical combat. Social play provides frequent and repetitious opportunities for stimulation of neuromotor pathways involved in wrestling and chasing at the time when those pathways are being selected and then myelinated. On the cognitive level, play provides repeated exposure to situations that can promote the development of assessment abilities necessary in conflict situations. Once the process of neural and intellectual maturation is complete, the frequency of play declines to very low levels.

HUMAN PLAY

Play is a frequent and predictable component in the development of human behavior. Bloch (1989) found that American children between two and six years of age spent 30 percent of daytime hours in some form of play. A comparable sampling of similar-aged children in rural Senegal found that approximately 33 percent of nonschool waking hours were spent in play.

The three forms of play discussed here for nonhuman primates (activity play, object play, and social play) are also part of the developmental repertoire of human children. Children's activity play involves many of the same acrobatic movements that we see in nonhuman primate play. Rough-and-tumble play and chase play by human children share many features with the social play of nonhuman primates, including similar sex differences (Humphries and Smith 1984; Parker 1984; Whiting and Edwards 1988b).

The nonverbal forms of play that overlap between nonhuman primates and humans seem to follow similar developmental courses. There are no quantitative studies of children's play that encompass the entire period from infancy to adolescence, but developmental trends can be derived by combining results from different studies (Pelligrini and Smith 1998). Activity play starts at the end of the first year and peaks at around four to five years of age (McGrew 1972; Smith and Connolly 1980), before the emergence of the first permanent molars (see table 10.4 in this volume for timing of molar eruption). Rates of rough-and-tumble play are high during the nursery-school years (McGrew 1972). Among school-age children, the time spent in rough-and-tumble play appears to peak at about 7 years of age and then to decline gradually from ages 7 to 14 (Humphreys and Smith 1984; Pelligrini and Smith 1998). This places the peak of rough-and-tumble play after the emergence of the first permanent molars. Rough-and-tumble play rates are still relatively high at age 11, when the second molars emerge, and they continue to decline to approximately 3 percent of the time at puberty, before the appearance of the third molars. This is comparable to the timing of social play for nonhuman primates, when scaled to molar development.

The same hypothesis that is proposed here for nonhuman primates—that play influences the development of behavioral and cognitive capacities via neural selection—can be extended to children's play. The arguments that object play contributes to cognitive development were derived from studies of human children, and this continues to be an active area of research in developmental psychology (Bates et al. 1979; Bornstein and O'Reilly 1993; Greenfield 1991; Levy 1978; McCune 1995; Piaget 1952). Neurologically, morphometric studies of developmental changes in cerebral cortex in humans have shown the same pattern of changes in synaptic density as described earlier for monkeys, in reference to the rate of maturation. For example, neuronal density and the number of synapses per neuron in visual cortex for human children peaks during infancy, at 8 to 10 months of age, and then declines steadily from 10 months to 10 years of age (Huttenlocher 1990). Limited data also suggest that in frontal cortex, synaptic density peaks a few months later and then declines more slowly, not reaching adult levels until 16 years of age (Huttenlocher 1979, 1990). Thus, changes in the rates of children's object and nonverbal social play with age correspond to changes in synaptic density.

The primary difference between play in nonhuman primates and human children is not in the form and timing of activity and rough-

and-tumble play relative to other developmental milestones but in the large number of additional play types that human children engage in (Parker 1984). The addition of language and symbolic reasoning is associated with a dramatic increase in play types, including make-believe games, board games, word games, and structured team games. These uniquely human forms of play do not emerge after the play types that are shared with other primates. Instead, each has its own developmental trajectory, and some appear before the peaks of activity and rough-and-tumble play. For example, human infants readily engage in games such as peek-a-boo when they are only a few months old. Parker (1984) proposes that these early social contingency games are the earliest manifestation of turn-taking, which plays a central role in language development. There is a close association between the complexity of early social and object play and the development of the mental representations involved in language for young children between 8 and 24 months of age (Bates et al. 1979; Bornstein and O'Reilly 1993; Greenfield 1991; McCune 1995). Play reflects these changes, but it is also likely that play actively influences the process of neural and cognitive development during this period (Bates and Elman, this volume; Deacon, this volume).

Make-believe and role-playing games become prominent in the preschool and early elementary years (Howes and Matheson 1992). Sex differences are often noted in the content of make-believe play at this age, with the play of boys modeling adult male activities, and girls' play, adult female roles (Bloch 1989). Piaget (1962) discussed make-believe games as manifestations of intellectual development, with early games reflecting preoperational thought patterns and later games exhibiting operational intelligence. Research in this area is focused primarily on the early appearance of language-based play forms (Howes and Matheson 1992), and quantitative data on changes in rates of these play types with age are scarce.

Games with formal rules replace make-believe games in frequency among older children and adolescents (Hughes 1991). Formal games encompass many kinds of physical, mental, social, and linguistic skills. As with play fighting among nonhuman primates, the frequency of engaging in most play games declines with age, but some uniquely human forms of play continue into adulthood and even into old age.

Comparison of human and nonhuman primate play suggests that there was a qualitative shift in capacities associated with language during human evolution that is reflected in the addition of play forms throughout development. Many aspects of the form and timing of nonverbal activity play, object play, and play fighting have been preserved, when scaled to maturation as assessed by molar development. But human play also includes features that are absent in the play of other primates. These additional play types reflect the extraordinary human capacities of language, operational understanding, and abstract reasoning. The preceding discussion of the functions of nonhuman primate play suggests not only that these uniquely human forms of play reflect human linguistic and cognitive capacities but that they have evolved specifically to promote the development of these abilities through neural selection during the extended period of brain plasticity and synaptic pruning. A precise analysis of ontogenetic timing of uniquely human forms of play is likely to reveal associations between the frequency of engaging in specific play types and periods of plasticity in relevant domains of neural and cognitive development.

PLAY IS NOT ESSENTIAL

Discussion of the function and evolution of play must take into account the fact that rates of play can vary widely according to circumstances. During times of stress or limited food availability, juveniles play less. The influence of reduced food supply on play has been documented for a variety of species in both field and controlled laboratory settings (Baldwin and Baldwin 1976; Lee 1984; Loy 1970; E. O. Smith 1978; Sommer and Mendoza-Granados 1995). The sensitivity of play rates to environmental circumstances supports the idea that play is an energetically costly activity, but it creates problems for the view that play is necessary for normal development. If play were essential for normal social and cognitive development, then we would expect it to be better buffered against expected variability in the environment. Baldwin and Baldwin (1974) proposed that normal social behavior can develop without the benefit of play but that individuals can enhance their ability to read subtle social cues and respond at a higher level of competence with social play experience. In one of the few studies to test the relationship between rates of play and success in competition,

Chalmers and Locke-Haydon (1984) demonstrated that infant common marmosets that played more often were more likely to take food from their mothers in a test for food competition and were more likely to anticipate the consequences of being in conflict by their behavior. This result supports the assertion that experience with social play at appropriate times in development can enhance competitive ability.

It is important to remember that play is not the only way to gain experience. Particularly in the development of object manipulation skills and their accompanying cognitive structures, it is possible to practice a skill without playing. Smith and Simon (1984) demonstrated that practice in a specific task was as effective as free play in children's performance in problem-solving tasks. Individuals with restricted opportunity for object play may be able to accomplish the developmental functions that are served by play through direct experience with handling objects in nonplay contexts. Object play would provide more opportunities for rapid assimilation of object constructs, but restriction of object play would not mean that adult cognitive levels could not be achieved.

Opportunities to stimulate neural pathways involved in fighting skills in the absence of social play would be more limited. Therefore, we would expect individuals with limited energy or with few appropriate play partners to make the best of the opportunities they do have. Biben (1986, 1989) demonstrated that juvenile squirrel monkeys varied the quality of play according to the characteristics of their play partners. When they played with partners who could be dominated, young squirrel monkeys were more likely to use a type of wrestling play in which one animal tried to pin the opponent to the ground. When playing with partners that could not be dominated, the monkeys switched to a nondirectional form of wrestling that used more suspensory postures and had no clear dominance roles. The form of play took advantage of opportunities for developing fighting skills when winning was likely and for physical flexibility training when it was not.

CONCLUSION

Much of the large body of literature on mammalian and non-human primate play argues function from the design features of play. From an extensive list of proposed functions, the most widely supported are the development of competence in motor coordination, fighting ability, and food-handling skills. A close examination of the ontogenetic timing of different forms of play has allowed us to delimit more precisely how play is likely to influence development to achieve competence in these areas.

Primate play is timed to coincide with periods when the central nervous system is relatively plastic and susceptible to environmental influence. Primate play rates decline throughout the juvenile period in a trajectory that is similar to the decline in synaptic density and unmyelinated pathways in the brain, a pattern that is consistent with the neural selection hypothesis that play has evolved to influence postnatal development of neural systems. The effects of play on strength and endurance are short-lived, but if play promotes motor training during the sensitive periods of cellular differentiation and synaptic selection, its influence on neuromotor capacities can be permanent.

The neural selection model of primate play is consistent with theories of sensitive periods for experiential feedback in the development of competence in the cognitive domain. All three play types involve cognitive processes that parallel neural development. Activity play requires judgment of distance and risk, object play propagates the development of familiarity with object properties and relationships, and social play involves communication, interpretation, and prediction of the actions of others. Play is timed to occur when cognitive understanding is not fully developed and when particular experiences have the capacity to generate permanent changes in cognitive capacities.

The types of movements and interactions involved in activity play, object play, and social play serve to repeatedly stimulate pathways involved in performing actions that will be useful in later life. Thus, play is a way to prepare the developing nervous system for capacities that will be needed in the future by selective retention of stimulated pathways. There are many socio-ecological differences between the behavior and activity profiles of infant, juvenile, and adult primates (Janson and vanSchaik 1993). Play ensures that neural development will not be limited to the kinds of experiences that are typical for non-playing infants and juveniles. Activity play prepares the nervous system for physical challenges in the environment, such as leaping across gaps in the canopy, that must be faced after infants can no longer rely on their mothers to assist or carry them. Object manipulation provides opportunities to experience the properties of objects that may be included in the diet only after adult canines have developed or during times of food shortage. Social play gives infants and juveniles a chance to experience behaviors that are important to adult reproductive success, such as fighting and sexual behavior, that would otherwise be absent from the immature animal's behavioral repertoire. Through play, a young monkey can move activities from the time when they will be used seriously to the time when the brain is most modifiable. Early stimulation of the appropriate pathways will then selectively augment the development of capacities that will influence later competence.

Careful consideration of the developmental timing of primate play leads to the conclusion that each of the different play types influences later capacities via the process of neural selection, by increasing relevant experience during the time when the neural and cognitive mechanisms involved are undergoing ontogenetic modification. Thus, primate play contributes to the development of fighting skills, object handling abilities, and motor performance by selectively enhancing the basic neuromotor structures that will later be available to perform these tasks.

Comparative analysis of play across the primate order suggests that the three basic forms of play (activity play, object play, and social play) are scaled to the period of immaturity. The longer the period of development, the longer the period of play, but there is no evidence for major changes in the order of appearance or relative timing of each play type between the prosimians, monkeys, and apes in reference to molar development. Comparable data on neural development and age-related changes in synaptic density are currently unavailable for prosimians and great apes, but there is a strong correlation between the length of the juvenile period and adult brain size across primate species (B. H. Smith 1989). It is reasonable to expect that the evolutionary extension of the timing of play activities reflects an extension in the period of synaptic pruning and neural plasticity. Among the higher primates, the longer period of neural plasticity creates more opportunities for fine tuning of neural networks through play. By evolving in concert with changes in the timing of maturation, the motivation to play serves to enhance relevant early experiences and thereby to contribute to phylogenetic differences in neurobiology and behavior.

Human play includes a large number of distinct play types that appear in addition to the basic nonhuman primate repertoire. Like the timing of play forms that are shared with the nonhuman primates, it is likely that the timing of these uniquely human forms of play reflects early neural plasticity in relevant motor, neural, and cognitive systems and that play has evolved to hone these skills in accordance with the neural selection model proposed here.

Note

I would like to thank the chairpersons and participants in the SAR seminar "The Evolution of Behavioral Ontogeny" for a stimulating week that inspired the ideas in this chapter. Sue Parker, in particular, made many helpful suggestions in the preparation of the manuscript. Collection of the behavioral data would not have been possible without the able assistance of Karin Blau and Jill Cullin and funding from NSF grants BNS 84-022292 and BNS-87-09765.

6

Evolutionary Development, Life Histories, and Brain Size

Finding Connections via a Multivariate Method

J. L. Gittleman, H.-K. Luh, C. G. Anderson, and S. E. Cates

In this chapter we attempt to draw connections among three issues: evolutionary development, life history traits, and brain size. Each has received considerable attention. Yet despite close interrelationships among them and the common thread of phylogenetic history running throughout them, relatively little work has been directed toward synthetic analysis (see also McKinney, this volume). There are many reasons for this, not the least of which is that each problem involves many variables that are not necessarily interconnected. For example, studies of life history traits traditionally involve aspects of ecology and population biology that are separate from evolutionary development or brain size. Two important advances now permit types of comparative analysis that will search for the right variables. First, the nature of phylogenetic reconstruction has improved dramatically with better analytical tools and more accurate molecular-morphological phylogenies. Second, new ways of using the comparative method with modern phylogenies allow rigorous testing of hypotheses about the direction and extent (heterochrony) of the evolution of traits (brain size and life history).

We first describe how and why different kinds of information in phylogenies dictate the types of evolutionary questions that can be asked. We then selectively review studies of evolutionary development (heterochrony), life histories, and brain size that suggest causal connections. Finally, we introduce a new comparative method for finding multivariate evolutionary patterns in these problem areas. Specifically, we apply our method to the hypothesis that brain size evolution in mammals is influenced mainly by age-specific mortality rates, which in turn affect variation in gestation length and litter size. Conceptually, we emphasize that for comparative studies involving many traits, it is time to recognize the true complexity of covariation in trait evolution and move away from reliance on univariate and bivariate analyses.

A warning is needed at the outset. Much of our chapter deals with difficult issues of how to analyze interrelationships among many traits. Why should anyone work through the tough equations involved with these issues, much less in a book devoted to "the evolution of behavioral development"? There are two reasons. First, as the chapters in this volume reflect, ontogenetic changes are complex. Not only are many variables involved (e.g., morphology, physiology, behavior, ecology), but each variable likely shifts (evolutionarily) in a unique temporal fashion (i.e., through heterochrony). It therefore becomes necessary to adopt a multivariate approach, at least at this early stage of finding out which variables are salient. Second, behavioral ontogeny is critical for analyzing how and why certain patterns are maintained or transformed during an organism's life history (see Shea, this volume). A promising exchange between life history theory and behavioral ontogeny has emerged in recent years (Bateson 1982; Gould 1977; Horn 1978; McKinney and Gittleman 1995; McKinney and McNamara 1991). We similarly view behavioral development as a critical component of life histories. For example, the development of feeding or communicative behaviors both influences and is influenced by life history traits such as size, number, and precocity of offspring. As more systematic comparative information becomes available on behavioral ontogeny, similar to that for traditional life history traits (e.g., gestation length, birth weight, litter size), the methods discussed here will amplify this important exchange.

WHAT IS A PHYLOGENY?

The limiting step in any study of evolutionary development, including studies of behavioral ontogeny, is the availability of reliable phylogenetic information. Without information about phylogeny—"the study of the evolutionary routes followed by particular organisms" (Desmond 1982:148)—there is no null hypothesis against which to examine evolution. This is why it is necessary at the outset to review a few of the assumptions and methods of phylogenetics.

Information conveyed through phylogenetic study is typically presented in some graphical form such as a phylogenetic tree, cladogram, or branching diagram, each of which has analytical variants (e.g., rooted or unrooted) (Eggleton and Vane-Wright 1994; Penny, Hendy, and Steel 1992). Such illustrations represent evolutionary transformation, but in a tricky sense: no observations are known for the actual transformations. Transformations are inferred in various ways, including the use of developmental (ontogenetic) characters. For example, by looking at the appearance of gill arches in embryos of vertebrates—after making some assumptions for distinguishing primitive and derived character change—we can trace ontogenetic transformations among fishes, amphibians, and mammals. A phylogeny is thus constructed, although we must bear in mind that a phylogenetic tree is only an estimate (hypothesis) of the actual historical events (evolutionary transformations).

A phylogenetic tree contains four elements that are useful for examining character evolution (fig. 6.1): roots, nodes, branches, tips. A root is a branching point at the base of a tree. Nodes represent putative ancestors at branching points. Branches connect internal nodes or, at the terminal branches of a tree, a node to a tip. And tips are the taxa of study, most often species, occurring at the terminal branches. With this information in hand, we can theoretically measure the overall relationship between a trait and the phylogenetic tree (i.e., "phylogenetic correlation"; see Gittleman and Luh 1992), evolutionary change in traits at precise phylogenetic time intervals (i.e., whether traits are evolving in certain ways at nodal splits; see Gittleman et al. 1996a), and the extent of trait evolution and branch lengths (i.e., rates of trait evolution; see Gittleman et al. 1996a). In essence, we

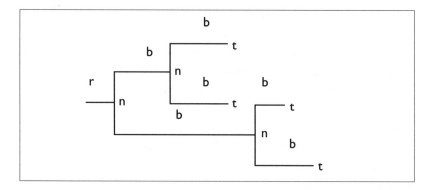

Figure 6.1.

Constituent parts of a phylogenetic tree. Key: b = branch, n = node, r = root, t = tip.

want to evaluate *relative* evolutionary patterns among traits (mortality, life histories, brain size) as they change phylogenetically.

In using phylogenetic information to examine character evolution, a distinction is often drawn between whether a phylogeny expresses a natural hierarchy or serial transformation (Patterson 1983). A Haeckelian phylogeny is a chain of direct ancestors that represent a branching process; a more von Baerian phylogeny is one that reflects serial ontogenetic transformation. Patterson argues that the latter type of phylogeny is, as expected, the most direct link between phylogeny and ontogeny. Indeed, this often leads to the view that there is no difference between phylogeny and ontogeny. But given that a phylogeny can reflect true historical changes among taxa, it makes no difference whether this is expressed as transformation or hierarchy. Both types of phylogenetic information pose hypotheses for character evolution, ontogenetic or otherwise; this is the context for using phylogenies to generate hypotheses about mammalian brain size and life history evolution.

Historically, phylogenies were based on morphological information such as skull shape or presence of gill slits. The problem with using these phylogenies to study trait evolution is that the phylogeny is built on the same traits we are using to test the evolutionary hypotheses: there is circularity in the use of traits. Phylogeny reconstruction has entered a new era in which tests of character evolution are independent of the phylogeny itself. Molecular phylogenies now provide an opportunity to examine trait evolution on the basis of relatively neutral characters (Gittleman et al. 1996b). In terms of studying interrelationships of life histories, brain size, or indeed the evolution of any trait, it is critical to ask whether the characters under study are also ensconced in the phylogeny that is being used. Such circularity is obviously problematic in dealing with ontogeny. Characters of size, growth, or morphological specialization are specifically linked to ontogenetic change *and* to the phylogeny. Now, with molecular phylogenies, this circularity can be broken.

In the present study, we use phylogenetic information based on both morphological and molecular information. This is because for the trait relationships and the taxa involved, complete molecular phylogenies are unavailable. Even so, with the data at hand we present a multivariate phylogenetic problem and a comparative method for tackling it.

THREE PHYLOGENETIC ISSUES

The following is a selective account of theoretical issues and empirical studies of heterochrony, life histories, and brain size that suggest interconnections among the three. The "selective" aspect specifically refers to our emphasizing issues that have a significant phylogenetic component.

Heterochrony

Of all the terms in evolutionary biology, few are used more often yet are so poorly understood (or defined) as *heterochrony*. We do not review here the many definitions or applications of heterochrony, or the controversies surrounding it, but refer the reader to thorough discussions elsewhere (Gould 1977; McKinney 1988a; McKinney and Gittleman 1995; McKinney and McNamara 1991). In general, heterochrony deals with changes in developmental timing, typically relative to the appearance or rate of development of a character during phylogeny. Many classification schemes have been developed for types of heterochrony, but they all pinpoint different kinds of change in terms of rate, onset, and offset. In these terms, McKinney and McNamara (1991) showed six kinds of heterochrony (fig. 6.2): neoteny (slower rate), acceleration (faster rate), postdisplacement (late onset), predisplacement (early onset), progenesis (early offset), and hypermorphosis (late offset). Patterns of growth rate, onset time, and offset time can be

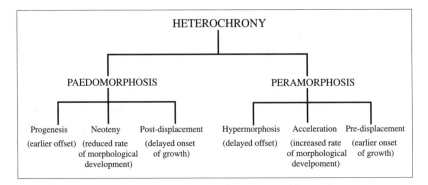

FIGURE 6.2.

The hierarchical classification of heterochrony. (Modified from McNamara 1986; see also McKinney and McNamara 1991.)

examined empirically between taxa with ancestor-descendent relationships (fig. 6.3). As argued earlier, employing phylogenetic information that is independent of the characters themselves provides a neutral model for testing heterochrony.

One problem in connecting the concept of heterochrony with explanations of differences in trait evolution (e.g., life histories, brain size) is in distinguishing it as a noun or verb: a developmental shift may be an empirical observation of a particular kind of ontogenetic change in a character or a mechanism for evolutionary change. In the present context it does not matter. Heterochrony can be applied to character evolution as a descriptive classification of empirical pattern; for example, brain size in chimpanzees reaches an adult form later than in gorillas (Shea 1983d). Heterochrony can also be used as an evolutionary mechanism; brain size in chimps is larger than in gorillas because it is neotenic.

Life Histories

Life history patterns represent various reproductive traits and the probabilities of survival at each age during life. Gestation length and litter size are typical life history traits in mammals. There has been a considerable resurgence in life history studies, as reflected in relatively new books on the subject (Charnov 1993; Roff 1992; Stearns 1992). Comparative studies across eutherian mammals consistently reveal the following trends among most life history traits (Berrigan et al. 1993;

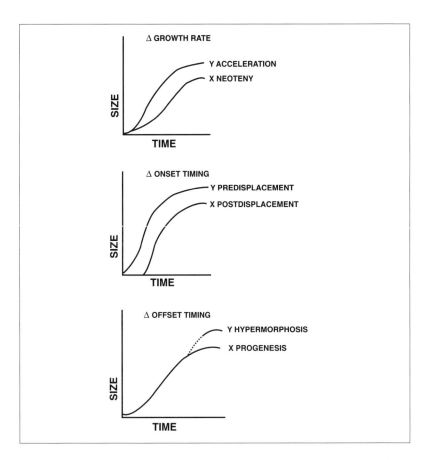

FIGURE 6.3.

The six major types of heterochrony in terms of size versus age plots. The ancestral trajectory would be located between Y and X. (Taken from McKinney 1988; see also McKinney and McNamara 1991.)

Gittleman 1993, 1994a; Harvey, Read, and Promislow 1989; Purvis and Harvey 1995): (1) life histories are significantly correlated with phylogeny, with the highest correlation occurring at the generic and family levels; (2) after accounting for phylogeny, life histories are correlated with allometric variables (brain size, body size), though there is considerable residual variation around lines of best fit such that size is considered only partially influential; (3) basal metabolic rate and various ecological factors (e.g., diet, habitat) are seemingly unimportant in explaining life history variation; and (4) age-specific mortality rates, specifically juvenile mortality (at less than one year of age) and adult mortality (at greater than three years of age), are significantly correlated with life histories and, more importantly, in the direction predicted by theory. That is, following Charnov's (1991, 1993) model, life histories compensate for environmental mortality patterns: high mortality is correlated with fast temporal variables (e.g., early age at sexual maturity, short gestation length, young age at weaning) and large litters, whereas low mortality rates are correlated with slow life histories and small litters (see Purvis and Harvey 1995 for a more rigorous test of Charnov's model). An extreme example of the first trend is observed in the cheetah *(Acinonyx jubatus)*, which has juvenile mortality of up to 95 percent and associated high reproductive output. Contrast this with the spotted hyena *(Crocuta crocuta)*, which has a juvenile mortality of less than 10 percent and a correspondingly slow life history (Gittleman 1993).

Clearly, the foregoing trends are somewhat preliminary in the sense that of the more than 4,600 mammal species, only a few hundred have been analyzed comparatively. Although relationships between mortality rates and life histories are now better understood from theoretical models and independent empirical study, the precise nature of how mortality influences life histories is unknown. Many intrinsic sources of mortality (e.g., individual resource allocation) and extrinsic sources (e.g., competition, low resource availability) are observed in nature, though few single-species field studies have systematically determined which sources are most important. Caro's (1994) and Laurenson, Wielebnowski, and Caro's (1995) recent analyses of the cheetah, showing that high juvenile mortality is due mainly to predation by lions *(Panthera leo)* and spotted hyenas, is an important example of what kinds of data are needed. At this stage, no comparative analysis has pinpointed whether a single source of mortality influences life history evolution more than others. Nevertheless, we now at least have an important explanatory variable—mortality rate—that appears to influence mammalian life histories.

Brain Size

Studies of brain size have always been and will continue to be controversial. In addition to political issues (see Gould 1981), there remain critically important questions about whether brain size relates to cell number, functional connections among neurons, or information-processing abilities. Such questions must be answered before comparative studies that claim significant behavioral-ecological correlates of brain size (e.g., Deacon 1990c) are fully accepted. Nevertheless, it seems premature to ignore comparative studies in generating hypotheses and, certainly, in detecting repeated and consistent trends that support functional associations of brain size across taxonomically diverse groups (e.g., associations between diet and perceptual complexity; see Gittleman 1994b; Harvey and Krebs 1990; Hofman 1983; Jerison 1973). We do not wish to advocate here a specific biological significance of brain size. Rather, we are interested in the relationship between life histories and brain size across mammals.

Pagel and Harvey (1988, 1990) analyzed species differences in neonatal brain size—a surrogate for adult brain size—across 116 eutherian mammal species from 13 orders. Independent of maternal size and metabolic rate, neonatal brain size was strongly correlated with gestation length and litter size. As examples of this pattern, the Delphinidae and Phocoenidae (dolphins and porpoises), the Procaviidae (hyraxes), the Chinchilladae (chinchillas), and various primate families have long gestations, small litters, and large neonatal brain sizes. By contrast, taxa with relatively small neonatal brain sizes, short gestations, and larger litters include the Felidae (cats), Canidae (dogs), Suidae (pigs), and Erinaceidae (hedgehogs). Thus, Pagel and Harvey argued that mammals develop brain sizes in association with life history strategies, which in turn relate to effects of environmental mortality patterns. We now turn to this multivariate hypothesis, which provides the context for our introduction of a multivariate comparative method.

A MULTIVARIATE PROBLEM

The preceding sections include two primary conclusions about mammalian evolution (fig. 6.4): (1) age-specific mortality rates influence life history traits, and (2) life history traits, particularly gestation length and litter size, influence the development of brain size. This begs the question of whether there exists a multivariate relationship involving mortality, life histories, and brain size. Indeed, Pagel and

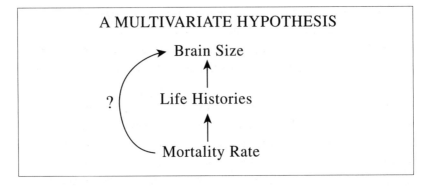

FIGURE 6.4.

A hypothesized multivariate relationship, as suggested by Pagel and Harvey (1988), wherein mortality rates influence life histories, which in turn affect brain size.

Harvey (1990:121) originally suggested the idea when they wrote: "If variation among [mammal] species in rates of mortality is the reason for the evolution of species' differences in gestation length, and ultimately of variation in neonatal brain size, then species with relatively high rates of mortality should have relatively short gestations and produce neonates with relatively small brains."

Pagel and Harvey used the best comparative methods of the time, but these did not incorporate actual phylogenetic (topological) information. More importantly, their work was restricted to bivariate analyses. Comparative study to date has been almost exclusively univariate and bivariate. This relates to many methodological reasons, a primary one being that most comparative data sets contain empty cells, which do not allow for multivariate approaches. This is not the case here, because life history and brain size data are available for a select group of mammal species. Rather, the problem is one of complexity: How do we develop a multivariate model that tests interrelationships of numerous traits, all of which covary with allometry and phylogeny and, indeed, among the traits themselves? Figure 6.5 gives a schematic illustration of a multivariate analysis of the relationship among mortality, life history, and brain size.

We now present an analytical model, more fully developed by Luh and colleagues (n.d.), that examines the multivariate hypothesis that mortality rate affects life history traits, which in turn affect brain size.

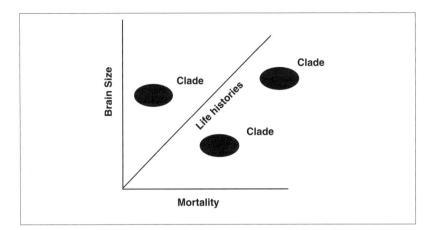

FIGURE 6.5.

An illustration of a multivariate problem that includes variation in a size variable (brain), life history variables (e.g., gestation length, litter size), and mortality rate. All of these traits are also changing phylogenetically (as shown by the positions of clades), which may be expressed in different patterns of heterochrony.

All of the species trait values and phylogenetic information used in our analysis are taken from Promislow and Harvey (1990) and are supplemented with more recent data drawn from the literature (the complete data file is available upon request). The data set contains eight traits (adult body size, adult brain size, gestation length, litter size, age of sexual maturity, period of lactation, juvenile mortality rate, and adult mortality rate) sampled from 48 mammal species. Because all of the variables involved in the analysis are significantly correlated with body size across mammals (Stearns 1992), it is necessary to remove allometric effects on each variable prior to employing the multivariate model. Thus, all examined variables are residuals after accounting for differences in size.

A potential problem in adopting a multivariate approach to these data is that our sample size does not fulfill the general rule in multivariate statistics that the sample should represent 10 times the number of variables being examined. Our multivariate approach, however, does not involve hypothesis testing in a statistical sense. Here, we are using multivariate model selection in a descriptive sense to construct an evolutionary model of the proposed variables.

FINDING CONNECTIONS: A MULTIVARIATE COMPARATIVE METHOD

An appropriate multiple linear regression model that fits the relationship of different traits across a broad range of taxa must incorporate two steps. First, regression is used to investigate one dependent variable (brain size) in relation to one or more independent variables (life histories, mortality rates). Second, phylogenetic autoregression is used to remove any spatial dependence (i.e., phylogenetic nodal sequences and branch lengths) in the functional relationships among the traits. Appendix 6.1 gives the estimates of unknown parameters in the model, as well as the regression coefficients and the phylogenetic autocorrelation coefficient.

Model Selection and Parameter Space

Using the multiple regression model with phylogenetically correlated error generates an interesting problem. Most comparative data are observational rather than experimental. Since the true form of the relationship among traits is therefore typically unknown, each combination of the variables is a viable evolutionary model for potential adaptive explanations. Which model best fits the data at hand? To answer this question, we need to (1) investigate the parameter space systematically (i.e., all possible combinations) and (2) develop robust numerical criteria for selecting the best explanatory variables.

The problem of model selection in multiple regression analysis can be considered a selection process on a fitness landscape (fig. 6.6). In this network, the vertices are regression models with different combinations of explanatory variables. Each model contains a fitness value representing how well the observed data fit this particular model. There are two approaches to exploring this parameter space. First, if we are interested only in the best model, we can simply compare the fitness values of all possible regression models and choose the one with the best performance. Second, if we are interested in how the models develop, we can look at the best one or best few models (i.e., combination of variables) at each subset level. For example, if the best models at three subset levels in figure 6.6 are (X1), (X1, X2), and (X1, X2, X3), then we can say that the influence of the three variables on the models is additive (or nested). However, if the best models at three subset lev-

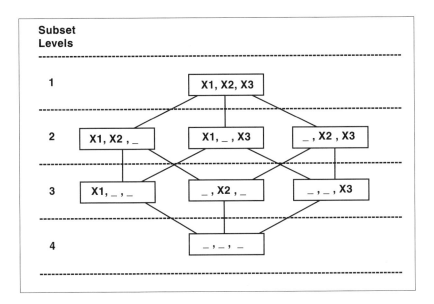

FIGURE 6.6.

The parameter space of multiple regression models involving three explanatory variables. Each vertex is a possible regression model.

els are (X1), (X2, X3), and (X1, X2, X3), this means that (X1, X2) and (X1, X3) have strong interactions and perform worse than (X2, X3) for fitting the models.

Model Selection Criteria: Information-Based Approach

Over the past decade, many information-based approaches for statistical model selection have been developed (see Burnham and Anderson 1992). The information-based criteria contain two components: lack of fit (precision of model) and penalty (complexity of model structure). The first component is a measure of the lack of fit between the data and the models. Like the residual sum of squares (RSS) used in a multiple regression model, the fit will improve as more parameters are added to the model. If the model selection is based on this criterion, it usually overfits complex data sets. In this case, the second component will penalize the model because of the increasing cost of complexity. Thus, information-based criteria provide a tradeoff between the precision and the complexity of a model structure.

Bozdogan's ICOMP (information-theoretic measure of complexity) criterion (1988, 1990) incorporates a measure of the complexity of the model into the penalty term. In terms of statistical modeling, ICOMP plays an important role in indicating the performance of a model; large values of ICOMP indicate a high interaction among the variables, and small values of ICOMP indicate that more information is captured from the estimation of the model. Therefore, in selecting the best combination of independent variables in a multivariate analysis, low ICOMP values are preferred. Appendix 6.2 gives the derivation of the ICOMP value.

Results

In order to test the importance of accounting for phylogenetic effects, we calculated ICOMP values for two model selection procedures: one that removed phylogenetic effects and one that did not. Examining all 127 possible models with the intercept plus six independent variables (gestation length, litter size, age of maturity, lactation period, juvenile mortality rate, adult mortality rate), we found that the ICOMP values ranged from 60.921 to 43.490 when the model did not account for phylogeny. When the model took phylogeny into account, the range of the ICOMP values was reduced to 53.308–39.611. This reduction indicates that removing phylogenetic effects in the model selection procedure results in a more representative model.

The optimal models—those with the smallest ICOMP values at different subset levels—are listed in table 6.1. We look first at the optimal models across subset levels. The variables lactation, adult mortality rate, and juvenile mortality rate are always chosen at subset levels greater than two. This indicates that these three variables are relatively more important than the other variables in contributing to variation in brain size.

The performance of the models becomes better (i.e., lower ICOMP values) as additional variables are added up, to the fourth level. Although the ICOMP values at levels above 4 indicate that those models are less optimal, they are still lower than those for the first two levels. The fact that the models with more than two variables yielded lower ICOMP values actually supports Pagel and Harvey's (1990) conclusion. That is, brain size evolution is not influenced by just a single trait variable but rather should be accounted for by multivariate interactions.

TABLE 6.1

The Optimal Model at Each Subset Level in the Multiple Linear Regression

Subset Level	Best Subset							ICOMP (F–1)
7	Intercept	Gestation length	Litter size	Age of maturity	Lactation	Juvenile mortality	Adult mortality	51.9793
6	Intercept	Litter size	Age of maturity	Lactation	Juvenile mortality	Adult mortality		51.4706
5	Litter Size	Age of maturity	Lactation	Adult mortality				51.2045
4	Litter Size	Lactation	Juvenile mortality	Adult mortality				50.8341
3	Lactation	Juvenile mortality	Adult mortality					50.9952
2	Lactation	Adult mortality						52.2164
1	Lactation							52.6351

Note: Each model consists of the independent variables that best account for the dependent variable, brain size. The optimal model is the one with the smallest ICOMP value at each subset level. There are 6 independent variables and therefore 7 subset levels, each including the following number of variable combinations (total permutations = 127): 7 = 1; 6 = 7; 5 = 21; 4 = 35; 3 = 35; 2 = 21; 1 = 7. In subset levels 7 and 6, the intercept represents a constant term in the multiple regression model when brain size is regressed against the remaining independent variables at each level.

173

Examining the combination of independent variables in the optimal model at each subset level reveals that these variables are hierarchically nested (a variable selected in a lower-level model is selected in all subsequent levels). Although this pattern may not directly support a causal relationship among mortality rates, life histories, and brain size, it does indicate that the explanatory power of these variables is additive.

DISCUSSION

Conventional comparative approaches often require two steps: (1) removal of phylogenetic effects, and (2) application of comparative statistics to test the underlying relationships among traits that are free from phylogenetic effects. In this chapter we have introduced a flexible framework to deal with these two steps *simultaneously*. We first postulated a regression model representing the correlated evolution among traits. Because the observed trait data are correlated with the phylogeny, the random error terms in the regression model are not independent of one another. Thus, the proposed model should incorporate the first-order phylogenetically autoregressive process with standard error components in order to deal with the phylogenetic dependencies in trait evolution. Therefore, we propose a new approach that combines regression models with phylogenetic comparative methods.

The multivariate aspect of our approach involves an information-based model selection procedure to determine the optimal combinations of explanatory (independent) variables. In the example given here, we show that brain size evolution is most closely associated with litter size, lactation period, juvenile mortality, and adult mortality. These factors are consistent with the multivariate hypotheses of Pagel and Harvey (1988). However, they also identified gestation length as an important factor in brain size evolution. A possible reason why lactation period and litter size are more salient than gestation length in the multivariate analysis is that body size was removed from all life histories prior to searching for trends. Across mammals, size is more highly correlated with gestation length than with litter size or lactation period. Thus, variability might be greater in these last two variables and therefore have more potential evolutionary influence on brain size than does gestation length.

Apart from allometry, variability in life history traits can be caused by changes in developmental timing—that is, heterochrony (McKinney and Gittleman 1995). Variation in mortality rates can also be affected by heterochrony through differences in the onset of senescence, while developmental states of neonates can relate to certain mortality factors such as disease and predation.

Future comparative analyses should assess whether mortality perhaps affects brain size via the number of offspring a mother has rather than the length of time needed to produce them. To answer this question, it is necessary to develop a multivariate approach for determining causality; certainly, an approach such as path analysis (see Cheverud et al. 1985; Wright 1968) that isolates the relative contribution that each variable has on an independent variable would be helpful. The difficulty will be to then incorporate phylogeny into the analysis. At this stage, the methods proposed here will be useful for identifying which variables are relevant in a multivariate model of trait evolution.

Note

We thank Mark Kot for discussion of phylogenetic analysis, Louis Gross and Pierre Legendré for discussion of spatial statistics and modeling, and the SAR advanced seminar group for comments on applications of multivariate comparative methods. We also thank the organizers of the seminar—Sue Parker, Jonas Langer, Mike "SJ" McKinney—for inviting this work, and the School of American Research for hosting the meeting. Our research is partially supported by the National Science Foundation, through grant BIR-9318160, and by the Howard Hughes Medical Institute Undergraduate Biological Science Education Program Grant 71195-539601 to the University of Tennessee. During the writing stage, JLG was also supported by a Professional Leave Award and sponsored by the Department of Zoology, Oxford University. Last but not least, we are grateful for help with the revised manuscript in the form of Jonas's loaning us his special pen.

APPENDIX 6.1: ESTIMATES OF UNKNOWN PARAMETERS, REGRESSION COEFFICIENTS, AND PHYLOGENETIC AUTOCORRELATION COEFFICIENT

Consider a multiple regression model defined by

$$\mathbf{y} = \chi\beta + \mathbf{e} \qquad (1)$$

where $\mathbf{y}$ is a given trait (dependent variable) for n species, χ is a matrix of independent trait variables associated with n species and q traits, β is a one-dimensional vector of regression coefficients, and $\mathbf{e}$ is a one-dimensional vector of errors (residuals) that are presumably related to a phylogeny. To account for this presumed phylogenetic pattern in the trait data, we now consider an autoregression model in which the correlated errors regress themselves along a given phylogeny:

$$\mathbf{e} = \rho\mathbf{W}\mathbf{e} + \mathbf{u} \qquad (2)$$

where ρ is a phylogenetic autocorrelation coefficient representing the level of phylogenetic autocorrelation, $\mathbf{W}$ is an n-by-n weighting matrix that represents a phylogenetic structure, and $\mathbf{u}$ is now a standard error term with a normal distribution:

$$E(u_i) = 0, E(u_i^2) = \sigma^2, E(u_i, u_j) = 0 \text{ for } i \neq j.$$

To estimate the unknown parameters in equations (1) and (2), we follow the maximum likelihood framework (Cliff and Ord 1981; Gittleman and Kot 1990). The likelihood function written in terms of $\mathbf{u}$ is

$$L = \frac{1}{\sigma^2 (2\pi)^{\frac{n}{2}}} \exp\left(-\frac{\mathbf{u'u}}{2\sigma^2}\right) \qquad (3)$$

and the log-likelihood function is

$$\log L = \text{constant} - \frac{n}{2}\log(2\pi) - \frac{n}{2}\log(\sigma^2) - \frac{1}{2\sigma^2}(\mathbf{u'u}). \qquad (4)$$

The maximum likelihood estimates for $\hat{\beta}$ and $\hat{\sigma}^2$ can be obtained by

$$\hat{\beta} = (\chi'\mathbf{A'A}\chi)^{-1}\chi'\mathbf{A'A}\mathbf{y} \qquad (5)$$

and

$$\hat{\sigma}^2 = \frac{1}{n}\left(\mathbf{y'A'A}\mathbf{y} - 2\hat{\beta}'\chi'\mathbf{A'A}\mathbf{y} + \hat{\beta}\chi'\mathbf{A'A}\chi\,\hat{\beta}\right), \qquad (6)$$

where

$$A = I - \rho W. \tag{7}$$

One immediate difficulty in obtaining the $\hat{\beta}$ and $\hat{\sigma}^2$ is that equations (5) and (6) still contain one unknown parameter, the autocorrelation coefficient ρ, which is embodied in A. The most straightforward method for searching for ρ is to minimize equation (4) via a direct search process within the bounds

$$\frac{1}{\lambda_{min}} < \hat{\rho} < \frac{1}{\lambda_{max}}, \tag{8}$$

where λ_{min} and λ_{max} are the smallest and largest eigenvalues of W, and W should be standardized by the row sums (Haining 1994).

APPENDIX 6.2: DERIVATION OF THE ICOMP VALUE

The information-theoretic measure of complexity (ICOMP) is defined as

$$\mathbf{ICOMP} = -2\left(\log L\left(\hat{\theta}\right)\right) + 2C_1\left(\hat{\Sigma}_{\text{model}}\right), \tag{9}$$

where $\log L(\hat{\theta})$ is the maximum log likelihood value when a maximum likelihood function is used to estimate the unknown parameters in a statistical model, and $\hat{\Sigma}$ is the estimated covariance matrix from a given model. C_1 represents a measure of complexity of the covariance matrix $\hat{\Sigma}$ and is defined by

$$C_1 = \frac{k}{2}\log\frac{\text{tr}\left(\hat{\Sigma}\right)}{k} - \frac{1}{2}\log\left|\hat{\Sigma}\right|, \tag{10}$$

where k is the dimension of the estimated covariance matrix, $\hat{\Sigma}$; $\text{tr}(\hat{\Sigma})$ denotes the trace of the covariance matrix; and $\left|\hat{\Sigma}\right|$ denotes the determinant of the covariance matrix.

In the case of multiple regression models with phylogenetically correlated errors, the inverse-Fisher information matrix (IFIM), also called the scored function, is used as a variance-covariance matrix of estimated parameters $\hat{\sigma}^2$, $\hat{\beta}$, and $\hat{\rho}$. IFIM measures the average accuracy or the precision of the parameter estimates of a statistical model. Following Upton and Fingleton (1985), we obtain the estimated IFIM:

$$\hat{F}^{-1} = \hat{V}\left(\hat{\sigma}^2, \hat{\rho}, \hat{\beta}\right)$$

$$= \hat{\sigma}^4 \begin{bmatrix} \dfrac{n}{2} & \hat{\sigma}^2\text{tr}(\mathbf{B}) & 0 \\ \hat{\sigma}^2\text{tr}(\mathbf{B}) & \hat{\sigma}^4\text{tr}(\mathbf{B}'\mathbf{B}) - \hat{\sigma}^4\alpha & 0 \\ 0 & 0 & \hat{\sigma}^2\left(\chi'\hat{\mathbf{A}}'\hat{\mathbf{A}}\chi\right) \end{bmatrix}^{-1} \tag{11}$$

where

$$\mathbf{B} = \mathbf{W}\hat{\mathbf{A}}^{-1}, \tag{12}$$

$$\alpha = -\sum_{i=1}^{n}\frac{\lambda_i}{\left(1 - \hat{\rho}\lambda_i\right)^2}, \tag{13}$$

and λ_i is the ith eigenvalue of the matrix $\mathbf{W}$. We substitute equation (11) into equations (10) and (9) so that ICOMP is:

$$\text{ICOMP} = n\log(2\pi) + n\log(\hat{\sigma}^2) + n - \log\left|\hat{\mathbf{A}}\right| + \frac{s}{2}\log\frac{\text{tr}(\hat{F}^{-1})}{s} - \frac{1}{2}\log\left|\hat{F}^{-1}\right| \quad (14)$$

where s is the dimension of IFIM.

7

Current Issues in the Investigation of Evolution by Heterochrony, with Emphasis on the Debate over Human Neoteny

Brian T. Shea

This volume attests to the persistent notion that the study of heterochrony may be able to provide key insights into our understanding of the morphological, life history, and behavioral evolution of humans. In spite of examples ranging from the long-known but still intriguing retention of larval features in neotenic salamanders to our growing knowledge of the molecular genetic changes underlying development and evolution in such powerful experimental systems as the nematode, *Caenorhabditis elegans* (e.g., Sommer and Sternberg 1996), more attention in the realm of heterochrony continues to be focused on the human lineage than on any other. This, as I noted elsewhere in a review of human neoteny and heterochrony (Shea 1989), is often the unfortunate, if hardly surprising, result of humans' being both the objects and the executors of such studies. Although few researchers know the details of Gould's rich empirical work on neoteny in Bermudian land snails of the genus *Poecilozonites* (Gould 1968, 1969), most are well aware of his claims regarding human neoteny, a recent example of which appeared in *Discover* magazine (Gould 1996a).

I have long maintained that the "case study" of human evolution should be tested against a traditional and well-informed framework of heterochronic processes and categories as developed by evolutionary biologists on the basis of a wide variety of organisms. We must not let either the natural human or the traditional anthropological tendency to view our lineage as "special" drive the debate over the role of heterochrony in human evolution. This is a mistake that has been made in many different realms (see Foley 1987 for discussion and examples).

In this chapter, therefore, I first address and try to clarify some key points extant in current general discussions of evolution via heterochrony. Secondarily I stress how misunderstandings may be at play in debates over human heterochrony and over neoteny in particular. I begin by advocating a more limited and precise definition of heterochrony, following a number of relatively recent authors (e.g., Hall 1992; Raff 1996; Raff and Wray 1989) who have legitimately claimed that the term as often defined—or at least operationalized—has come to mean simply "evolutionary change." I proceed to focus on the nature of the characters considered by various studies of morphological heterochrony to have been transformed, and I point out some differences between studies of what we might call "qualitative" and "quantitative" heterochrony. These issues lead into a consideration of some current debates over the operationalization of the heterochronic framework, the role of traditional studies of growth in time, and recent critiques of the comparative use of allometry in studies of heterochrony. The nature of the acceptable "criterion of standardization" in heterochrony, and the inescapable relativity of standardized shape transformations, is considered in detail. In each case, some implications for our understanding of current debates over human neoteny are provided. A section reviewing two case studies of heterochronic transformation in "pygmies," human and chimpanzee, allows for an application of some of the previous issues to empirical bases. Finally, I offer some summary comments about the human neoteny debate and attempt to identify the types of new data and perspectives that might help to advance this debate out of the rather unproductive semantic and obfuscatory wrangling that has characterized it since Gould (1977) resurrected the notion of human evolution via neoteny.

HETEROCHRONY: DEFINITIONS, CONSTRAINTS, AND RELEVANCE

Stephen Jay Gould's *Ontogeny and Phylogeny* (1977), which helped to rekindle interest in evolutionary change via heterochrony, treated in considerable detail the various usages and definitions of heterochrony in the hands of naturalists such as Haeckel, de Beer, and many others over the decades. Gould defined heterochrony as "changes in the relative time of appearance and rate of development for characters already present in ancestors" (1977:2). Although the "characters" mapped in heterochronic analyses have traditionally been morphological, there is no a priori reason why other types of characters (e.g., behavioral) could not be productively analyzed from this perspective.

It is important to note that this definition of heterochrony explicitly associates the concept with character homology and continuity between ancestor and descendant and with the displacement of such characters in relative developmental time. Characters may change size or shape in certain respects or be displaced in developmental time, but the continuity is recognizable and no *fundamental* novel transformation has occurred. This has to be something of a relative assessment, particularly since a developmentalist perspective might lead one ultimately to reject the validity of traditional "characters" as such and to replace them with dynamic developmental processes (e.g., Goodwin 1984; Sattler 1994; Shea 1990). Nevertheless, I advocate that we not continue to employ a broad definition of heterochrony, roughly equivalent to "changes in the rates or timing of ancestral developmental patterns," even though I have published similar definitions myself (e.g., Shea 1989:70). My reason for now explicitly favoring Gould's more restrictive definition is that otherwise heterochrony becomes equated basically with "evolution," since all transformations must emerge through some sort of change in ancestral developmental patterns—how could it be otherwise in a theory of "descent with modification" (Darwin 1859)? This is true even for the evolution of characters that we depict as "novel."

My preference for the more restrictive and historically accurate definition admittedly limits the analysis of morphological heterochrony to the relative temporal shifting of ancestral characters. Yet I believe this most closely follows the original and most cogent criterion for the recognition and description of heterochronic change: "Evolution occurs when ontogeny is altered in one of two ways: when *new* characters are introduced at any stage of development with varying effects upon subsequent stages, or when characters *already present* undergo changes in developmental timing. Together, these two processes exhaust the formal content of phyletic change; *the second process is heterochrony*" (Gould 1977:4, italics added).

Operationalizing this definition of heterochrony is not without its difficulties and drawbacks. The narrow definition does tend to inculcate notions of characters as static entities, a view that in some ways is the antithesis of a developmentalist perspective (e.g., Goodwin 1984; Sattler 1994), particularly of the emerging molecular genetic emphasis (e.g., Raff and Wray 1989). Moreover, how do we define "characters" and, especially, the boundaries between different characters? What is the relationship between characters in traditional studies of qualitative heterochrony and the "shapes" that comprise the focus of quantitative heterochrony? How much change in a character is permitted before we recognize the presence of a new character? In some cases a character may have changed enough to have passed beyond some consensus threshold to a configuration universally recognized as "new," even though the novelty is but a reconfiguration of ancestral pieces and/or is underlain by a relatively small developmental shift in terms of genetic and epigenetic bases. The turtle carapace, a key innovation traditionally viewed as the basis of an ordinal radiation, may provide just such an example (Burke 1989).

On a different taxonomic and morphological scale, I am inclined to view the derived pattern of morphological reorganization of the skeletal and soft-tissue anatomy of the upper respiratory and vocal tract in human evolution as such a "new" feature. As Laitman, Crelin, and colleagues have demonstrated in a series of papers (e.g., Laitman and Crelin 1976, 1980; Laitman and Heimbuch 1982; Laitman, Heimbuch, and Crelin 1978, 1979; Lieberman 1984), novel localized growth patterns likely underlie the derived reconfiguration of the modern human laryngeal region, which, along with other cranial and neural modifications, effectively allows for the production of human speech (fig. 7.1). I would advocate recognition of this composite suite of upper respiratory

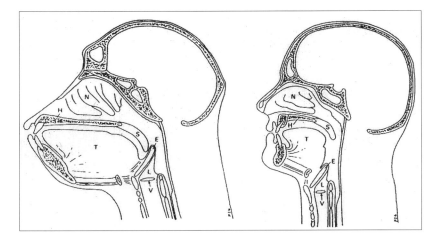

FIGURE 7.1.

A midsagittal section through the head and neck of an adult chimpanzee (left) and an adult human during normal respiration. Note the lower position of the larynx in the human; infant humans and young chimpanzees exhibit a configuration that resembles the adult chimpanzee, not the adult human. Key: E, epiglottis; L, laryngeal cavity; H, hard palate; N, nasal cavity; S, soft palate; T, tongue; V, vocal fold. (Redrawn from Laitman and Heimbuch 1982.)

changes as a new feature or character, even though it involves only a repositioning and changes in the relative sizes and shapes of "parts" and "features" already present in the ancestral state.

From the foregoing I draw a simple conclusion about discussions of heterochrony, and particularly neoteny, in human evolution. Heterochrony is probably not going to contribute any vital perspective or approach when we deal with features, characters, or character states that are arguably novel or fundamentally altered. It follows that heterochrony will be more relevant and successful when we address relatively minor transformations of timing and allometric patterning among closely related and similar morphs than when we address more fundamental and profound morphological transformations. Indeed, it is in the former kinds of comparisons that the parallels between ontogeny and phylogeny are most compelling, which is exactly what originally generated much of the historical interest in heterochrony. Thomson (1988:92) recognized this well when he wrote:

> Phenotypic features set in place at late morphogenetic stages
> basically involve modulations of pattern control mechanisms,
> because the basic elements of organogenesis have already been
> set in place during early pattern formation stages....Changes
> caused at this stage would seem to be the basis of differences
> among most closely related species of birds and mammals....
> In fact, a large array of phenotypic differences involving size
> and proportions due to the operation of allometric and hetero-
> chronic mechanisms are caused at the late morphogenesis
> stage.

One issue for the biological anthropologist to keep in mind, then, in discussions of human evolution via heterochrony is whether the key features being discussed constitute something new or something old merely shifted in relative ontogenetic timing. I would argue that most of the features or characters of relevance in our highly autapomorphic lineage approach the former characterization more closely than the latter—but more on this later.

MORPHOLOGICAL HETEROCHRONY: IT'S ABOUT SHAPES, DISPLACED IN DEVELOPMENTAL TIME

I have opted here for a relatively narrow and historically accurate definition that restricts morphological heterochrony to displacements in descendants of features already present in ancestors (Gould 1977). Some of the recent literature invoking heterochronic perspectives requires that we carefully consider exactly what constitutes such "features"—that is, just what is being shifted in ancestral-descendant heterochronic transformations? The answer, quite simply, is something that "has shape," either implicitly (as in many "qualitative" heterochronic transformations, such as neotenic paedomorphosis in salamanders that retain larval gills and tail structures as sexually mature, full-sized adults [Gould 1977]) or explicitly (as in morphometric and allometric heterochrony [Alberch et al. 1979; Gould 1977]).

We must be clear that shape is inherently multidimensional and relative; it is proportion, form, pattern, morphology. It cannot be unidimensional—as, for example, a humerus length or body weight. According to Gould's (1977) clock models or Alberch and colleagues'

(1979) formalizations, "shape" cannot be represented by a single dimension, though it may be a single number, such as a ratio or a regression slope, relating two or more variables. (Interestingly, "size" in these frameworks is represented by a single variable such as humerus length or body weight or by some multivariate composite ranging from a geometric mean to a score on a first principal component.) In sum, I advocate that in discussions of morphology, we adhere closely to Gould's (1977:251) notion of heterochrony: "Heterochrony is defined as the evolutionary displacement of a specific feature (a 'shape') relative to a common standard of size, age or developmental stage." The centrality of focus on shape change is further revealed by the fact that Gould (1977) did not consider proportioned gigantism or dwarfism, in which size and possibly age vectors are shifted but the descendant shape vector is unaltered at common developmental stages, to lie within the realm of heterochrony (though he acknowledged similarities between these geometric transformations and such heterochronic shifts).

I stress this point in the hope of both clarifying my specific use of features or characters in this chapter and highlighting an unfortunate recent trend in the heterochronic literature that obfuscates discussion of specific cases and general approaches. Simple plots of single dimensions against time in ancestors and descendants (or two or more groups in any comparative framework) do not in themselves represent heterochronic transformations. Such a plot is shown in figure 7.2, and a few examples of such recent analyses include comparisons by Richtsmeier and Lele (1993), German and Myers (1989), McKinney and McNamara (1991), and Godfrey and Sutherland (1995a, 1995b, 1996). These are simply classic plots of growth in time, and although there is every reason to give them our focused attention and evolutionary explanations, they do not, strictly speaking, fall within the realm of morphological heterochrony, either historically or conceptually.

Moreover, even traditional plots of shape versus age can be misleading if we fail to recognize that what we seek in traditional analyses of heterochrony is normally a standardization by developmental (physiological) stage, and not a standardization by absolute chronology. This is not merely a pedantic claim for historical accuracy and methodological purity. Because features of form and morphology are

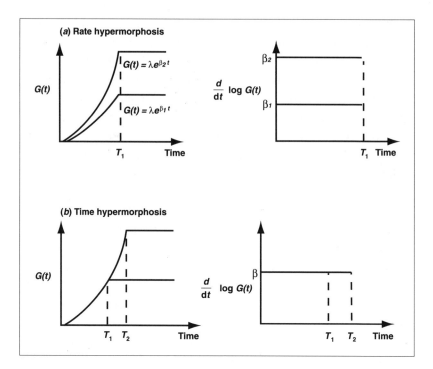

FIGURE 7.2.

A schematic depiction of heterochronic patterns of rate and time hypermorphosis as perturbations in simple curves of growth in time for individual dimensions. G(t) represents growth as a function of time. (Redrawn from Richtsmeier and Lele 1993.)

inherently multidimensional and relative, there is no necessary relationship between a transformation in the overall feature or proportion and any change in the individual dimensions of the feature. For instance, we may speak of the derived increase in *relative* brain size in human evolution that resulted from a tripling or more of *absolute* brain size as paedomorphic because the ancestral juvenile shape (brain/body ratio, or slope) is high relative to the adult configuration. However, we can envision an evolutionary transformation that is identical on the heterochronic scale (i.e., paedomorphic) but that in this case entails no change or even a decrease in absolute brain size, so long as it is accompanied by an even greater decrease in absolute body size, again yielding a *relatively* greater, or juvenilized, brain size. This argument is not so fanciful when we realize that many of the cases of relatively high brain/body ratios in mammalian and primate evolution are unrelated to actual encephalization events in phylogenetic transformations. Instead, they are due to rapid body size decreases, in many cases mediated by developmental controls that do not much affect overall brain size (Gould 1975; Jerison 1973; Riska and Atchley 1985; Shea 1983d, 1988, 1992a, 1992b). This is what gives the talapoin monkey (Shea 1992a) and the human pygmy (Shea and Gomez 1988) their relatively enlarged, juvenilized brain/body ratios. In related fashion, it is the marked increase in the absolute volume of the brain and various of its key functional components that leads many authorities to conclude that much more in terms of true evolutionary novelty than a simple neotenic heterochronic displacement of ancestral features underlies the evolution of brain size in our own lineage (e.g., Deacon 1990a, this volume; McKinney 1998; McKinney and McNamara 1991; Shea 1989).

The point to emphasize for the present discussion is that shifts in standard plots of absolute brain size (in cubic centimeters or grams) against time in these respective transformations or comparisons would yield quite different pictures from the heterochronic ones of relative brain size. Of course we want to know this information; it is vital, in some ways more fundamental and revealing than the relative (allometric and heterochronic) comparisons themselves. I certainly do not advocate that we not undertake these comparisons; I merely stress that the choice of comparison depends on one's question and that morphological heterochrony is about form and shape. Using the framework and language of heterochrony to describe univariate changes only obfuscates the statements made. Some of the confusion injected into discussions of the role of allometry in heterochrony by Godfrey and Sutherland (1995a, 1995b, 1996; e.g., see 1995a:47, 49, 53; 1995b: 420–22; 1996:27, 36) emerges from such problematic plotting of individual dimensions against age.

In sum, I call here for an excising of traditional growth-in-time plots from the conventional realm of heterochrony. This in no way denigrates or elevates the evolutionary significance of either type of comparison. It merely facilitates effective testing and communication of the morphological shape concepts central to analyses of heterochrony.

THE ROLE OF ALLOMETRY IN STUDIES
OF HETEROCHRONY

The role that allometric analyses should play in the investigation of heterochrony is currently a topic of some debate. This debate has been exacerbated recently by the claims of Godfrey and Sutherland (1995a, 1995b, 1996) that the approach of "allometric heterochrony" has generated erroneous interpretations in the realm of evolutionary heterochrony in general and human neoteny in particular. Here I attempt to clarify some central concepts in allometry and heterochrony raised by their critiques.

"Size" and "Shape" in Allometry

The first issue to consider is how traditional approaches to allometry within the framework of "allometric heterochrony" deal with the surprisingly complex and contentious morphometric concepts of "size" and "shape." Consider the two typical allometric plots illustrated in figure 7.3. How do we "translate" such bivariate plots into our vernacular notions of "size" and "shape"? Size is often depicted as the relevant point along the x axis, especially when body weight, height, or some other such global measure is used. Yet even in simple bivariate allometric plots, and especially those of the variety of figure 7.3 (right panel), in which the axes are completely interchangeable, size is probably best thought of as a variable that is "latent" and here composite, in the same sense as a first principal-component score. Size might thus be best depicted as x plus y, or an average, or a geometric mean. In practice, either x or y alone will generally be suitable, since the trajectory is linear and monotonic.

This seems to have (mis)led some into thinking that the allometrician implies that the y axis therefore captures "shape," since the x axis is often used for "size," and various schematic frameworks in the literature on allometry and heterochrony depict plots of "shape" (y) versus "size" (x). Nothing could be more incorrect. As I laid out in detail earlier, shape cannot be a single dimension; it is inherently multidimensional. Shape is unquestionably latent in allometric plots (indeed, this is their purpose) and is explicitly derived for each particular allometric plot as either the dimensionless x/y ratio at each point along the trajectory or the dimensionless slope of the regression line relating the two

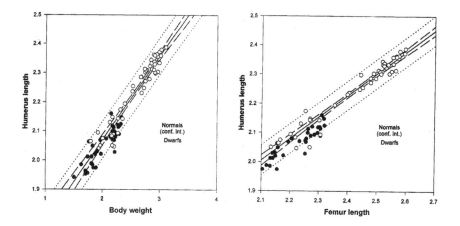

FIGURE 7.3.

Unpublished data comparing little (lit/lit) dwarf mice (filled circles) with their normal lit-termate controls, illustrating typical plots of ontogenetic allometry (relative growth). The panel on the left illustrates a log-log plot of an individual bone dimension (length of the humerus in mm) against body weight (g); that on the right shows a plot of one bone dimen-sion (humerus length) against another (femur length). The 95-percent regression confidence intervals and prediction intervals are shown with dashed and dotted lines, respectively.

variables (from a particular origin or y/x value). A given allometric plot such as that in figure 7.3 (right panel) therefore encapsulates a trajectory of "shape" versus "size" that is unique to that $y{:}x$ allometric relationship and can be replotted as y/x versus x or as $x + y$. It can also be characterized as the slope k, the coefficient of allometry.

With shape and size thus defined and depicted, the dissociation in descendants of patterns of ancestral covariation between two variables capturing the "shape" and "size" of a "feature" can certainly be read as divergences in growth allometries, or departures from ontogenetic scaling. Of course such depictions of shape and size relate only to those two dimensions and translate in no direct way if we substitute another dimension for one of the original ones or utilize two new dimensions. Also, if we define size and shape differently (and there are myriad pos-sibilities—just check the voluminous morphometrics literature), we can generate frameworks that will not map precisely to this allometric one (cf. Godfrey and Sutherland's [1995a, 1995b, 1996] notions of "growth" and "development"). But the allometric framework is internally consistent, traditionally extremely successful, and moreover based in our current understanding of real growth processes via cell division (see Huxley 1932 and discussions of the developmental basis of ontogenetic scaling and allometric divergences in Shea 1992b).

Chronological and Developmental Time in Allometry

Next I must stress that the typical allometric plot contains no direct information about either absolute age or relative developmental stage. It is not meant to; relative growth emerged on the theoretical premise and empirical evidence that the passage of time was not necessarily causal in terms of the control of growth, but rather that fundamental growth substances and gene products underlay the growth and covariance of structures (e.g., Huxley 1932). Information about absolute and relative time in allometric plots is at best available *indirectly*—for example, when we happen to know that successive points along a trajectory represent age groups or standard and homologous developmental stages such as a dental eruption sequence (though even then this implies nothing about absolute age save some monotonic increase), or that the terminal points on the plot represent sexually mature adults. Directly stated, this is why allometry does not simply equal heterochrony; it tells us nothing directly about age or stage, though it tells precisely what we need to know about size and shape (as defined earlier). Therefore, allometry can be but one key component of an analysis of heterochrony.

The foregoing review should have been unnecessary. Yet we still read with somewhat distressing regularity about how allometric approaches are misleading or inadequate because they do not magically reveal the underlying curves of growth in time for the individual variables (e.g., Godfrey and Sutherland 1995a) or because a divergent allometric trajectory does not definitively localize which variable or variables have changed relative to underlying curves of growth in time (e.g., Blackstone 1987; German and Myers 1989). Similar conceptual difficulties are at the base of most of the complaints by Godfrey and Sutherland (1995a, 1995b, 1996) that allometric analyses do not directly translate into what they wish to know about either variable growth-in-time curves or "true rates of heterochronic shape change" as they define these (commonly relative to chronological age rather than to developmental stage). I deal with these issues in greater detail later in the section on the relativity of heterochronic categories.

Let me summarize here by restating that allometric plots indeed do not directly yield information on timing (chronological or developmental), but neither do single-variable, growth-in-time plots directly yield information on either shapes or developmental timing or allometric covariance. Nor do plots of shape versus age directly yield information on developmental timing or the localization of change to the numerator and/or denominator of the composite shape variable. None of this is surprising. One should choose an approach (allometric, growth-in-time, etc.) designed to test a specific hypothesis, not argue for some universal guideline.

Allometric Dissociations in Neoteny and Acceleration

The concepts of size, shape, age, and developmental timing have been introduced, particularly as they relate to the role of allometry in broader analyses of heterochrony. I turn now to a focus on comparing multiple (presumed ancestor-descendant) allometries as part of the study of heterochrony in an effort to clarify some recent debates. In their somewhat confusing and redundant triumvirate of critiques, Godfrey and Sutherland (1995a, 1995b, 1996) attacked previous studies in the realm of "allometric heterochrony," emphasizing one critical point above all others. This point essentially reduces to the claim that the comparative analysis of growth allometries has led to profound misunderstandings regarding heterochronic transformations because it has generated incorrect criteria for assessing neoteny and acceleration. In critically discussing a general framework presented by McKinney (1988b) and the specific example of the pygmy chimpanzee raised by me (Shea 1983b, 1984), Godfrey and Sutherland (1995a, 1995b, 1996) repeatedly emphasized the fundamental misunderstanding they believed they had uncovered: "In the current literature on heterochrony, growth allometries are regularly being misinterpreted, with regard to both the heterochronic processes and the heterochronic products they may or may not imply. Progress in the field of heterochrony, the study of the evolution of ontogeny, depends on our correcting these analytical errors" (1995b:429).

Specifically, the key error that they believed they had recognized and that generated their lengthy critiques was that the shape retardation characteristic of (paedomorphic) neoteny in fact requires allometric dissociations in divergent directions in the descendant, depending on whether the ancestral ontogeny was positively or negatively allometric (a parallel situation occurs with [peramorphic] acceleration, of course). They claimed that the erroneous notion that neoteny "uniformly lowers the slopes of growth allometries" was either explicitly or implicitly present in my work and that of McKinney and others utilizing an allometric framework, to such an extent that "this particular assumption is built into the diagnostic formalism of 'allometric heterochrony'" (Godfrey and Sutherland 1996:28).

Now consider the specific framework actually articulated in my review of human neoteny and heterochrony (Shea 1989), fully a half-dozen years prior to Godfrey and Sutherland's papers (1995a, 1995b, 1996). In describing the changes in allometric patterning required to meet an expectation of neotenic paedomorphosis, I emphasized the following (Shea 1989:73):

> The changes in trajectories that are linear in a logarithmic and allometric sense would be straightforward. Positively allometric trajectories yielding relative enlargement in adults of the ancestral group would exhibit lower slopes so that descendant adults would be characterized by relatively smaller such structures, just as in juveniles of the ancestors....Negatively allometric trajectories yield relative diminution of structures in the adult, so our descendant adults would exhibit a relative enlargement of such features compared to ancestral adults, thus resembling ancestral juveniles.

Later in the same paper, in stressing that such divergent allometric transformations were in fact unlikely to occur at all on any pervasive or global level, let alone as a relatively simple genetic or developmental perturbation, I further explicitly noted (Shea 1989:96) that "such widespread neotenic retardations appear unlikely in theory since trajectories of positive allometry require slope decreases and/or downward transpositions to yield paedomorphosis, while trajectories of negative allometry require slope increases and/or upward transpositions."

The key here is to realize that although in one sense a decrease in (or downward transposition of) a positively allometric slope is opposite from an increase in (or upward transposition of) a negatively allometric slope, the *direction* of shape change is the same—that is, it retards the ancestral pattern of stronger shape change, be it positively or negatively allometric. Another way of getting this point across is to stress that further decreasing the absolute value of a negatively allometric slope—for example, going from 0.67 to 0.33—actually accelerates the shape change, here the relative diminution of x to y as growth proceeds and size increases.

Godfrey and Sutherland (1995a, 1995b, 1996) were either unaccountably cursory or less than ingenuous in their review of previous work in the realm of allometric heterochrony, particularly as it applies to the case of human neoteny. They presented as novel and incisive a point clearly made a half-dozen years prior to their critiques, and they implied that there was a general theoretical flaw in the approach rather than simply highlighting specific cases that admittedly presented only a partial picture by depicting positively allometric patterns alone (e.g., McKinney and McNamara 1991; see also Futuyma 1986:415, who depicted neoteny as a reduced slope on an allometric plot explicitly linked to an ancestral trajectory of positive allometry, and see below for clarification of the pygmy chimpanzee case discussed by me). The point they belabored regarding McKinney's (1986, 1988b) framework is correct (as I myself pointed out but did not elaborate on in my review of McKinney and McNamara's [1991] book [Shea 1993]). Since this is a relatively minor point, however, and since the correct framework was long ago explicitly detailed (Gould 1977; Shea 1989; and myriad other references and applications), I am tempted to describe Godfrey and Sutherland's (1995a, 1995b, 1996) energetic efforts as "blazing a path down a well-worn road." Their challenge in raising this point as implicit support for Gould's (1977) notion of human evolution via neotenic paedomorphosis was to document specific cases in which such misconceptions had led to the rejection of particular features as not supporting neotenic paedomorphosis when in fact the appropriate corrective reveals that they do provide such evidence. Godfrey and Sutherland

(1995a, 1995b, 1996) provided neither specific correctives nor new empirical evidence, and I know of no cases in which the misapplication of "allometric asymmetry" has resulted in such rejection of evidence supporting neotenic paedomorphosis.

It would be incomplete to imply that the preceding represents the only criticism of allometric heterochrony made by Godfrey and Sutherland. In fact, they offered numerous other criticisms, but all of them represent well-known and long-appreciated "truisms" about this approach (unless otherwise noted, these points appear in all three of Godfrey and Sutherland's papers [1995a, 1995b 1996]). These criticisms include the notion that allometric comparisons are inherently relative (though the authors show less appreciation that this is also true of all heterochronic assessments, as I discuss later); that size cannot be taken as a direct surrogate for age without sacrificing some information or making some assumptions; that a given allometric pattern for one bivariate comparison does not necessarily imply the same pattern for all such measured comparisons (Godfrey and Sutherland 1995b:422); that allometric plots do not always yield straight lines, and underlying plots of growth in time do not always follow precisely the same growth curve; that a definition of shape retardation relative to different standards, such as versus size, age, or developmental stage, can generate different heterochronic assessments; that a pervasive patterning of allometric trajectories, such as ontogenetic scaling, found between closely related species does not guarantee that other, unexamined allometries may not necessarily accord with this general pattern (Godfrey and Sutherland 1995a:56–57); that allometries which exhibit a particular pattern, such as parallel transpositioning, between two groups over the observed and measured range may actually converge, cross, or bend "when more of ontogeny is available" (Godfrey and Sutherland 1995a:57); that a particular heterochronic or allometric change in one body region may be different from those in others, producing mosaic heterochronies; that allometric trajectories are not always "simple" and linear; and that what is true for an allometric plot of y versus x (e.g., linearity) is not necessarily true when those features are plotted against time or some other feature. It would be a tedious and unproductive task to list the many places in the literature on allometry and heterochrony where these points had already been made and further discussed.

Godfrey, King, and Sutherland's (1998) most recent attack on the use of comparative growth allometries focuses on the rather unsurprising occurrence of isometric trajectories. They label cases of the ontogenetic scaling of such trajectories as "trivial" in the analysis of allometric heterochrony. This entirely misses the point that comparative studies of ontogenetic allometry are designed to determine whether ancestral patterns of multivariate feature covariation are changed, or dissociated, in descendant forms, regardless of their particular slope values in selected, specific bivariate comparisons. It is Godfrey's undue focus on isometric transformations, rather than the fundamental issue of the maintenance versus dissociation of ancestral patterns of covariance, that is wanting here.

Finally, then, we come to a point for summation and assessment. Can and should allometry be utilized to describe size/shape relationships and test predictions within the traditional heterochrony framework? Gould (1968:83) explicitly advocated the use of allometry as part of the assessment of heterochrony. In his 1977 book he explicitly developed an allometric guideline for recognizing heterochrony (see his fig. 30); in describing this schematic allometric plot of log organ size (y) versus log body size (x), he wrote (1977:238–39):

> [L]et us consider the simplest dissociation of size and shape. If we can measure the correlation of size and shape in an ancestor, the change of this correlation in descendants will indicate whether heterochrony has occurred and, if it has, will measure both its direction and magnitude. Allometry is the study of relationships between size and shape (Gould 1966). For more than 50 years, since Julian Huxley generalized its earlier use in the study of brain-body relationships, allometric work has relied largely upon the power function: $y = bx^k$. From this common formula, we can extract a simple measure of heterochrony.

Gould then described this allometric test for the recognition of heterochrony using the organ/body-size plot. He concluded by noting, "We simply plot the descendant's ontogeny on the graph and read the effects of heterochrony directly" (Gould 1977:240).

In addition, one might note that nearly all of the accumulated empirical evidence generated by Gould as evidence of heterochrony (see 1966, 1975, and 1977 for general reviews and many additional references) was substantiated using traditional allometric comparisons. One need only think of the classic plots of brain/body allometry (e.g., Gould 1977:373) or hypermorphosis in Irish elk (e.g., Gould 1977:342) for examples. And Godfrey and Sutherland (1996:40), after myriad implications that the general framework of allometric heterochrony is theoretically bankrupt, come full circle to acknowledge that the framework does apparently make sensible and directly testable predictions for neoteny and acceleration (having already acknowledged that the allometric framework is easily applied to cases such as rate and/or time hypo- and hypermorphosis, where descendants follow ancestral patterns of size/shape—in the allometric sense—covariation). They in fact ultimately specify the precise allometric predictions for various categories of heterochrony themselves (Godfrey and Sutherland 1996:40), in the process repeating frameworks long known and undermining their own critical perspectives.

It appears that allometry will continue to play a key role as *one* central component in heterochronic analyses. That a complete analysis of heterochrony also requires data on age and developmental stage and may permit any number of alternative (i.e., nonallometric) approaches to assessing "size" and "shape" is simply a statement that *allometry does not equal heterochrony.* No one who has ever built from the voluminous literature stretching from Huxley (1932) to Gould (1977) and on through more recent workers has ever suggested that it does.

ISSUES OF STANDARDIZATION AND RELATIVITY IN THE INVESTIGATION OF HETEROCHRONY

A key issue at the base of much unnecessary misunderstanding and debate over purportedly incorrect or inconsistent uses of heterochronic terminology is the failure to either (1) utilize the same (or most appropriate, depending on the question) criterion of standardization or (2) acknowledge the inevitable (and desirable) relativity of heterochronic categories and processes, regardless of the criterion of standardization used. Put more simply, heterochronic categories necessarily vary depending on both whether and how we assess the four primary inputs into the comparative analyses—that is, size, shape, age (absolute), and relative developmental stage (Gould 1977). Shape must be included, for it is transformation in shape that we assess relative to one or all of the remaining three inputs, any of which may in practice be unavailable to us.

But each of these variables can be depicted in different ways or using various surrogates. Size can be defined any number of ways, as we well know from ongoing debates in the literatures of allometry and morphometrics. The same is true for shape, and since shape is a relative depiction, necessarily involving at least two variables, obviously the variables we choose will directly affect our characterization of shape. Age is most appropriately an absolute estimate on a continuous scale (such as years), but various scales are used, and this also affects heterochronic output. The key criterion of standardization in studies of heterochrony—developmental stage—is normally some depiction of relative developmental time such as physiological maturation, but the myriad choices available (including sexual maturity, a dental eruption stage, time of weaning, epiphyseal fusion, etc.) clearly indicate an element of inbuilt relativity. A heterochronic classification for an ancestor-descendant comparison at one developmental stage may actually differ from the classification for a comparison at another such developmental stage. This point has been made cogently by Klingenberg (1998).

The configuration of heterochronic input used will depend on both the question being asked and the data available. No problems need emerge so long as one clearly states the criterion of standardization and how size, shape, and time have been determined and compared. Obfuscation arises only when these guidelines are not followed or when there is a failure to comprehend that the assessments must be relative—that is, when one researcher accuses another of using the "wrong" standard or scale and thus of coming up with the "incorrect" heterochronic transformation. Perhaps the most blatant example of this type of misrepresentation is Godfrey and Sutherland's paper "Flawed Inference: Why Size-Based Tests of Heterochronic Processes Do Not Work" (1995a). What they demonstrate is essentially that such analyses yield results *different* from those in which shape is plotted against age, an issue long understood and, ironically, departing from heterochrony's traditional focus in that it places undue emphasis on a time standard of chronological age rather than on developmental stage. In other words, age-based tests of heterochrony "do not work" relative to size- or stage-based tests. It cannot be otherwise.

The "Best Case": Standardization by Developmental Stage

Let me begin with what I consider a "model" situation in that data for all possible inputs are available in this hypothetical heterochronic comparison. I take this situation directly from Gould's (1977) discussion of his "clock models," and I follow his overall scheme as closely as possible. Simply put, a study of heterochrony asks whether a descendant "shape" has changed (has been displaced) from the ancestral condition at a common "time" standard. *Clearly, that time is most appropriately a developmental stage, not an absolute age.* This is where the "time" aspect of heterochrony (Gk. *heteros:* other, different; *chronos:* time) comes from, not from plots of shape or trait values versus chronological age, as is often mistakenly thought. Alberch and colleagues' much-cited paper (1979) may be partially responsible for this misunderstanding, since a number of their bivariate plots include age on the x axis, and their three-dimensional plots provide no easy way to compare ancestor-to-descendant values on the three axes across common developmental stages. (This is the primary advantage of Gould's [1977] clock model, with its explicit four inputs, though it suffers from being restricted to a mere "snapshot" in time.) Hall and Miyake (1995) provide further fruitful discussion.

What about "size" and "age," the other inputs? These are merely used to further subclassify types of changes in which features are displaced and shape is altered. One classic example might be the case of distinguishing between paedomorphosis as generated by an early cessation of growth in time (what Gould [1977] called progenesis) and paedomorphosis as generated by a dissociation and retardation of ancestral shape change (what Gould called neoteny). In a further type of subclassification, I argued for a distinction between paedomorphosis as generated by a reduction in age at a common developmental stage, such as an early overall cessation of growth in time (Gould's progenesis; my *time hypomorphosis*), and paedomorphosis as generated by a simple terminal reduction in overall body size, with no change in age, at the common developmental stage (no category for Gould; my *rate hypomorphosis*). Parallel changes underlie rate and time hypermorphosis (Shea 1983a). The object here is not merely the proliferation of daunting terminology but rather a more accurate localization of transformations as an aid in our attempts to reconstruct selective scenarios, identify possible foci of correlated change, understand genetic and developmental covariance, and so forth. Indeed, one of Gould's (1977) primary goals and accomplishments was to illustrate how identical morphological transformations could in some cases relate to selection for rapid maturation and reproductive turnover (thus size/shape change as a correlated by-product) and in others be linked to selection on the morphological configurations themselves—quite a different evolutionary proposition.

There are at least two reasons why developmental stage is preferable to age (absolute time, chronology) as the criterion of standardization in studies of heterochrony. The first is that comparative biologists have long preferred notions of "time" that are more intrinsic to the biology of the organisms being compared (see Dettlaff, Ignatieva, and Vassetsky 1987; Hall 1992:200; Hall and Miyake 1995; Jones 1988; Reiss 1989); this is also why in many cases size is actually preferred to chronological age in mapping developmental age/stage. The second and no less compelling reason is that one of the primary goals of heterochronic analysis is to assess whether chronological age itself has changed relative to our common developmental stage (this was a major thrust of Gould's 1977 treatise, with its myriad examples of selection on duration of growth and age at sexual maturity). This obviously cannot be done if we root our comparisons along a common age scale.

Gould's (1977) clock models were explicitly constructed to follow traditional investigations of morphological heterochrony and to assess shifts in the three vectors of shape, size, and age at the chosen common developmental standard. Gould (1977:260–62) was quite clear about this:

> Developmental stage is clearly the most satisfactory criterion
> for comparing ancestors and descendants. It is also the crite-
> rion of classical literature on ontogeny and phylogeny, since
> these accounts treat the transference of specific shapes
> between juvenile and adult developmental stages of ancestors

and descendants. Other standardizations are "restricted" because their representation on the clock precludes an account by developmental stage, while the "clock" for developmental stage permits us to infer results for all other standardizations. Each restricted standardization can lead to pitfalls or causally ambiguous results.

Having established that the optimal criterion of standardization is developmental stage, as opposed to size or age, let us now consider a range of "incomplete" models.

Missing Data and Incomplete Frameworks

The ambiguity of results alluded to in the previous quotation from Gould can be manifested in various ways. It is well known in cases where size has been taken as a proxy for age or developmental stage (e.g., see Shea 1983a). Less well appreciated, at least by some (Godfrey and Sutherland 1995a, 1995b, 1996), is that plots of features (shapes) against age are equally ambiguous. This should be obvious, since we can think of many cases in which a variety of species compared at comparable ages are quite divergent in developmental stage. That is why studies such as Watts's (1990) comparative investigations of development in various primate taxa utilize intrinsic developmental time criteria such as ossification or epiphyseal fusion rather than absolute age.

Too many examples to cite here have depicted plots of shapes versus age in ancestors and descendants as definitive evidence of heterochronic change because of the shift in the y axis value (e.g., shape or some ratio) at a common age. But at a common developmental stage, age may in fact differ, and thus there may be no change in shape, or heterochrony, in such cases (or at least it is not deducible from such plots). Therefore, in plots that depict morphogenesis (developmental shape change) relative to age (e.g., Godfrey and Sutherland 1996:36), we cannot even be certain whether heterochrony (true shape transformation) has occurred, since at truly common developmental stages the shape variable may actually turn out to have identical values, indicating a geometric transformation. The same type of argument could be made about allometric divergences of slope or position: in themselves, such divergences provide no definitive evidence of heterochronic shape transformation if we do not know where common developmental stages fall along the respective allometric trajectories.

The interested reader should carefully navigate Gould's (1977: 260–62) cogent discussion of "incomplete" analyses. He methodically lays out the information optimally present in the best-case, complete framework that is explicitly lost when one or the other input is unavailable and/or is used as a surrogate for developmental stage. Earlier I argued that traditional growth-in-time plots are irrelevant and misleading as heterochronic analyses because the y axis is a unidimensional feature that has no shape or form as such. With this attack on the undesirability of utilizing (exclusively) chronological time or age as the x axis, I hope to have effectively laid to rest this needlessly confusing aspect of some recent heterochronic literature.

Does any of this general discussion have significant implications for the specific ongoing debates over human neoteny? In several ways, perhaps. First, I have repeatedly stressed (e.g., Shea 1988, 1989) that the received wisdom (e.g., Montagu 1981 and many others) that the markedly extended duration in absolute time of humans' developmental periods (e.g., age at sexual maturity, longevity) is a priori evidence of neoteny is misleading. Rather, the appropriate criterion for neoteny is retardation of morphogenesis (shape change) *at common developmental standards*, regardless of what transformations have occurred in absolute age (though Gould [1977] clearly depicted and defined neoteny as also yielding no change in size at the common developmental stage). Having carefully articulated the entire system of analysis, Gould (1977) was somewhat disingenuous in discussing the human case, in my opinion. He was thereby ultimately driven to assume some necessary link between retarded morphogenesis and prolonged growth in time (Shea 1989:80–85, 89–90), a linkage that actually undermines other aspects of the heterochronic framework and has no basis in our growing knowledge of molecular genetics and growth control. In any case, the focus on time and chronology in human development is a symptom of the ambiguity noted previously.

A second ambiguity arises when some insist on plotting shapes (however determined) against age for the criterion of assessment of shape retardation or acceleration. Godfrey and Sutherland (1995a, 1995b, 1996; e.g., see 1996:36–37, fig. 14) repeatedly make this mistake in an attempt to salvage Gould's (1977) interpretation of human neoteny. But of course shape so assessed will be "retarded" in a species that has significantly prolonged the absolute chronological duration of its life history periods—that is, in which the *x* axis is stretched out relative to underlying physiological criteria of developmental time. It cannot be otherwise, but what we instead need to know is whether descendant human features are in a juvenilized state at common developmental stages. This sort of approach, unhappily, also leads the already jargon-laden heterochronic framework to such difficult labels as "peramorphic paedomorphosis" (Godfrey and Sutherland 1996). (Though I hate to fuel the fire, I must point out that the appropriate tag even within their reconstructed framework would be "neotenic peramorphosis.") There is ultimately nothing "wrong" in plotting shape versus age and examining ancestral perturbations, following the same argument about the inherent relativity of categories that I present next (and so, too, with other incomplete models, such as plotting shape versus size—or their comparable allometric surrogates). It is merely not the appropriate test for the question of heterochronic paedomorphosis via neoteny in the *complete* framework, and thus it misleads when this distinction is not realized.

Inherent Relativity

The relativity of heterochronic shape transformations just described occurs because our assessments of shape change (and thus of heterochronic category) may be altered if we substitute one developmental standard (such as size or age) for the traditionally favored developmental stage. But there is a second potential input to such relativity that may occur independently of, or in concert with, that just noted. It occurs when we change the variable being used to represent any of the three primary inputs of shape, size, and age or that reflecting developmental stage. Obviously, a substitution of one type of developmental marker (e.g., second molar eruption) for another (e.g., basicranial synchondrosis fusion) can alter actual assessments of paedomorphosis or peramorphosis and can alter them differentially depending on the variables depicting shape. If paedomorphosis is observed for a given shape transformation based on a particular size, age, or stage estimate, isometry or peramorphosis may result relative to some other such estimate. This is precisely the same as noting that variables positively allometric to one other variable may not necessarily be positively allometric to all variables, or that variables ontogenetically scaled relative to basicranial length in African apes are not necessarily ontogenetically scaled relative to some other (e.g., postcranial) dimensions (Hartwig-Scherer and Martin 1992). A more recent "discovery" of the true-by-definition framed as a critical rectification is that "'size/shape dissociation' does not require change, from ancestor to descendant, in the slopes and intercepts of *all* growth allometries" (Godfrey and Sutherland 1995b:422).

A final aspect of the inherent relativity of heterochronic categories relates to possible complex changes along a chosen continuum of the particular developmental stage (or substituted age or size). For example, suppose we are using Gould's (1977) clock models to assess heterochronic shape (plus age and size) change at the common developmental standard of full dental eruption. Depending on the complexity and directionality of both the size/shape trajectories and the spacing of the developmental stage markers, we may find a totally different heterochronic transformation at another point—say, deciduous M1 eruption. A specific example of this is provided by Ruvkun and colleagues' (1991) study of dominant gain-of-function mutations and the misregulation of the *Caenorhabditis elegans* heterochronic gene lin-14; this results in the repetition/retention of larval stage 1 morphology into sexual maturity (and therefore, a form of paedomorphosis via neoteny), although morphogenesis is only delayed, not blocked, since the mutant worms eventually do develop the normal adult alae and cuticle morphology. Thus, if we choose some later developmental stage in adulthood, subsequent to sexual maturity, to make our comparison, there would be no heterochronic change in morphology at all. This further emphasizes that Gould's clock models represent something of a "snapshot" view of ongoing processes and results that can change dramatically throughout development (Klingenberg 1998). We need a common developmental standard, but we need also to recognize its inherent relativity.

Nothing illustrates this issue better than the huge amount of ink spilled debating the "right" heterochronic transformation characterizing human brain evolution. Gould (1977), following many previous researchers, described the relatively large brain of modern humans (relative to skull size, body size, etc.) as paedomorphic because it resembled ancestral juvenile rather than adult (brain/body) proportions at the common developmental stage of sexual maturity. Others, however, have argued that human brain/body enlargement instead reflects hypermorphosis or peramorphosis (e.g., McKinney 1998; McKinney and McNamara 1991). How can this be? Must not someone be fundamentally wrong?

Unfortunately for the absolutist, both are correct. This is simply because of the oft-noted complex allometry of brain/body scaling, with its early period of strong positive allometry followed by a prolonged period of strong negative allometry. When we compare humans with nonhuman primates at a common developmental marker shortly after birth (e.g., Gould 1977:374, fig. 65), we can conclude that relative brain size is peramorphic via (time) hypermorphosis (Shea 1983a), for the positively allometric slope is extended to new size/shape ranges, yielding an even higher brain/body ratio in the derived human (at this point a relatively lower ratio would be paedomorphic, reflective of earlier stages on the trajectory). But when we compare humans with nonhuman primates at the common developmental stage of sexual maturity, after the prolonged period of *negative* brain/body allometry that begins early in postnatal life, a relatively larger brain is a *paedomorphic*, not a peramorphic, feature (see Shea 1989 and Klingenberg 1998 for additional discussion). One developmental stage cannot be deemed inherently superior or more accurate than the other (or than myriad alternatives). Both heterochronic perspectives and sets of results are equally valid, and both tell us something important about the complex ontogeny of brain/body scaling and the evolutionary transformation in early humans. A full understanding of the heterochronic framework acknowledges, welcomes, and controls for the relativity inherent in this approach.

HETEROCHRONY IN PYGMIES—HUMAN AND CHIMPANZEE

The morphological differences between adult human pygmies and nonpygmies may be profitably viewed from the perspective of heterochrony. I have previously reviewed the morphological and developmental bases for this claim (e.g., Shea 1988, 1990, 1992b; Shea

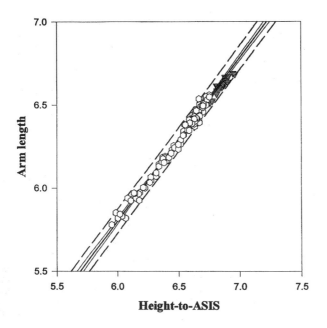

FIGURE 7.4.

A representative example of ontogenetic scaling in body proportions of Efe pygmies (ontogenetic sequence, hexagons) and non-Efe African adults (triangles) of larger size. The 95-percent regression confidence intervals and prediction intervals are shown with dashed and dotted lines, respectively. ASIS refers to the anterior superior iliac spine. (Redrawn from Shea and Bailey 1996.)

and Bailey 1996; Shea and Gomez 1988) and will not detail that evidence here again. Suffice it to say that this work has demonstrated that (1) significant differences in skeletal proportions and relative tooth and brain size exist between adult pygmies and nonpygmies, and (2) those differences carefully examined to this point appear to result from global truncation in pygmies of common patterns of ontogenetic allometry (relative growth) shared with nonpygmies.

In this occurrence of microevolutionary size reduction we have a model case of heterochronic transformation as described in this chapter. In other words, the characters that change are proportions (points along allometric continua), so this is truly a morphological, shape transformation (fig. 7.4). The changes are not major novelties but rather are altered proportions in the same basic "features," merely shifted in relative timing in the ancestral-to-descendent transformation. Finally, the morphological transformation involves an entire network of covariant characters (the redundant allometric truncations), which we even understand in terms of the developmental growth factors and controls (if not genes) involved (see Shea 1992b and Shea and Bailey 1996 for discussion).

This is not, by the way, to imply that *all* morphological distinctions between human pygmies and nonpygmies are necessarily explicable in terms of this relatively simple allometric and heterochronic transformation, but merely that many of them appear to be. Godfrey and Sutherland's (1995a:56–57) implication to the contrary, including their claim for "the empirical ubiquity of ontogenetic scaling among closely related sister taxa," is thus clearly misguided and certainly not a position promulgated by those long working in the area of allometry and heterochrony. I have repeatedly shown and stressed that such ontogenetic allometric dissociations exist for the African apes and various other primate comparisons. Indeed, one of the most effective uses of ontogenetic scaling lies in developing a "default baseline" against which to recognize special adaptations, or apomorphic function via altered covariance structure (see Shea 1981:180–81 and Shea and Bailey 1996:315–17 for discussion; see Shea 1996 for a recent review of primate examples). Godfrey, King, and Sutherland (1998) seem to miss this key point in their recent consideration of ontogenetic scaling and comparative allometric approaches.

The pygmy chimpanzee, or bonobo (*Pan paniscus*), provides a second case of allometric and heterochronic transformation, this time on the macroevolutionary level, at least according to current consensus taxonomy. I have previously detailed the allometric and heterochronic bases of the transformation from the common chimpanzee to the pygmy chimpanzee, or its converse (Shea 1983a, 1983b, 1983c, 1984). First, although there is indeed considerable overlap in adult body size between pygmy and common chimpanzees, and this overlap is differentially expressed depending on the subspecies of *Pan troglodytes* being discussed, average weights for the two species definitively place *Pan paniscus* as the smaller. Even in comparisons in which average weights may not differ—that is, *P. t. schweinfurthii* versus *P. paniscus*—the latter is skeletally dwarfed (e.g., Jungers and Susman 1984), though differentially so in disparate body regions. I have previously shown that patterns of ontogenetic allometric covariance within the major body regions of the skull (especially the face), trunk, and limbs are predominantly shared between the two species. Additionally, the pygmy chimpanzee is differentially dwarfed along these common allometric trajectories, depending on the part of the body being discussed. The face is most strongly dwarfed, the trunk and upper limbs somewhat less so, and the hind limbs not really at all (fig. 7.5). This composite means that the pattern of allometric concordance noted *within* these body regions is broken or dissociated when we make comparisons *between* the regions—for example, a plot of palate length versus femur length. In my original discussions of heterochrony and allometry as they apply to *Pan paniscus*, I linked the paedomorphosis of facial morphology first to an overall marked reduction in facial size (with concomitant allometric changes) and, second, to possible selection for reduced dentofacial dimorphism in the context of the pygmy chimpanzee's social system (e.g., Shea 1983b:522, 1984:121–23).

Godfrey and Sutherland (1996) have critically discussed this example at some length. There are two issues that require clarification here, the first dealing specifically with the morphological transformation in these chimpanzees and the second with its more general implications for heterochronic analysis and even the broad assessment of human neoteny. Godfrey and Sutherland (1996:21, 35) contend that my assessment of cranial paedomorphosis in the pygmy chimpanzee was made relative to trunk length. Yet a plot of skull length versus trunk length (Shea 1983b:fig. 1) in fact yields a negatively allometric slope for both species' ontogenies, with *Pan paniscus* exhibiting the lower slope. How can this be, and does this not imply a *systematically erroneous application of an improper criterion* for recognizing heterochronic change?

In fact, this turns out to be a "non-issue" and an example of the relativity discussed earlier. My original figure 1, panel *c* (Shea 1983b), here reproduced as figure 7.5, was generated simply to demonstrate the reduced growth of the skull (and particularly the face) relative to various measures of overall size (here reflected by trunk length) in the pygmy chimpanzee. There will of course be some variance in the specific allometric coefficients as we choose among different potential size surrogates. Given the reduced overall skull (and particularly face) size

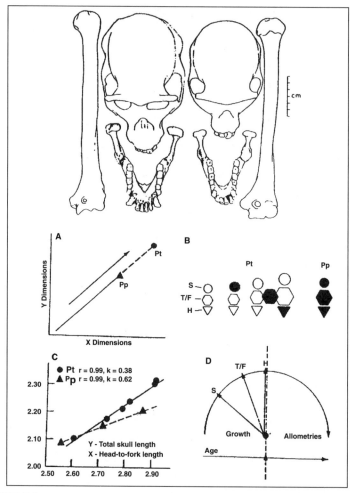

FIGURE 7.5.

A depiction of cranial (facial) paedomorphosis via neoteny in the pygmy chimpanzee (Pan paniscus) relative to the common chimpanzee (Pan troglodytes). The schematic at the top illustrates the reduced size of the skull (and especially face) relative to humerus length. Panel A schematically depicts the pervasive ontogenetic scaling of growth allometries within the face (and within most other body regions) and the paedomorphosis consequently result- ing from truncation in terminal size in Pan paniscus. Panels B and D depict the differen- tial size reduction by body region in a comparison of Pan paniscus with Pan troglodytes; the skull (face) is most strongly dwarfed, the hind limbs barely or not at all. Panel C shows a plot of total skull length versus trunk length for both species (note the incorrect reversal of the slope, k, values for the two species). (Redrawn from Shea 1989.)

and the pervasive ontogenetic scaling of the features *within* the face, we can then first note that the facial skeleton of *Pan paniscus* is paedomorphic relative to that of *Pan troglodytes*. We also know from previous discussions that the criterion for neotenic (as opposed to hypomorphic or progenetic [Shea 1983a]) paedomorphosis varies depending on whether the ancestral ontogeny is positively or negatively allometric (Shea 1989:73, 96). Here a discussion of the "skull" as a whole admittedly becomes problematic, since the cranial vault grows during postnatal ontogeny with negative allometry (the brain having ceased its period of early marked growth), while the face grows with positive allometry. This general mammalian pattern (e.g., Gould 1975) is also true of the chimpanzees (Shea 1984), though this is always relative to particular baselines. So we connote this pattern as a general one, knowing full well that one can determinedly peruse an array of bodily dimensions and come up with some number that will show the chimpanzee face to grow only isometrically or perhaps even with negative allometry, or the vault to grow isometrically or perhaps even with positive allometry. When the skull is divided into roughly the splanchnocranium and the neurocranium, the face of the pygmy chimpanzee is appropriately seen as resulting from paedomorphosis via neoteny (contra Godfrey and Sutherland 1996), because dimensions such as palate length and many others in the face exhibit positive allometry relative to numerous general size estimators during ontogeny and a downward shift (in position or slope) in the pygmy chimpanzee as compared with the common chimpanzee. The braincase would be determined to result from paedomorphosis via neoteny if the pygmy chimpanzee's trajectory of postnatal negative allometry were transposed or shifted above that of the common chimpanzee. This is clearly the case for many vault dimensions scaled against basicranial size and also probably for many, if not all, dimensions of the postcranium. And we may complicate this example even further, while continuing to emphasize the previous statements regarding the relativity of heterochronic categories, by pointing out that the paedomorphic face of the pygmy chimpanzee results from rate hypomorphosis and not neotenic dissociation at all, if we restrict our comparative (*x* axis) allometric baseline to the skull base or some other such measure of overall facial size (Shea 1983a, 1983b, 1983c, 1984).

Those interested in a careful assessment of the full comparative data base and allometric and heterochronic comparisons are referred to previously published material (Shea 1981, 1982, 1983a, 1983b, 1983c, 1984, 1985). A careful reading of the relevant passages shows that the neotenic paedomorphic features being discussed relate to the face, not the braincase. For instance, I linked facial and gnathic neotenic paedomorphosis to reduced sexual dimorphism (face, canines, etc.) in the context of observed behavioral features of *Pan paniscus* such as high male-female affinity and less sexual differentiation of social structure (e.g., Kano 1980; Kuroda 1979; but see White 1996). Whatever the validity of this linkage, it is clearly the face and not the skull as a whole that is the target of morphological discussion. Finally, I explicitly linked the discussion of neotenic paedomorphosis in the pygmy chimpanzee to the facial component of the skull in my review of human neoteny (Shea 1989:92).

The confusion and debate over the example of the pygmy chimpanzee therefore vanishes once one acknowledges the long-appreciated component of relativity inherent in all heterochronic assessments (whether based on allometry or not). The direct and implicit claims of Godfrey and Sutherland (1995a, 1995b, 1996) that they have exposed fundamental flaws in the theory and methodology of allometric heterochrony are unfounded. Moreover, few will be convinced by an argument that even legitimate minor criticisms of previous empirical work on other taxa in any way strengthen the unrelated case for human neoteny as advanced by Gould and others. Indeed, Godfrey and Sutherland (1995a, 1995b, 1996) offer no new data to strengthen such claims. The debate awaits the careful presentation of new data that would significantly support the hypothesis in terms of both morphological expression and underlying genetic and developmental control (see Shea 1989).

CONCLUSION

Evolutionary change via heterochrony involves the derived displacement in developmental time of morphologies present in ancestors, yielding the classic parallels between ontogeny and phylogeny. Heterochronic change does not encapsulate all developmental perturbations underlying morpological transformations and in fact is typically not very relevant to the evolution of significant novelties and major transformations. This bears directly on the rekindled debates over the role of neoteny in human evolution, since the major derived transformations of the brain and skull (behavior), the upper respiratory apparatus (speech), and the postcranial skeleton (bipedalism) reflect novelties that are not easily interpreted as due to the mere shifting in developmental time of features present in great apes. Indeed, it is possible to interpret some selected morphological features in modern humans as indicative of neotenic paedomorphosis (Shea 1989); evolution, whether by heterochrony or other processes, is seldom an all-or-none proposition, as contributors to this debate have realized since at least the time of Bolk (1926). But as I have detailed previously (Shea 1989), there are no compelling morphological, developmental, or genetic data to suggest that these key features in human evolution are linked or that they result from any coordinated heterochronic transformation. The data of allometric heterochrony, the source of some contentious debates in the literature that I hope I have helped clarify, also support this conclusion by making any globally coordinated suite of neotenic retardations unlikely in theory (Shea 1989:96, n.d.) and undocumented empirically (Shea 1989).

All evolutionary transformations must be explicable in terms of underlying developmental and genetic changes, and therefore I have no doubt than an understanding of the interrelationships between ontogeny and phylogeny (e.g., Gould 1977; Hall 1992; Raff 1996) will be fundamental to unraveling the mysteries of human evolution. To this point, however, the evidence seems not to implicate any simple heterochronic transformation, nor any coordinated neotenic paedomorphosis in particular, as the predominant change involved.

8

The Heterochronic Evolution of Primate Cognitive Development

Jonas Langer

Neoteny and terminal addition—or terminal extension, the more precise term introduced by McKinney in this volume—are well-known alternative hypotheses about the evolution of ontogeny in general and the evolution of cognitive development in particular (Gould 1977, 1984; McKinney and McNamara 1991). My focus here is on primate cognition. The mechanism that I propose for its development is dissociated heterochrony, resulting in mosaic evolution.

Neoteny implies the infantilization of human cognitive ontogeny caused by delayed development. Terminal extension implies adultification of human cognitive ontogeny caused by accelerated development. So humans are either "retarded" or "precocious" apes.

By either account, cognitive ontogeny concords with its phylogeny, subject to two modifications. Developmental velocity increases or decreases. But significantly, by both accounts the extent of cognitive ontogeny increases (dare one say "progresses"?) in phylogeny. "Directionality" of cognitive evolution and development is inherent in both hypotheses. I will return to this important issue later, but I want to stipulate from the outset that directional processes are statistical, not deterministic.

Although terminal extension is closer to the mark than is neoteny, neither accounts for central data on primates' comparative cognitive development. Most importantly, I believe that an adequate evolutionary theory must account for the increasing structural reorganization in early ontogenesis that marks the cognitive development of successor species in phylogeny. Such a theory has at least three advantages. First, it fits the known data on primate cognitive and linguistic development. Second, it implies that cognitive development is a process of cascading change. Cognitive development is increasingly reorganized in (many, perhaps most) successor phylogenetic species and in successor ontogenetic stages. There is much phylogenetic divergence and variation in cognitive developmental organization. This leads to the third advantage. The kind of theory I advocate is consistent with the observation that both the evolution and history of cognition are open systems. Humans hardly represent the last stage in either the evolution of intelligence or the history of ideas.

The vital impulse for further cognitive change, according to this view (which harks back to Baldwin [1896, 1902, 1915]), is ontogeny. The ontogenies of descendant primate species are marked by increasing and more powerful constructions of cognitive possibilities or variations. Thus, according to this hypothesis, ontogeny provides the permanent cognitive possibilities or variations for future cognitive evolution and history.

To get a workable handle on comparative structural reorganization in primate cognitive development, I focus on key formal features of its ontogeny. These include similarities and differences in onset and offset ages, velocity or rate, extent, sequencing, and organization of monkey, chimpanzee, and human cognitive development. Primate cognitive development includes foundational physical cognition, such as knowledge about causality and objects; logical cognition, such as classificatory categorizing; arithmetic cognition, such as exchange operations, especially numerical substituting; and language, if any.

Much of the available data come from cross-species comparisons with my findings on young human children. I first sketch essential features of the research methods I devised to generate the data and then turn to the similarities and differences in formal features of primate cognitive development.

RESEARCH METHODS

The research began with the study of 6- to 60-month-old children's spontaneous constructive interactions with 4 to 12 objects (Langer 1980, 1986). The range of objects spanned geometric shapes to realistic things such as cups (see figs. 8.1–8.4). Class structures were embodied by some of the object sets presented—for example, multiplicative classes such as a yellow and a green cylinder and a yellow and a green triangular column (see fig. 8.1). But there was nothing in the procedures that required subjects to do anything about the objects' class structures. No instructions were given and no problems were presented to the subjects. Children were simply allowed to play freely with the objects as they wished, because my goal was to study their constructive intelligence.

With human children, this initial nonverbal and nondirective method was followed by progressively provoked conditions. To illustrate, in one condition designed to provoke classifying, children were presented with two alignments of four objects. For example, one alignment might comprise three square rings and one circular ring while the other comprised three circular rings and one square ring. By the age of 21 months, some infants began to correct the classificatory "mistakes" presented to them (Langer 1986), and by 36 months all children did (Langer n.d.c; Sugarman 1983).

This range of methods was used to study human children's spontaneous and provoked constructions of logical, arithmetic, and physical cognitions. Many of the findings on humans that I will review have been replicated with both Aymara and Quechua Indian children in Peru, aged 8 to 21 months (Jacobsen 1984). The Indian children were raised in deeply impoverished conditions in comparison with the mainly Caucasian, middle-class San Francisco Bay–area children in my samples. Many of the findings have also been replicated with 6- to 30-month-old infants exposed in utero to crack cocaine (Ahl 1993). No differences were found in onset age, velocity, sequence, extent, or organization of cognitive development during infancy in these different human samples.

Most comparisons of primate cognitive development in the next sections are based on these studies of human children and on parallel studies of capuchins *(Cebus apella)*, macaques *(Macaca fascicularis)*, and chimpanzees *(Pan troglodytes* and *Pan paniscus)* using the nonverbal and nondirective methods developed to study human children's spontaneous cognitive constructions. My colleagues and I have yet to use the provoked methods with nonhuman primates.

INVARIANT INITIAL ELEMENTS OF COGNITION

Perhaps the most important foundational similarity (and difference, as we shall see later) among primates in their cognitive and linguistic development lies in their composing of sets. All primates we have studied so far compose sets of objects as elements for their cognition (see figs. 8.1–8.4). They compose these sets by bringing two or more objects into contact with or close proximity to each other—that is, no more than 5 centimeters apart.

This is a fundamental similarity. Combinativity structures, including especially the composing of sets, are foundational to constructing cognition and language (Langer 1980, 1986; Piaget 1972). Thus, combinativity is a central general-purpose structure. It includes composing, decomposing, and recomposing operations. (Here I focus only on composing for the sake of brevity.) These operations construct fundamental elements such as sets and series. They are foundational and fundamental because without them little if any cognition and language is possible. To illustrate the generality of these combinativity structures, consider an aspect of composing. At least two objects must be composed with each other if (1) they are to be classified as identical or different or (2) a tool is to be used as a causal instrument to an end—for example, if one object is used to hit another. Similarly, at least two symbols must be composed with each other if they are to form a minimal grammatical expression.

Combinativity operations, such as composing sets, are also foundational and fundamental because a set of cognitions can have no structure unless (1) the elements of the set are themselves constant and (2) one or more operations, functions, and/or relations performed on the set are defined (Langer 1990). To illustrate partially, the structuring of determinate equivalence and inequivalence relations by one-to-one correspondence or by substitution requires, at a minimum, construct-ing invariant objects of comparison. For example, using one-to-one cor-respondence to determine numerical equality or inequality requires, at a minimum, the composing of two sets in which the members of each set are stabilized (conserved) at least momentarily. Human infants begin to do this during their second year by composing stable sets of objects in temporal overlap and spatial proximity. Common chim-panzees also begin to do so, but not until their fifth year. Capuchins and macaques do not do so (at least up to their fourth year, which is the oldest age at which we have tested them).

INVARIANT INITIAL LOGICOMATHEMATICAL AND PHYSICAL COGNITIONS

It has long been recognized that all primates develop foundational physical cognitions such as object permanence and causal instrumen-tality (e.g., Kohler 1917; Parker and Gibson 1979). We discovered that human infants and juvenile chimpanzees and monkeys also develop (1) logical operations such as classifying by the identity of objects (Langer 1980, 1986; Langer et al. 1998; Spinozzi 1993; Spinozzi and Langer 1999; Spinozzi and Natale 1979; Spinozzi et al. 1998, 1999) and (2) arithmetic operations such as substituting objects in sets to produce quantitative equality (Langer 1980, 1986; Poti' 1997; Poti' and Antinucci 1989; Poti' et al. 1999).

These findings support our originalist hypothesis that logicomath-ematical cognition is a primary and initial primate development, just as physical cognition is (Langer 1990, 1996). These findings are inconsis-tent with the derivationist hypothesis that grammatical language is pre-requisite to the development of logicomathematical cognition (e.g., Braine 1990; Carnap 1960; Falmagne 1990; MacNamara 1986; Vygotsky 1962) and therefore uniquely human.

Instead, I have proposed that cognition provides axiomatic proper-ties necessary for any grammatical symbolic system, including language (Langer 1982a, 1983, 1986, n.d.a). Cognitive operations are necessary in order to begin to produce and comprehend the arbitrary conven-tions used to make symbolic combinations stand for and communicate semantic referents in syntactic forms. Cognition enables organisms to begin to generate syntactic constructions in which linguistic elements are progressively combinable and interchangeable, yet semantically meaningful.

The basic idea is that cognitive operations provide the axiomatic rewrite rules without which grammatical language is impossible. To illustrate, protogrammatical sentences are impossible without at least rudimentary substituting of elements between two sentential compositions. We have found wide variation in primate species' ability to substitute objects between two sets. It develops toward the end of the second year in human infants, the age at which they begin to develop protogrammatical language. It barely originates by the fifth year in common chimpanzees. It does not develop in capuchins and macaques, at least up to their fourth year. So grammatical language is developed by human children, and protogrammatical language can be learned by chimpanzees but not by monkeys.

This is a first and obvious example of the divergences in primate species' early development that have cascading consequences for subsequent development. This phylogenetic proposal about the cascading consequences of the diverging early development of different primate species is consistent with Deacon's hypothesis (this volume) that "a localized shift in the differentiation clock" during embryogenesis produced cascading later changes in brain size, body segmentation, and developmental duration (Finlay and Darlington 1995).

INVARIANT ONSET AGE OF PHYSICAL COGNITION

Human infants begin early to construct knowledge about the existence and causal relations of objects in space and time. The earliest symptoms are newborns' sensorimotor activities, such as tracking objects and hand sucking. These activities maintain contact with objects, thereby constituting stage 1 of Piaget's (1952, 1954) 6-stage sequence of object permanence development during infancy. They also require (1) exerting effort ("work" or energy) and (2) taking into account spatiotemporal contact in order to maintain effective causal relations—thereby constituting the stage 1 efficacy and phenomenalism of Piaget's (1952, 1954) 6-stage sequence of causal means-ends development during infancy.

More robust symptoms of infants' constructing of physical cognition are found from about ages 3 to 7 months. Now infants begin to anticipate correctly the reappearance of disappearing objects, which constitutes stage 3 object permanence (Piaget 1952, 1954). They also begin to use objects purposefully as causal intermediaries or instruments, which constitutes stage 3 causality (Piaget 1952, 1954) or first-order causality (Langer 1980).

Comparative research has paid little attention to the onset age of physical cognition. I have been able to discover only that the earliest symptoms of stage 1 object permanence begin to be manifested during their first week by macaques *(Macaca fuscata* and *M. fascicularis)* (Parker 1977; Poti' 1989), during their second week by *Cebus appella* (Spinozzi 1989), and during their fifth week by *Gorilla gorilla gorilla* (Redshaw 1978; Spinozzi and Natale 1989). Though limited, the data suggest little or no difference between human and nonhuman primates in the onset age for developing physical cognition. A fairly secure estimate would put onset age in the neonatal to early infancy range in all anthropoid primates.

INVARIANT SEQUENCES OF DEVELOPMENT WITHIN COGNITIVE DOMAINS

The within-domain developmental stage sequences are universal and invariant, with one partial exception detailed later. The order of stage development is conserved, including no stage skipping or reversal, in all primate species studied so far. This generalization applies to all cognitive domains studied so far.

Universal invariance has been found for the most extensively studied developmental stage sequence of physical cognition, object permanence. Because it provides the most reliable data, it will serve as my example. Sequential invariance has been found in at least a variety of monkey species (i.e., capuchins, macaques, and squirrel monkeys), in gorillas, in chimpanzees, and in humans (e.g., Antinucci 1989; Doré and Dumas 1987; Doré and Goulet 1998; Piaget 1954; Redshaw 1978). Indeed, the universality of the invariant object permanence stage sequence extends to the other mammal species that have been studied so far, cats and dogs (Doré and Goulet 1998).

We have begun to investigate whether within-domain stage sequences in logicomathematical cognitions are also universal in primate species. The data indicate universality with one partial exception

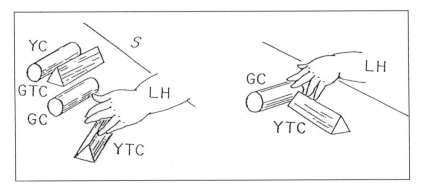

FIGURE 8.1.

A six-month-old subject composing a set consisting of a green cylinder and a yellow triangular column using the left hand.

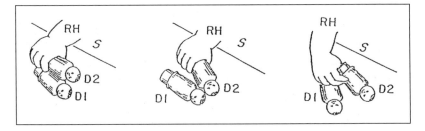

FIGURE 8.2.

A subject composing a set consisting of two dolls using the right hand.

in the domain of logical classification. Recall that we presented objects that embodied class structures—for example, four cylinders and four square columns. The sequence of classifying is invariant in humans (Langer 1980, 1986; Langer et al. 1998) and common chimpanzees (Spinozzi 1993; Spinozzi and Langer 1999; Spinozzi et al. 1998).

It is simplest to illustrate the sequence shared by common chimpanzees and humans by summarizing its development in humans only. First, human infants construct class-consistent single categories. They consistently compose single categories of *different* objects at age 6 months (fig. 8.1). This changes to *random* composing at ages 8 and 10 months, when they are equally likely to compose single sets of different, identical, or similar objects. The next change is the onset of composi-

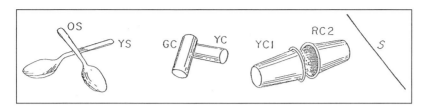

FIGURE 8.3.

Three compositions of similar objects: an orange and a yellow spoon; a green and a yellow cylinder; and a yellow and a red cup.

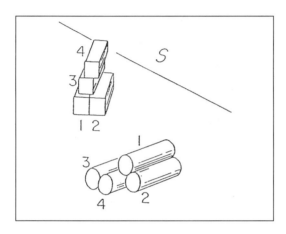

FIGURE 8.4.

Second-order classifying by a 21-month-old subject.

tion of single categories of *identical* objects at age 12 months (fig. 8.2). Studies using somewhat varying procedures and analyses found comparable single-category classifying by identities at age 12 months (Nelson 1973; Riccuiti 1965; Starkey 1981; Sugarman 1983). The major change after this is the onset of composition of *similar* objects into single sets at age 15 months (fig. 8.3). Similarity is defined as identity in some properties, such as form, combined with differences in others, such as color.

In this entire developmental sequence, infants compose only one category with one class property at a time. I therefore call this *first-order classifying*. This structure is what differentiates these four sequential developments in first-order classifying from *second-order classifying*. Second-order classifying originates at age 18 months. Now infants begin consistently to compose two temporally overlapping categories. The objects composing each category are identical, but the objects in the two categories are different (fig. 8.4). Thus, second-order classifying means composing two categories and two class properties at the same time.

This entire sequence, including the stages of first-order classifying and the transition to second-order classifying, is similar in young common chimpanzees and humans. It is different, however, in young monkeys. They develop only first-order, not second-order, classifying. Capuchins, for example, begin with mainly random classifying at age 16 months, change to classifying mainly by differences by age 36 months, and shift to classifying by identities and similarities by age 48 months (Spinozzi and Natale 1989). Indeed, capuchins and macaques differ from each other, as well as from common chimpanzees and humans, in their sequences of developing first-order classifying.

VARIANT EXTENT OF DEVELOPING ELEMENTS OF COGNITION

During their first three years, human infants' combinativity operations already construct ever more powerful elements of cognition (Langer 1980, 1986, 1998). Two measures permit central comparisons between the elements composed by human infants and those composed by young nonhuman primates (Antinucci 1989; Poti' 1997; Spinozzi 1993; Spinozzi et al. 1999). I outline these findings in turn.

With age, human infants include more objects in the sets they compose. For example, at age 30 months, 14 percent of their sets comprise 8 objects each. Similarly, the number of objects composed into sets increases with age in common chimpanzees. Up to age 5 years, the limit is about 5 objects. Thus, while already breaking out of the limits of the law of small numbers, defined as no more than 3 or 4 units, young chimpanzees seem to be restricted to the smallest intermediate numbers. Minimal increases are found in capuchins and macaques during their first 4 years. With age, the set sizes increase from 2 to 3 objects. They do not exceed the limits of small numbers.

During their first year, human infants construct only 1 set at a time. By the end of their first year they begin to construct 2 sets at a time, and by the end of their second year they begin to construct 3 or 4 contemporaneous sets. Thus, whereas only 10 percent of their compositions at age 12 months comprise multiple contemporaneous sets, more than half do so by age 36 months. Young common chimpanzees also begin to construct contemporaneous sets, but up until age 5 years they are limited to constructing 2 sets at a time. And their rate of production is comparatively small: contemporaneous sets account for only 20 percent of their compositions. In contrast, with rare exceptions capuchins and macaques do not compose contemporaneous sets in their first 4 years.

VARIANT EXTENT OF DEVELOPING COGNITION

Young nonhuman primates do not progress to the level of cognitive development that human infants have already attained by their third year (Langer 1986, n.d.c). This fundamental comparative developmental difference is the result, in part, of divergence in different species' construction of the elements of their cognition. As already noted, set construction by composing objects together is a major source of cognitive elements for primates. Thus, the elements of cognition that primates construct constrain the level of intellectual operations they can attain. Disparities in the level of combinativity operations are central to the cascading divergence in primate species' cognitive development that begins early in life.

Up to at least the age of four years, capuchins and macaques are limited to constructing single sets of up to three objects. Human infants already begin to exceed these limits in their second year by constructing two contemporaneous sets of increasingly numerous objects. The comparative consequence is that capuchins and macaques are locked into developing no more than relatively simple cognitions, whereas progressive possibilities open up for children to map new and more advanced cognitions. For instance, as we have already seen, young capuchins and macaques are limited to constructing single-category classifications, whereas human infants begin to construct two-category classifications by the age of 18 months.

Young chimpanzees, like human infants and unlike young monkeys, construct two contemporaneous sets as elements of their cognition. Unlike monkeys, they are therefore not limited to developing first-order cognitions, such as single-category classifications. Instead, like human infants, young chimpanzees begin to develop second-order cognitions, such as two-category classifications, but not until their fifth year.

Up to the age of five years, and unlike human infants, common chimpanzees are limited to composing two contemporaneous sets. As a consequence, they construct no more than two-category classifications (Spinozzi 1993). In contrast, human infants in their second year begin to compose multiple contemporaneous sets, and during their third year they begin to develop three-category classifications.

This is a vital difference in the cognitive development attainable by chimpanzees and humans. It determines whether hierarchically integrated cognition is possible. Three-category classifying opens up the possibility of hierarchization, whereas two-category classifying permits nothing more than linear cognition. Minimally, hierarchic inclusion requires two complementary subordinate classes integrated by one superordinate class. Thus, during their third year human infants open up the possibility of hierarchization. Chimpanzees as old as age five years do not. They remain limited to linear cognition.

Another vital difference between humans and chimpanzees in their potential cognitive development is that human infants begin to map their cognitions onto each other. Only transitional mappings of cognition onto cognitions are constructed by bonobos up to at least adolescence and by common chimpanzees up to adulthood (Poti' 1997; Poti' et al. 1999). To illustrate, toward the end of their second year, human infants begin to compose two sets of objects in spatial and numerical one-to-one correspondence. Then they exchange equal numbers of objects *between* the two sets such that they preserve the spatial and numerical correspondence between the two sets. These infants map exchanges onto their correspondence mappings, thereby producing equivalence-upon-equivalence relations. In comparable conditions, chimpanzees at most exchange only equal numbers of objects *within* one of two corresponding sets they have constructed, thereby preserving their equivalence. Only the cognition of human infants, among young primates, becomes fully recursive. Recursiveness is a key to changing the rules of cognitive development (Langer 1986, 1994a, 1996). It further opens up possibilities for transforming linear into hierarchic cognition.

Recursive development drives progressive change in the relation between the forms and contents of cognition. This opens up possibilities for transforming forms (structures) into contents (elements) of cognition. Thus, initial simple linear cognitions (e.g., minimal classifying) become potential elements of more advanced hierarchic cognitions (e.g., comprehensive taxonomizing). In this view, recursive cognitive development is a precondition for the formation of all reflective cognition that requires hierarchization, including abstract reflection (Piaget 1977) and metacognition (Cassirer 1953). Linear cognition is insufficient.

With the formation of hierarchic cognition, the referents of human infants' intellectual operations are no longer limited to objects. Cognition is no longer limited to the concrete. Progressively, the referents of infants' cognitions are becoming relations, such as numerical equivalence and causal dependency, that are the products of other intellectual operations mapped onto objects. By mapping cognitions onto relations, infants' intelligence is becoming abstract and reflective or thoughtful. Then meaning, and certainly the significance of their representational logicomathematical and physical concepts, can begin to become abstract and reflective (Cassirer 1953). Reasons why or explanations for phenomena can begin to be constructed.

Reasoned explanation is an advanced cognitive development that requires a base of hierarchic conceptual integration. Conceptual integration is not truly possible without the hypotheticodeductive formal operations that are uniquely human and originate in early adolescence (Inhelder and Piaget 1958; Langer 1969, 1994b). Formal operational development continues through young adulthood, up to the age of about 30 years (Kuhn et al. 1977; Langer 1982b). Humans' extended cognitive development parallels our prolonged brain maturation, which extends into adolescence and young adulthood. This includes prolonged glial cell growth, myelination of axons, synaptogenesis, and, perhaps most importantly, dendritic growth in the cortex (Gibson 1990, 1991; Paus et al. 1999; Purves 1988).

The findings, then, support a phylogenetic model of progressively increasing cognitive peramorphic "overdevelopment" or terminal extension. Human cognition is overdeveloped in comparison with chimpanzee cognition, which is overdeveloped in comparison with monkey cognition. This model of cognitive phylogeny is consistent with the phylogenetic model of neural "overdevelopment" of brain size and complexity proposed by Gibson (1990, 1991).

VARIANT VELOCITY OF COGNITIVE DEVELOPMENT

The rate of developing logicomathematical cognition is accelerated in human ontogeny. The development of classification is prototypical. For instance, we saw that capuchins do not complete their development of first-order categorizing until the age of 4 years. In comparison, it is already developed by 15 months in humans. So, too, although common chimpanzees' development extends to rudimentary second-order categorizing, it does not originate until the age of 4.5 years. In comparison, it originates at 1.5 years in humans. This pattern of accelerated logicomathematical cognitive development in humans fits with the theory of terminal extension and not neoteny.

In comparing the velocity of developing logicomathematical cognition, I have used absolute age as the scale. Humans' velocity, then, is approximately three times as rapid as that of common chimpanzees. This underestimates the actual difference because the more appropriate comparative scale is relative age, which takes into account each species' life span. To illustrate, humans live about twice as long as common chimpanzees. Thus, humans' velocity of developing logicomathematical cognition is actually more like six times as rapid as that of common chimpanzees.

Nonhuman primates develop their physical cognition much more rapidly than their logicomathematical cognition (Langer 1993). This is vividly illustrated by the physical cognition for which the greatest amount of comparative developmental data is available, object permanence. A representative finding is that young capuchins develop to their most advanced level of object permanence, stage 5 (presentational or here-and-now object knowledge) in Piaget's 6-stage sequence, by the age of 7 months (Natale 1989). In contrast, young capuchins do not develop to their most advanced level of logicomathematical cognition, such as single-category classifying by identity and similarity, until their fourth year. So the velocity of capuchins' development of physical cognition is greatly accelerated in comparison with their logicomathematical cognition.

Though developing much more rapidly than their logicomathematical cognition, capuchins' physical cognition still develops more slowly than humans' physical cognition. We need only recall that human infants develop stage 5 object permanence by around the age of 10 months (e.g., Paraskevopoulous and Hunt 1971; Piaget 1954). Factoring in humans' life span, approximately three times longer than that of capuchins, reveals that the developmental velocity is about doubled in humans.

Indeed, this is the general pattern for all physical cognitions (including space and causality as well as objects) for which comparative developmental data are available (which includes data on macaques, gorillas, and chimpanzees as well as capuchins; Parker and McKinney 1999). The development of physical cognition is accelerated in human versus nonhuman primate ontogeny. Thus, although this acceleration is less rapid than that of logicomathematical cognition, the development of physical cognition is certainly not neotenous in primate evolution. Instead, it confirms that there is much terminal extension in the evolution of cognitive development.

Further confirmation comes from comparing the velocity of developing social cognition among primates. So far, the comparative data mainly concern the development of imitation and symbolization based upon Piaget's (1952) 6-stage sequences of these behaviors but extended to aspects of mirror self-recognition and theory of mind (Parker and McKinney 1999). The pattern again is acceleration in human versus nonhuman primate ontogeny.

All the available comparative developmental data, then, point to the same conclusion: Terminal extension is a general law of the evolution of primate cognitive development. Neoteny is not. All forms or domains of logicomathematical, physical, and social cognition studied so far develop more rapidly and are terminally extended in human ontogeny.

VARIANT SEQUENCE OF DEVELOPMENT BETWEEN COGNITIVE DOMAINS

We are discovering striking divergences in the organization of developmental sequencing between physical and logicomathematical cognitive domains in primates. This suggests divergent evolution (fig. 8.5).

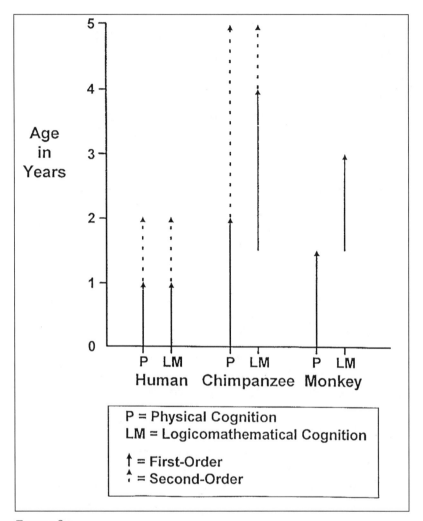

FIGURE 8.5.

Comparative cognitive development in humans, chimpanzees, and monkeys: vectorial trajectories of developmental onset age, velocity, sequence, and organization (but not extent or offset age).

Physical and logicomathematical cognition develop in parallel in human children. The onset age is the same—very early infancy and probably the neonatal period—and they develop in synchrony. To illustrate, infants construct first-order classificatory and causal relations during their first year (Langer 1980) and second-order classificatory and causal relations during their second year (Langer 1986). Neither type of cognition begins or ends before the other during childhood. Consequently, both forms of cognition are open to similar environmental influences and to each other's influence.

We find the other extreme in capuchins and macaques, namely, almost total asynchrony between their development of physical and logicomathematical cognition. To illustrate, capuchins complete their development of object permanence (up to Piaget's stage 5) during their first year (Natale 1989) and begin to develop classifying only during their second year (Poti' and Antinucci 1989; Spinozzi and Natale 1989). Because they are out of developmental phase with each other, capuchins' and macaques' physical and logicomathematical cognitions are unlikely to be open to similar environmental influences and to each other's influence.

In common chimpanzees' ontogeny, physical and logicomathematical cognition follow partially overlapping developmental trajectories. Though already well under way, common chimpanzees' development of physical cognition is not completed before the onset of logicomathematical cognition. We can therefore expect that these two cognitive domains may eventually begin to be partially open to similar environmental influences and to each other's influence, but not until relatively late in ontogenesis in comparison with humans.

From the start of human ontogeny, physical and logicomathematical cognition follow contemporaneous developmental trajectories that become progressively interdependent. Synchronic developmental trajectories permit direct interaction or information flow between cognitive domains. Mutual and reciprocal influence between logicomathematical and physical cognition is readily achievable because humans develop them simultaneously and in parallel. Thus, we have found that even in infancy, logicomathematical cognition introduces elements of necessity and certainty into physical cognition (Langer 1985). At the same time, physical cognition introduces elements of contingency and uncertainty into logical cognition.

In primate evolution, the unilinear growth trajectory of physical cognition *followed by* logicomathematical cognition evolved into multilinear growth trajectories of physical and logicomathematical cognition *at the same time*. The sequential trajectory of physical followed by logicomathematical cognition in the ontogeny of capuchins and macaques became "folded over" into concurrent trajectories—first into partially multilinear development midway in chimpanzee ontogeny and eventually into fully multilinear development from the start in human ontogeny (fig. 8.5).

Phylogenetic displacement in the ontogenetic onset and offset ages of cognitive development, along with the velocity of one cognitive developmental trajectory relative to another within the same organism, causes a disruption in the repetition of phylogeny in ontogeny. Such heterochronic displacement involves a dislocation of the phylogenetic order of succession. It produces a change in the onset and offset ages and the velocity of ancestral processes. The onset and velocity may be accelerated or retarded. But importantly, in humans as compared with nonhuman primates, the onset age and velocity of development of logicomathematical cognition are accelerated and the offset age is extended. And although the offset age of physical cognition is extended, only its developmental velocity is accelerated in humans relative to nonhuman primates. The onset age is not precocious.

The comparative data on the organization of and sequencing between cognitive domains in humans, chimpanzees, and monkeys are consistent with the hypothesis that heterochrony is a mechanism of the evolution of primate cognition. According to this hypothesis, hetero-chronic displacement is a mechanism by which consecutively develop-ing ancestral cognitive structures were transformed in phylogenesis into simultaneously developing descendant cognitive structures in human ontogenesis. Heterochrony produced the reorganization of nonaligned ancestral cognitive structures in capuchins and macaques into the partly aligned descendant structures in chimpanzees and the fully aligned descendant structural development of cognition in human infancy.

This reorganization opened up multiple cascading possibilities for full information flow between logicomathematical (e.g., classificatory) and physical (e.g., causal) cognition in human infancy (e.g., making it possible to form a "logic of experimentation"). These cognitive domains are predominantly segregated from each other in time and therefore in information flow in the early development of capuchins and macaques. They are partially segregated from each other in time and information flow in the early development of common chimpanzees.

The possibilities opened up for further development vary accordingly and, I propose, reciprocally constrain the direction of progressive cognitive ontogeny in primate phylogeny. As we have seen, cognitive development is already quite substantial during the youth of capuchins and macaques. Their asynchronic early cognitive development, however, hampers much further progress with age. The partially synchronic and relatively advanced early cognitive development of common chimpanzees multiplies the possibilities for substantial if still limited further progress with age. Human synchronic and still more extensive early cognitive development opens up comparatively unlimited, permanent, unique, and cascading possibilities for further intellectual progress, such as a history of science (see Langer 1969:198–200 for five criteria in determining progressive cognitive development).

Its ontogenetic consequence is terminal extension of human cognitive development from (1) its initial sensorimotor stage, including first- and second-order cognition, through (2) one-way functions or intuitive preoperations to (3) two-way functions or concrete operations, followed by (4) hypotheticodeductive formal operations (Langer 1994b). The cascading cognitive scope of these consecutively developing stages is marked by two central features. First, successor stages construct increasing quantities of progressively complex knowledge. Second, successor stages integrate knowledge constructed during previous stages by recursive hierarchization into progressively power-ful forms.

The corresponding increase in duration of the stages—sensorimotor from ages 0 to 2 years, preoperations from 2 to 6 years, concrete operations from 6 to 12 years, and formal operations from 12 to 30 years—thus becomes predictable. This does not reflect any slowing down in the velocity of development. Instead, the increasing duration of each successor stage in human ontogeny reflects the cascading terminal expansion in constructing cognitive possibilities and necessities (Piaget 1987a, 1987b).

CONCLUSION

These findings on key formal features of comparative cognitive development in primates lead to five core propositions about its heterochronic evolution.

First, the velocity of developing logicomathematical, physical, and social cognition is accelerated in juvenile humans versus juvenile nonhuman primates. Juvenile humans' cognitive developmental acceleration is consistent with their relatively precocial brain development, as compared with the relative deceleration of the rest of their physiological development (see table 10.3 in this volume).

Second, comparative intellectual acceleration is most pronounced in juvenile humans' development of logicomathematical cognition. The acceleration is less for their development of physical cognition. Conversely, comparative retardation is most pronounced in juvenile nonhuman primates' development of logicomathematical cognition and is less for their development of physical cognition.

Third, these diverging patterns of co-occurrence *within* the cognitive ontogeny of descendant primate species are supplemented by diverging patterns of co-occurrence *between* their cognitive, physiological, and noncognitive behavioral ontogeny. The velocity for much of juvenile humans' physiological maturation, such as tooth eruption—with the exception of brain development—and for much of their noncognitive behavioral development, such as locomotion and dependency, is retarded in comparison with that of juvenile nonhuman primates. Comparatively, then, juvenile human cognition develops more rapidly than does physiological maturation (excepting the brain) and noncognitive behavior, whereas juvenile nonhuman primate physiology and noncognitive behavior develops more rapidly than cognition. Hence, evolution has provided humans more time and more speed for their cognition and brain to develop during their juvenile period.

Fourth, diverging patterns of co-occurrence within cognitive ontogeny and between cognitive, physiological (excepting the brain), and noncognitive behavioral ontogeny during the juvenile periods of descendant primate species are at the heart of the mosaic organizational heterochronies producing cascading terminal extension of human cognitive development.

Fifth, terminal extension of brain development may well have been produced by a localized temporal displacement during human embryogenesis (Deacon, this volume; Finlay and Darlington 1995). Its cognitive developmental consequences, however, do not await adulthood. Cascading cognitive constructions originate during childhood.

Displacement of the origins of cascading cognitive development to the juvenile period, I have hypothesized (Langer 1998), led to the unique human transformation of ahistorical preformal knowledge into formal knowledge that gave rise to the history of ideas.

The descent of cognitive development in primate phylogeny is heterochronic (Langer n.d.*b*). Heterochronic evolution extended humans' ontogenetic window of opportunity for progressive cognitive development. Its ontogenetic consequence par excellence is the development of formal thought. Humans' cognitive developmental specialization in formal reasoning to explain their Umwelt, or world as it appears to themselves, underpins the construction of the cultural history of ideas.

9

Cultural Apprenticeship and Cultural Change

Tool Learning and Imitation in Chimpanzees and Humans

Patricia M. Greenfield, Ashley E. Maynard,

Christopher Boehm, and Emily Yut Schmidtling

Often biological evolution is contrasted with cultural process, but the dichotomy is problematic because ultimately the capacity to learn culture is naturally selected. This capacity is critical for humans, and it is tied deeply to the child's potential for cultural learning, which for our species involves the re-creation of cultural traditions in each generation (Lock 1980). This natural ability to acquire (and transmit) culture is actualized through cultural apprenticeship, the subject of this chapter.

Apprenticeship processes are an important key to behavioral ontogeny in a species that must, in the course of a single lifetime, both acquire extensive cultural knowledge from the preceding generation and transmit it to the next. A few other animals participate similarly in socially learned group traditions (e.g., McGrew 1992), but there remains an important question (see Tomasello 1989, 1994): Do species such as chimpanzees learn their cultural behaviors in the same way that humans do, and in particular, are they able to imitate?

One way to explore the imitation controversy is to focus on behavioral ontogeny, which involves coordination between skill development and apprenticeship processes. We do so in this chapter by comparing apprenticeship behavior in humans (acquisition of traditional craft skills) with apprenticeship in chimpanzees (acquisition of tool-use skills in feeding). Our working hypothesis is that chimpanzees may well be capable of imitative learning, both in captivity and in the wild, and that the best way to explore this important question is to compare apprenticeship behaviors of the two species under natural conditions. After considering chimpanzee learning processes, we move on to those of Maya weavers and draw evolutionary conclusions from the comparative studies.

Cultural apprenticeship can be conceptually decomposed into processes of learning and teaching; these take place both between and within generations (Greenfield and Lave 1982; Lave and Wenger 1991; Parker 1996a; Rogoff 1990). Because an important aspect of human culture and its evolution lies in technology, the apprenticeship processes required to learn and transmit technological knowledge are the focus of this chapter. In what follows, we explore apprenticeship in a human tool system (Greenfield 1999; Greenfield, Maynard, and Childs 1997; Maynard, Greenfield, and Childs 1999) and in a chimpanzee tool system (Yut 1994). For this purpose, we utilize two unique data sources: (1) a cross-sectional and historical (long-term diachronic) video study of human tool apprenticeship and (2) a cross-sectional and longitudinal (short-term diachronic) video study of chimpanzee tool apprenticeship. The human tool system is weaving in Zinacantán, a Maya community in Chiapas, Mexico; this is the first microanalytical study of naturally occurring transmission of a human tool system (Childs and Greenfield 1980; Greenfield 1999). The chimpanzee tool system is termite fishing in Gombe National Park, Tanzania (video archives of the Jane Goodall Research Center, University of Southern California, used by Yut 1994), and this is the first microanalytical study of naturally occurring transmission of a chimpanzee tool system. This chapter constitutes the first time these two data sources have been put together for comparative purposes. We supplement our own analysis of the acquisition of chimpanzee tool use with nonhuman primate data from other sources.

The evolution of powerful means of learning and teaching creates culture by providing a way to transmit and transform knowledge from generation to generation. This is the hallmark of human culture (Bruner 1972; Parker and Russon 1996). The evolution of powerful means of learning and teaching opens the door to behavioral flexibility on the individual level and to cultural change on the social level. In other words, *the phylogenetic evolution of learning and teaching mechanisms creates a biologically based potential for cultural change.*

In recent years, much has been written about the existence of learning biases; these predispose human learning in particular cultural directions such as language and complex social relationships (Boyd and Richerson 1985; Fiske et al. 1998; Pinker 1994a; Trevarthen 1980). Tools and technology constitute another such direction. The focus of this chapter, however, is less on predispositions to learn particular content than on predispositions to utilize particular learning and teaching mechanisms. The role of teaching mechanisms has been relatively neglected in approaches to both the evolution of culture and cultural change; these mechanisms will be central to the discussion that follows.

Pérusse and colleagues (1994:328–29, their emphasis) have pointed out that

> teaching biases (evolved tendencies to convey adaptive infor-
> mation to offspring) are potentially more powerful co-evolu-
> tionary forces than learning biases (evolved tendencies to
> acquire adaptive information) because they capitalize on an
> *epigenetic* component in parents: the extensive knowledge that
> comes from experience gained during the long transition to
> adulthood. If they evolved, such biases could thus have given
> rise to an almost-endless production of adaptive *and* diverse
> cultural variants, as the biases, in addition to being naturally
> selected, would necessarily be context-dependent in their oper-
> ation. In effect, teaching biases might help reconcile the appar-
> ent contradiction between the adaptiveness and the diversity of
> human culture. A necessary condition for teaching biases to
> evolve, of course, is that parental rearing be under genetic
> influence.

The authors then verified this condition empirically (Pérusse et al. 1994). Using a large sample of adult twins, they did a behavior genetics study in which the target behavior was parental child-rearing behavior. Their results established that parenting behavior, in general, is heritable. This finding opens up the possibility that parental teaching style, in particular, is also heritable. Pérusse and colleagues (1994:328) argued for the natural selection of parental teaching techniques: "Genes promoting the transmission by parents of adaptive versus neutral or maladaptive stimuli would likely generate the acquisition of adaptive information by children and, hence, be naturally selected through the process of kin selection."

Knowledge transmission is a defining feature of culture. Indeed, Parker and Russon (1996:432) have defined cultures as "representations of knowledge socially transmitted within and between generations in groups and populations within a species which may aid them in adapting to local conditions (ecological, demographic, or social)." Within the framework of this definition, all that is necessary for culture is some form of social transmission of knowledge—that is, communication. We include in our definition of communication *unintentional* as well as *intentional* communication. For example, a model observed and imitated by another is often an unintentional form of communication. Individuals of the same species must communicate in order to transmit behavioral information—that is, information that affects the behavior of another animal.

Bonner (1980) traced the development of culture through extant animal species, stressing the need for some form of communication, not for a full-blown language capacity. In Bonner's phylogenetic analysis, learning in animals preceded simple forms of teaching, which preceded more complex forms of instruction. Bonner stressed that complex forms of teaching are more recent and are basic to cultural *change* in the human species. These more complex forms are not necessary, however, for cultural *transmission.*

Central to his ideas about cultural evolution is that certain kinds of information are best transmitted by means of social behavior, because genetic transmission would require a complex code and would be nearly impossible. If the transmission of these kinds of behavioral information is adaptive, then there is strong selection pressure for effective social transmission; genes that make this kind of transmission possible will be favorably selected.

In the first part of the chapter, we focus on and compare the tool use and teaching-learning techniques used by chimpanzees and humans. We will also look at within-species variability in cultural traditions and transmission across both space and time. By making a cross-species comparison of teaching-learning processes in humans and chimpanzees, we can learn two kinds of things relevant to the evolution of tool apprenticeship.

First, we can identify similarities in the teaching-learning processes of the two species. These similarities produce knowledge of a possible common evolutionary foundation for teaching-learning processes in the two species before their phylogenetic divergence five million years ago. We present data primarily from chimpanzees *(Pan troglodytes)*, but from time to time we bring in findings from bonobos *(Pan paniscus)*, a species that diverged from *Pan troglodytes* about two million years ago (Caccone and Powell 1989). *Pan troglodytes, Pan paniscus,* and *Homo sapiens* are considered to be sibling species. Our evolutionary case is strongest where we have data from all three sibling species.

Second, we can identify differences in the teaching-learning processes of *Pan* and *Homo.* These differences are evidence for possible evolutionary change in the ontogeny of cultural behavior in the last five million years, after the phylogenetic divergence of chimpanzees and humans.

In the second part of the chapter, we explore the utility of evolved processes of human apprenticeship for adapting to ecocultural change. We focus on weaving, a complex technology, and the means by which skilled use of this technology is transmitted and transformed from generation to generation. The particular ecocultural change on which we focus is the economic transition from agriculture to commerce. Insofar as this change entails changes in learning and teaching, we conclude that certain types of apprenticeship processes will be adaptive and may even be selected for under particular ecological conditions.

THE EVOLUTION OF APPRENTICESHIP AND TECHNOLOGY

Japanese primatologists, starting in the 1950s, began to find evidence in macaques for two key features of human culture, social transmission of information and tool use (Itani and Nishimura 1973; Kawai 1965; Kawamura 1959; Parker and Russon 1996). (For purposes of this chapter, we define a tool as an object that is purposely fashioned to accomplish a task involving at least one other object.) The Japanese research suggested that cross-specific studies could help in reconstructing the evolutionary foundations of human culture.

The notion of culture in nonhuman primates (Parker and Russon 1996) soon spread beyond Japan (Kummer 1971) and became current with works such as those by Wrangham, Goodall, and Uehara (1983), Goodall (1986), McGrew (1992), Wrangham and colleagues (1994), Boesch and colleagues (1994), and Boesch (1996). More specifically, this body of research (recently synthesized by Whiten et al. 1999) showed that chimpanzees have the rudiments of cultural knowledge, the rudiments of cultural variability, the rudiments of intergenerational learning processes for cultural transmission, and the rudiments of cultural change.

Apprenticeship in Tool Use: A Human–Chimpanzee Comparison

Scaffolding. Scaffolding is an interactional process by which an older, more skilled member of the species supplies help to a younger, less skilled one, thus enabling the younger to accomplish a task he or she could not complete independently (Wood, Bruner, and Ross 1976). At a later point in this developmental sequence, the role of the older guide is internalized (Vygotsky 1978) or appropriated (Saxe 1991) by the younger learner, who is now able to carry out the same task independently.

Although Wood, Bruner, and Ross (1976) conducted their research with human children, scaffolding is not unique to humans. Chimpanzee mothers in the Taï forest of Ivory Coast provide a scaffold for their young who are learning to use a hammer and anvil to crack nuts (Boesch 1991, 1993). For example, when a mother goes to gather more nuts, she might leave a nut positioned in the hollow of a tree-root anvil with a hammer stone on top of it. She has prepared and positioned all the necessary materials for nut cracking; all the infant has to do is pound the nut. Chimpanzees without infants have never been observed leaving intact nuts behind when they go to gather more. Mothers leave hammer and/or nut on or near their anvils while collecting additional nuts significantly more often for infants aged three and above than they do for younger infants. Interestingly, three corresponds to the age when chimpanzees first become interested in nuts and in using hammer tools (Boesch 1991). Thus, leaving nuts and hammers behind is not a generalized chimpanzee behavior but is (1) a specific maternal behavior and (2) concentrated in the chronological age window when chimpanzees are maturationally ready to acquire the skill of cracking nuts with hammer and anvil. Thus, this maternal behavior appears to be directed toward the goal of facilitating tool apprenticeship in infant chimpanzees.

Whereas this type of maternal scaffolding is common, there is another, extremely rare type of scaffolding that is much more dramatic because it involves collaborative learning (Tomasello, Kruger, and Ratner 1993). (Later we discuss the potentially important role of infrequent or rare behaviors in the evolutionary process.) In this example, Salomé is cracking nuts with her son Sartre. "After successfully opening a nut, Sartre replaced it haphazardly on the anvil in order to try to gain access to the second kernel. But before he could strike it, Salomé took the piece of nut in her hand, cleaned the anvil, and replaced the piece carefully in the correct position. Then, with Salomé observing him, he successfully opened it and ate the second kernel" (Boesch 1993:177).[1] Collaborative learning is Tomasello, Kruger, and Ratner's (1993) third and last level of cultural learning. Their view is that only humans have this ability (cf. Reynolds 1993). This example, however, albeit a rare one, suggests a preadaptation for collaborative learning that may stretch back five million years.

Analogs to these examples of scaffolding and collaborative learning in chimpanzee tool apprenticeship are found among humans. For example, Zinacantec Maya mothers can be observed preparing the materials that their daughters will need in weaving; an example of this phenomenon is shown in the sequence of actions and interactions in figure 9.1. The woman on the right in the first video frame (fig. 9.1A) is using a knife to prepare a stick to be used as a bobbin in the weaving. She hands this stick to one of her daughters (fig. 9.1B), who then uses it as she helps her younger sister learn to weave (fig. 9.1C). The human example is more socially complex than any example of scaffolding observed in chimpanzees: the human mother is dealing simultaneously

Figure 9.1.

Video frames showing the coordinated use of a tool in weaving apprenticeship. A: Mother (hands at right of frame) makes a stick used in the weaving process. B: She hands the stick to her older daughter. C (opposite): The older daughter uses the stick as she helps her younger sister (Katal 1) learn to weave. (First name plus number identifies individual subjects in our complete database of video and other records.) Nabenchauk, Zinacantán, Chiapas, Mexico, 1970. Video by Patricia Greenfield.

C

with two other people, whereas the chimp mothers are dealing with only one other animal. Nonetheless, in these examples both human and chimpanzee mothers are clearly anticipating the needs of their children for aid in successfully completing their respective tool-based tasks. In addition, Salomé, the chimpanzee, appears to take action to prevent an error on her child's part.

Another aspect of scaffolding has also been observed among Taï chimpanzees. They go through the nut-cracking motions more slowly when one of their offspring is present and watching than when a child is not present to watch (Boesch 1991). In Parker's (1996a) typology of teaching techniques, this aspect of scaffolding is called demonstration teaching.

Demonstration teaching is an important part of weaving apprenticeship for the Zinacantec Maya. In the sequence depicted in figure 9.2, demonstration teaching is combined with its complement, observational learning. In the first video frame (fig. 9.2A), a young girl is weaving at a backstrap loom. As she comes to a difficult part of the process, her mother takes over. In the second frame (fig. 9.2B), the mother has begun to weave while her daughter observes. This sequence shows how demonstration can be a part of a scaffolding process: In taking over the weaving, the teacher not only serves as a

Figure 9.2.

Video frames showing demonstration teaching. A: A young girl (Katal 1) weaves on her own as her mother stands nearby. B: The mother then steps in and takes over the weaving while the girl observes. Nabenchauk, Zinacantán, Chiapas, Mexico, 1970. Video by Patricia Greenfield.

model for the learner but also helps the learner get through a difficult part of the weaving process. The learner is also given an opportunity for observational learning.

Keep in mind, as we make our cross-species comparison of tool apprenticeship, that weaving technology is much more complex than termiting technology. Using Piaget's scheme of cognitive development, we could place the former at the level of concrete operations (Greenfield 2000) (completed by many children around age 10), and the latter at the level of sensorimotor intelligence (completed by children around age 2).

Observation and Imitation. In contrast to the Taï chimpanzees (and humans), chimpanzees from Gombe National Park in Tanzania do not use scaffolding to transmit the use of sticks to fish for termites, nor do they scaffold dipping for driver ants, another tool-based activity at Gombe. Gombe chimpanzees use no special techniques whose goal is to help learners achieve mastery of tool use. Instead, learners have opportunities to observe older chimpanzees as the older animals use stems or vines to fish for termites or use sticks to dip for ants in everyday practice. The older animals at Gombe, unlike those at Taï, have never been observed to carry out a deliberate demonstration for the purpose of helping a young chimp to learn. Correlatively, unlike nut cracking at Taï, nut cracking is not part of Gombe culture (Whiten et al. 1999). We conclude that not only is there cross-cultural variability in chimpanzee cultural behavior, but there may also be cross-cultural variability in chimpanzee mechanisms of cultural transmission.

Chimpanzees are exposed to models of tool use, but do they watch and imitate these models? Our observations (as well as those of Custance, Whiten, and Bard [1995] and Boesch [1996]) run contrary to what one would predict from Tomasello's position that chimpanzees can *emulate goals* but cannot *imitate means* of tool use (Tomasello 1989; Tomasello et al. 1987) unless they have received human enculturation (Tomasello, Savage-Rumbaugh, and Kruger 1993). In contrast to Tomasello and colleagues, we studied what chimpanzees *observe*, rather than what they *do*, as a window into the question of observational learning and imitation. Using the Jane Goodall Research Center's video archives at the University of Southern California (USC), we

Figure 9.3.

Frames from a video of a two-year-old juvenile male chimpanzee observing the means by which an adult reaches the goal of feeding on termites. The numerals indicate that the sequence takes 3.85 seconds. A: In the first frame, the infant's head is oriented toward watching the mother's hands as she engages in the fishing process. B: As the mother's hand brings the fishing tool to her mouth, the infant's head adjusts upward as its gaze is directed at the mother's hands and mouth. C (opposite): The infant follows through on these adjustments and raises its head further as its eyes continue to track in the direction of the mother's hands and mouth. Video frames courtesy of the Jane Goodall Research Center, University of Southern California.

248

C

found that young chimpanzees pay close attention to the means as well as the end when they observe older chimpanzees engaged in fishing for termites. In the sequence shown in figure 9.3, for example, the juvenile's gaze follows the upward trajectory of the termiting tool (fig. 9.3A–C).

We identified 32 instances in which we were able to code the specific focus of attention of the subjects observing an experienced adult's termite fishing. In half of the observations, subjects paid attention to the entire trajectory—that is, to the adult's getting the termites with the stem or vine and transferring them to the mouth (Yut 1994). Indeed, it was relatively infrequent for subjects to focus on the adult's eating the termites (goal) without first observing the adult get the termites on the tool and then visually following the trajectory to the mouth—that is, observing the means (Yut, Greenfield, and Boehm 1995). This close *observation* of means is a prerequisite for the *imitation* of means. In a natural task context, this finding casts doubt on the idea that chimps are generally limited to emulation (replication of ends only) without being able to imitate the means that lead to the end.

These natural behaviors contradict the assertion of Nagell, Olguin, and Tomasello (1993) that captive chimpanzees paid attention to general functional relations in a laboratory task but not to the actual, demonstrated methods of tool use. This is not entirely surprising, because we are analyzing a natural chimpanzee task rather than an artificial human one. In accord with Piaget's (1962) perspective emphasizing the importance of cognitive understanding of the observed model that is to be imitated, we would expect accurate imitation of means to take place in an activity that is part of the lifeway (or culture) of a species or group (and therefore is well understood), as termite fishing is for the Gombe chimpanzees (cf. Boesch 1993). Boesch (1996) provides parallel examples of imitation from chimpanzees in the Taï forest of the Ivory Coast.

Tomasello's first criterion for imitation—rote copying of means—requires some discussion. This is a rather behavioristic approach to the topic of imitation. It stands in stark contrast to Piaget's (1962) cognitive approach to imitation, in which he emphasized not replication but the cognitive transformation of the model in accord with the developmental stage and schema of the imitator. In other words, in an act of imitation, a child (or an adult) will replicate the features of the model as he or she understands them.

For Piaget, therefore, unlike Tomasello, the production of an exact replica could never be a criterion for an imitation. Cognitive transformation of models was also part of Russon's (1996) definition of true imitation and was frequently observed in rehabilitant orangutans. Miles, Mitchell, and Harper (1996) also observed transformation of models in the imitative behaviors of a captive orangutan. Whiten (n.d.) made a similar point in noting that emulation of a model's ends, in combination with creative transformation of the model's means, may actually constitute more intelligent behavior than the (rote) imitation idealized by Tomasello.

Chimpanzees, however, are indeed capable of replicating a model's means by imitating them. Even accepting Tomasello's criteria for true imitation, we have observed that juvenile chimpanzees translate their close observation of adult tool use into the replication of specific features of the model's activity—for example, a juvenile moves to an adult's termite mound after watching successful termiting there. There is such an instance in the Jane Goodall Research Center video archives at USC used by Yut (1994) in her quantitative study of developmental changes in visual attention to tool activity. In one extended action sequence (fig. 9.4), the young chimpanzee's intentional activity clearly involves taking the same perspective on action as the original model; this is Tomasello's second criterion for true imitation. In this sequence, a juvenile observes an older chimpanzee fishing for termites at a particular location (fig. 9.4A). After the older chimpanzee has left the location, carrying his termiting tool with him, the younger one, swinging from a tree (fig. 9.4B), takes the older animal's tool from his mouth. The younger chimp then returns to the original location (fig. 9.4C) and begins to use the older chimp's tool to fish for termites (fig. 9.4D). The use of the older chimpanzee's tool, location, and activity constitutes the replication of specific features of the elder chimpanzee's methodology. From a learning perspective, this imitation functions, at the very least, to provide practice in fishing for termites. Because goal-directed activity is involved, this imitative sequence also involves some perspective taking: the young chimpanzee assumes the same perspective on the activity that the older chimpanzee had taken a few minutes earlier. As Russon (1996) has found, the imitation lies in the organization of individual elements, not merely in isolated behaviors.

We have observed similar imitative sequences in Gombe with infant and juvenile chimpanzees and their mothers. An infant or juvenile chimpanzee will often grab the mother's abandoned fishing tool when she gets up to leave; the young chimpanzee will then use the tool to fish for termites, often with no success. In one such example, the mother's termite hole and tool were far more productive than the hole and tool the infant had been using earlier. After taking over the mother's hole, the infant began to get a few insects, having assumed a position and posture similar to the mother's. The infant was torn, however, between staying close to its mother, who was moving away, and continuing to fish. The infant therefore whimpered whenever the mother began to move; in this fashion, the infant kept the mother near and continued to fish, using the mother's stick and termite hole with a moderate degree of productivity for about 45 minutes.

In captivity, we have a striking case of adult chimpanzees learning a complicated behavior not found in nature, by immediate imitation. An observation by Desmond Morris (1962) in a captive colony of chimpanzees makes a direct connection between observing and imitating. Morris gave six chimpanzees in the "Chimpanzee Den" of the London

FIGURE 9.4.

Video frames from a sequence of imitative activity. A: A two-year-old male juvenile chimpanzee observes an adult male using a stem or vine to fish for termites. About seven minutes later, the adult moves a few steps away from the fishing site, holding the tool in his mouth. B: The juvenile, now playing in a tree, takes the tool from the adult's mouth.
C: Less than a minute later, the juvenile descends with the stolen tool in his mouth and heads toward the location where he had observed the adult fishing for termites earlier.
D: The juvenile then uses the tool to fish for termites there. The whole sequence takes a little less than eight minutes. Video frames courtesy of the Jane Goodall Research Center, University of Southern California.

Zoo their first drawing experience. With the first three chimps, each tested alone, Morris provided the chimp with needed instructions in how to hold the pencil. He described what subsequently happened with the last three chimpanzees in this situation:

> But then, when the fourth one, Fifi, came out, to my astonishment she grabbed the pencil from me and started to work without any hesitation. The thought struck me that this must be the

result of imitation and I turned quickly round to see a dense cluster of young chimps all hanging from the wire of their rest-room at the spot which gave them the clearest view of the draw-ing Fifi was making. I had been so absorbed in watching the drawings emerge on to the paper, that I had not realized that the Chimpanzee Den was much quieter than usual. I was told afterwards that, behind my back, the silent huddle had hung throughout the tests, intently watching every move, as if their very lives depended on it.

Despite this, the fact that Fifi was the leader of the group made me suspicious. It could just have been that, being the leader, she always took action in a new situation, without waiting for directions. But the fifth chimp, Jubi, answered this doubt. She was small and the least assertive of the group. But, never-theless, she had seen enough. She did not actually take the pen-cil from me, but when I handed it to her she straightaway started to draw. (Morris 1962:39)

Chimpanzees are also capable of replicating a model's sequence of func-tional acts. Whiten (1998) created an experimental test of chimpanzee skill in replicating behavioral sequences. A human model showed humanly enculturated chimpanzees how to open artificial fruit. In line with the greater intelligence of emulating (i.e., copying the goal) rather than imitating (i.e., copying the means) under certain circum-stances, the chimpanzees began by imitating model behaviors that were unnecessary to the task; these dropped out over a number of trials (Whiten n.d.). Simultaneously, as subjects experienced repeated cycles of demonstrations and opportunities to perform the task, their behav-ior gradually converged on the *sequential pattern of the model,* including only those steps that were necessary to accomplish the task (Whiten 1998). Learning over time to replicate a task-relevant sequence of acts demonstrated by a model could be a powerful mechanism in the evolu-tion of cultural transmission. In sum, this array of findings indicates that observational learning (Bandura 1977; Goodall 1986) and imita-tion are important for chimpanzees, as they are for humans. In both species, they play an important role in cultural transmission.

The Role of Observation and Imitation in Development and Cultural Learning

The research of Bloom, Lightbown, and Hood (1974) on the role of imitation in language development suggests that imitation of models is most frequent with respect to a skill that is already in the process of being acquired but has not yet been mastered. The example of imitation in termiting reported earlier conforms to this principle: it occurred at the age of two years, when young chimpanzees first start attempting to use objects to fish for termites, a skill that is not completely mastered until the age of five or six years (Yut 1994). Apparently, this principle is general in higher primates; Russon and Galdikas (1995) found that orangutans concentrate their imitation on skills that are on the leading edge of their capacities (Parker and Russon 1996).

As in the human skill of weaving (Childs and Greenfield 1980; Greenfield 1984), close observation of tool use among chimpanzees is a group phenomenon and continues beyond childhood, as is evidenced in the Jane Goodall Research Center archives at USC (Yut, Greenfield, and Boehm 1995). Figure 9.5 depicts close observation of tool use among chimpanzees and humans, respectively. In figure 9.5A, a six-year-old juvenile female chimpanzee is using a tool to fish for termites as two other chimpanzees look on. In figure 9.5B, a young girl is learning to weave as her mother and younger brother stand near to watch.

Other Mechanisms in Chimpanzee Learning and Intergenerational Transmission

Chimpanzees use still other learning and transmission techniques. For example, the Gombe young learn termiting by playful experimentation with objects—specifically, by playing around with the kinds of objects that will later be used as termiting tools (Goodall 1995). Playing with tool-like objects (branches, vines, twigs, stems, and sticks) begins before one year of age (Yut 1994). In figure 9.6, a very young chimpanzee is using a twig, apparently to "termite" in her mother's fur. This playful activity models the real activity of fishing for termites.

A second example of playful experimentation at tool use is particularly creative. On videotape at the Jane Goodall Research Center at

Figure 9.5.

A: *Two chimpanzees observe another chimpanzee, at far left of frame, probe to fish for termites. Video frame courtesy of the Jane Goodall Research Center, University of Southern California.* B: *Mother and brother observe Rosy 206 learning to weave. Nabenchauk, Zinacantán, Chiapas, Mexico, 1991. Photo courtesy of Lauren Greenfield.*

FIGURE 9.6.

A two-year-old female juvenile plays with a twig. She has been putting the twig in her mouth, attempting to fish with it, and poking her mother's back with it. In this frame, she is dragging the twig across her mother's back. Jane Goodall remarked that "it looks as if she were fishing [for termites] in her mother's fur" (personal communication, 1995). Video frame courtesy of the Jane Goodall Research Center, University of Southern California.

USC, there is footage of Fifi's infant son as he plays with a small branch with leaves attached, next to a shallow stream. His juvenile sister, Flossi, takes the branch from him and strips a few leaves off, as adults do in making termite probes. She then probes the water for a considerable time, peering into the water as she does so. After repeated reviewing of this footage, it was noted that there were water bugs moving around in the bottom of the stream, which was perhaps six inches deep. That this is indeed playful experimentation is emphasized by the fact that "fishing" for water bugs is never done by adults and has never been otherwise observed in infants or juveniles.

Playful experimentation is an important aspect of the development of technological knowledge by young humans (Piaget 1952). We have observed it in the acquisition of weaving skill when very young girls engage in play weaving on a toy loom (Greenfield 1999). Like play termiting, play weaving models the activity of real weaving. Of course, the play loom of humans is a much more complex tool than a termiting tool used by chimpanzees. But this significant fact does not diminish the important parallelism between the use of playful experimentation in each case.

There is a fourth mechanism for acquiring tool knowledge that is seen at Gombe: independent practice, which is related to opportunity teaching (Caro and Hauser 1992). Independent practice may be stimulated by earlier observation of models, as in Russon's (1996) observations of recurrent rehearsal in rehabilitant orangutans. Our analysis of the Gombe tapes indicates that independent practice occurs for termite fishing between the ages of two and six years, by which time an approximation of adult competence is achieved (Yut 1994).

Similarly, we saw independent practice in videotaped data of young Zinacantec Maya girls learning to weave (Greenfield 1984, 1999). As among the chimpanzees, it was concentrated at the more advanced levels of learning. Again, the technology is much more complex in the human case, but the apprenticeship process and its developmental function are similar.

Cross-species comparison indicates that the human evolutionary legacy includes another important feature of culture: the use of both external and internal representational tools to guide action and to teach. Recently, Savage-Rumbaugh and colleagues (1996) made a startling discovery: in the forests of the Congo Republic (formerly Zaire), bonobo chimpanzees *(Pan paniscus)* use specially prepared branches as pointers to indicate which of two possible paths have been taken. More research on bonobos in the wild is needed to confirm these first observations and to indicate how widespread this practice is. But if these early observations are confirmed, we can note parallels between the bonobo use of branches as pointers and the printed pattern books used by modern-day Zinacantec Maya teenagers and women for embroidery and weaving. In both cases, an external representation indicates a motor path, though in the human case the path is much more complex and is traveled with the hands rather than the feet (Greenfield 1999).

Note that these modes of representation are all indexical or iconic; they are not arbitrary symbols. An index is a representational sign whose interpretation depends on the surrounding context. For example, the act of pointing creates an indexical sign whose meaning depends on what is being pointed to—that is, it depends on context. An icon is a representational sign that resembles its referent; an example would be a photographic image of something, which of course resembles that thing. A symbol is a sign that is arbitrary in the sense that it does not physically resemble its referent. The words of human language generally have this characteristic. Piaget (1962) found that in the first two years of life, these three forms of representation developed in the order index, symbol, and sign.

In the case of bonobo trail marking (Savage-Rumbaugh et al. 1996), the bonobo branches are in context; they are therefore indices of a path. The branches can also be thought of as an image of a path (iconic). The patterns used by the Zinacantec Maya for weaving and embroidery also utilize the iconic mode of representation.

In his work at Gombe, Plooij (1978) provided important insights into the developmental process by which communicative gestures, an internal form of representation, are innovated and developed in chimpanzees in the wild. He documented a developmental process of conventionalization in which mother-child pairs turned interaction into communicative gesture. In figure 9.7 we see how the act of a mother in raising her baby chimpanzee's arm to groom him is subsequently transformed into the baby's raising his arms as an indexical signal to request mother to groom him. Each act of conventionalization creates an innovation in communication for that particular pair of animals.

Boesch (1996) has provided additional evidence for the creation of social conventions in chimpanzee communication in the wild. For example, he noted instances in which the same action had a different communicative meaning in different groups of chimpanzees. And conventionalization has been a long-standing theme in Tomasello's (1989, 1994) longitudinal study of communication in a captive colony.

The genus *Pan* can also use conventionalized symbols to teach. In a videotaped example, Kanzi, a bonobo who uses a lexigram board to communicate with people, attempts to show a younger bonobo, Tamuli, how to slap, hug, and groom him (Ikeo, Jones, and Niio 1993; Tomlinson and Jones 1993). The researcher, Sue Savage-Rumbaugh, asks Tamuli, who has been raised only by his mother and does not understand English, to slap Kanzi. Kanzi then slaps Tamuli. When

Figure 9.7.

The conventionalization of a gesture, a request for grooming. Left: A mother chimp raises her baby's arm in order to groom it. Right: Later in development, the young chimpanzee raises its arm to request grooming. Reproduced with permission from Plooij 1978:118–19, courtesy of Academic Press.

Tamuli still does not slap Kanzi, Kanzi takes Tamuli's arm and attempts to get Tamuli to slap him. When Savage-Rumbaugh asks Tamuli to groom Kanzi, Kanzi takes Tamuli's arm and hand and shows Tamuli how to groom him. This is a robust example of one chimpanzee's using a communicative gesture to teach another chimpanzee how to do something. Visalberghi and Fragaszy (1996) independently analyzed this scene in the same way.

Not only a bonobo but also a language-trained chimpanzee has been observed to engage in deliberate teaching of symbolic communication (American Sign Language). For a number of years, Washoe and the other signing chimps from the original Gardner study have been collected in a group with no experimental intervention whatsoever, the purpose being to see whether they would use sign language spontaneously in their own interactions. Not only did they use signs frequently in communicating, but also what amounted to an imposed cultural tradition was transmitted to an infant introduced to the group. Some of this transmission was based on unassisted observational learning, but at times deliberate teaching was involved. Fouts, Hirsch, and Fouts (1982) reported a number of occasions in which the chimpanzee Washoe deliberately tried to teach new signs to her adopted infant, Loulis, through a combination of molding—placing the subject's hands in the correct position—and demonstration (Goodall 1986). In the following example, note how Loulis's observational behavior complements Washoe's deliberate teaching: "Washoe was observed to sign *food* repeatedly in an excited fashion when a human was bringing her some food. Loulis was sitting next to her watching. Washoe stopped signing and took Loulis's hand in hers and molded it into the food sign configuration and touched it to his mouth several times" (Fouts, Hirsch, and Fouts 1982:183).

Another example features scaffolded demonstration in which Washoe gradually withdraws the supportive scaffold, just as human mothers do (e.g., Childs and Greenfield 1980): "When Loulis was first introduced to Washoe, Washoe would sign *come* to Loulis and then physically retrieve him. Three days later, she would sign *come* and approach him but not retrieve him, and finally, 5 days later she would sign *come* while looking and orienting towards him without approaching him" (Fouts, Hirsch, and Fouts 1982:183–84). "It is as though because Washoe herself was taught, so she is able to teach" (Goodall 1986: 25).

These examples from *Pan troglodytes* and *Pan paniscus*, because they occurred spontaneously and without coaching, show a preadaptive potential for active teaching. Goodall makes the important point that what may have actualized this preadaptive potential was the animals' experience of having been themselves taught. In other words, environmental stimulation (and functionality) is crucial for transforming a preadaptation into an adaptation. Although a humanly devised language environment is not the "natural" one for chimpanzees, their reaction to it, including the spontaneous teaching of signs, may constitute a model of the way in which the preadaptation for language was transformed into language itself under particular environmental conditions in the course of human evolution.

Note that one aspect of this behavioral adaptation is the possibility of a delayed effect: Washoe was taught signs many years before she taught them to Loulis. Because Washoe's human caregivers had molded her hands to teach her signs many years earlier, her molding of Loulis's hand might even be considered delayed imitation. Indeed, many years had gone by without Washoe's being taught any new signs. This kind of delay might characterize the human mechanism for socialization and culture transmission: one teaches the next generation many years after being taught a skill or piece of knowledge oneself. Because being taught sign language is not normally a part of the chimpanzee environment, the effects of this experience on later teaching suggest a developmental relationship between how one is taught as a child and how one teaches one's children. This may be a key to cultural transmission among humans.

We have seen that chimpanzees in the wild manifest seven of the most basic mechanisms used by humans for the intergenerational transmission and innovation of cultural tools: (1) scaffolding, (2) playful experimentation with objects, (3) observational learning and imitation, (4) independent practice, (5) conventionalization, (6) the use of internal (gestural) and external representations to teach and to guide action, and (7) collaborative learning. Some may become more developed under the influence of human enculturation. Indeed, use of gestures to teach symbol meanings may be an example of such development. We must not, however, underestimate the mechanisms of naturally occurring chimpanzee enculturation. The presence of these mechanisms in both chimpanzees and humans opens up the possibility that these acculturative processes began their evolution in our common ancestor at least five million years ago.

Implications of Cross-Species Contrasts for the Evolution of Human Pedagogy in the Last Five Million Years.

There is one pedagogical complex used by humans but never observed in chimpanzees; it is the use of arbitrary symbolic means to teach a technological skill. This is a strong candidate for the evolution of learning and teaching modalities after humans split from chimpanzees five million years ago. As an example, the arbitrary symbolic means of language is very important in Zinacantec weaving apprenticeship (Childs and Greenfield 1980). Teachers (usually a learner's mother) use language to tell the learner what to do, but not to explain the process. They scaffold their language according to the learner's level of experience. For example, directives are most frequent at the earliest stages of learning, whereas descriptive statements become more frequent with more experienced learners. Language is combined with other modes of teaching, such as demonstrations. The sophisticated use of language, especially commands combined with gestures, in weaving apprenticeship is a skill apes have not evolved.

It may be that linguistic teaching is required by more complex human tool systems such as weaving. Linguistic teaching is clearly unnecessary for simpler tool systems, such as those used to fish for termites or crack nuts. An important hypothesis arises: that technology and teaching (apprenticeship) have co-evolved.

Cross-Cultural Variability and Change in Communicating, Teaching, and Learning among Chimpanzees

There is cross-cultural variability in the symbolic communication of wild chimpanzees. The chimpanzees in Mahale have a different set of communicative gestures from those of the chimpanzees in neighboring Gombe (Wrangham 1995b)—two distinctive nonverbal "dialects" (Goodall 1986:143–45). McGrew (1992) also found cultural variation in the extended arm grasp of chimpanzees.

As we have seen, the different chimpanzee groups do not use methods of intergenerational socialization and learning equally. Just as there is variability in the tools that are used by geographically separated groups in a single species *(Pan troglodytes)*, there is also intergroup variability in the learning-teaching mechanisms by which tool knowledge is transmitted and re-created in each new generation.

Cultural variability carries within it the seeds of cultural change. Clearly, as members of *Pan troglodytes* spread to different ecological niches in historical and evolutionary time, different cultural traditions of tools and communication developed (Whiten et al. 1999). The emigrant chimps must have undergone historical change in cultural traditions (Matsuzawa and Yamakoshi 1996).

Conventionalization, too, carries within it the seeds of cultural change. When a mother-child pair conventionalizes a new signal to use between them (Plooij 1978), cultural innovation on the personal level has taken place. The signal is ripe to be shared with and transmitted to others. And the learning techniques of playful experimentation and independent practice contain within them the potential that individual or dyadic learning can create novelty, thus creating the possibility of cultural change on a broader group level.

Equally important are observational learning from models and symbolic communication. These are mechanisms for the conservative transmission of cultural knowledge from generation to generation or for the spread of tradition from one group to another (McGrew 1992). In the Bassa Islands, Hannah and McGrew (1987) noted the spread of tool use upon the arrival of an adult female chimpanzee who used stones to crack nuts. Where there had previously been no tool use in the host community of chimpanzees, 9 of the 13 chimpanzees began to use stones to crack nuts after observing the behavior of the introduced chimpanzee.

Matsuzawa and Yamakoshi (1996; Yamakoshi and Matsuzawa 1993) have modeled this type of cultural change experimentally by introducing coula nuts into the chimpanzee colony at Bossou, Guinea, in West Africa. An adult female, Yo, most likely a migrant from a colony in which stone tools were used to crack coula nuts, immediately began to use tools to crack and eat the nuts. "When Yo cracked the coula nuts, a group of juveniles gathered around and peered at her while she was cracking and eating the strange nuts. The next day, an unrelated 6 $\frac{1}{2}$-year-old male named Vui cracked open a coula nut without any practice. Four days later, a six-year-old female named Pili did the same. The two juvenile chimpanzees cracked the nuts and sniffed the kernel and chewed and spat it out" (Matsuzawa 1994:364). Although these were the only two juveniles to crack nuts, the utilization of observation and imitation to spread novel behavior is clear. An important point is that the innovative tool use model was copied only by juveniles. Among humans, learning from innovative models is also concentrated in the "juvenile" period (Greenfield 1999).

Tomasello (1989) observed an example of cultural innovation in communication in a captive colony. This occurred when newly intro-duced wood chips became the basis for a communicative gesture; all of the chimpanzees in the study group began to throw the wood chips at each other in order to initiate play.

Because of the variety of learning mechanisms present in at least a crude form in modern chimpanzee life, the evolutionary foundation for both cultural continuity and cultural change in human groups could have been present in our common ancestor five million years ago, ready to be elaborated in the ensuing millions of years.

One striking change that probably took place in the last five million years is in the cumulative quality of human cultures. Unlike chimpanzee cultures, they cannot be reinvented in the space of a single generation by inter- and intragenerational interaction. Boyd and Richerson (1996) point out that whereas the basic mechanisms of cultural learning are present in nonhuman primates, cultural accumulation is hard to get started because it does not have adaptive advantages until it is quite widespread; at that point, it is easy to keep going. In essence, according to Boyd and Richerson (1996), cumulation is a quantitative, not a qualitative, difference between humans and great apes. The implication is that the common ancestor of *Homo* and *Pan* had the basic psychological mechanisms for cultural transmission. We hypothesize that these mechanisms were then strongly selected for in the hominid line after its divergence from the great apes. This selective process would then have resulted in cultural accumulation.

Contrasting Implications of Behavioral Frequency in Psychology and Evolution

Sometimes the argument is made that chimpanzee behaviors for transmitting cultural knowledge are irrelevant to human culture because they are relatively infrequent in comparison with such behaviors in humans (Tomasello 1994). Frequency is very important in the discipline of psychology, which puts great weight on mean and modal frequencies of behaviors; correlatively, psychology minimizes the significance of infrequent behaviors.

The reverse, however, is true for evolutionary research and evolutionary theory. In evolution, infrequent phenomena often hold the keys to an evolutionary process. This is so because natural selection operates to make adaptive but infrequent characters (including behavioral traits) more frequent in the course of phylogenetic development. Consequently, when we see behavioral traits in common between sibling species, but frequent in one species and infrequent in the other, the most parsimonious explanation is that the behavior was present but infrequent, fragile, and primitive in the common ancestor of the two species; it proved adaptive in one phylogenetic line when it appeared; and it was therefore selected for and became more common, more elaborated, and more robust in the course of that line's evolution. In sum, even rare data are useful in identifying preadaptive potential.

In comparative research, therefore, it is important to acknowledge infrequent phenomena for their potential evolutionary significance. For example, there is potential evolutionary significance in the minority of conventionalized gestures that spread to the chimpanzee group as a whole (Tomasello 1989).

If modern-day chimpanzees share with humans the basics of tools and the basics of intergenerational transmission of tool knowledge, then it is more than likely that the ancestor that preceded the phylogenetic split between hominids and chimpanzees also knew how to use tools and how to induct the young into this knowledge. In this pedagogical knowledge lies the evolutionary foundations not only of human culture and human cultural transmission but also of human cultural change.

CULTURAL CHANGE AND TOOL APPRENTICESHIP IN HUMANS

Given the evolution of this wide range of apprenticeship mechanisms, the possibility arises that different mechanisms would be useful under different ecocultural conditions. A change in the importance of various apprenticeship processes could therefore be a key to successful adaptation to ecocultural change. This hypothesis was tested through a long-term study of weaving apprenticeship conducted by Greenfield and Childs in the Maya community of Zinacantán. We explored the relations between sociohistorical transformations and apprenticeship processes in a direct way: by following a group of families over two generations, studying their processes of weaving apprenticeship before and after processes of important ecological change.

The study site of Zinacantán is a community in which the basis of the economy is in transition, from agricultural subsistence to commercial entrepreneurship and cash. In 1969 and 1970, Greenfield and Childs (1977; Childs and Greenfield 1980) conducted studies of culture, learning, and cognitive development in Nabenchauk, a

Zinacantec hamlet. At that time, a subsistence lifestyle based on corn and beans was almost universal in the community, although corn was also sold as a cash crop. One focus of our research was on the apprenticeship processes involved in the important cultural technology of weaving, the most complex skill in the culture and one that is acquired by virtually all Zinacantec girls. In 1991 we returned to a community that had expanded its economic basis to commerce and in which cash had taken on a much greater importance. We repeated our study of weaving apprenticeship with the next generation of girls (Greenfield 1999; Greenfield, Maynard, and Childs 1997).

Our major theoretical proposition was that not only do cultures change over historical time, but the importance of particular processes of cultural learning and cultural transmission also changes. More specifically, a somewhat different set of teaching and learning processes may be emphasized when cultures are in a more stable state compared with when they are in a more dynamic state. Correlatively, particular processes of apprenticeship should be highlighted in connection with the dominant ecocultural system of a particular time and place.

Insofar as the process of socialization prepares the next generation to participate in society, it should change when the conditions faced by that next generation differ from the environment in which their parents grew up. Socialization is intrinsically future oriented—it prepares children for an adulthood that still lies in the future. A key question, however, is this: Under conditions of change, do parents merely re-create the apprenticeship process they underwent as children (as Washoe did), or is there a capacity to develop new methods and processes as societal conditions—in this case, economic conditions—change? Pérusse and coworkers (1994:334) hypothesized that evolution has selected for the (unconscious) rule of parental teaching (which has its roots in childhood teaching [Maynard 1999]): "Teach what is most adaptive for your children." Is this the case in situations of sociocultural change?

Weaving was the focus for studying processes of informal education, teaching, and learning in a society in which education does not traditionally take place in school. Weaving, considered to be the essence of Zinacantec womanhood, is the means by which most clothing is made. In 1969 and 1970, woven artifacts, like other parts of the culture, were stable and unchanging, limited by tradition. Woven patterns were limited to two red-and-white striped configurations, one multicolor stripe, and one gray-and-white basket-weave pattern.

On the basis of our research in 1969 and 1970 (Childs and Greenfield 1980; Greenfield 1984; Greenfield, Brazelton, and Childs 1989; Greenfield and Childs 1977, 1991), we concluded that the implicit goal of Zinacantec education and socialization was the intergenerational replication of tradition: learning to weave meant learning to weave about four specific patterns. According to our findings, the particular way in which weaving was taught fostered this goal. The learning process was a relatively error-free one in which the teacher, usually the mother, sensitively provided help, a model for observation, and verbal direction in accord with the developmental level of the learner. The mother as a teacher provided a scaffold of help that allowed the learner to complete a weaving she could not have done by herself. There were no failures; every girl learned to weave. Because the process was highly structured by the older generation and allowed no room for learner experimentation and discovery, the method of informal education (or apprenticeship) was well adapted for the continuation of tradition, the status quo. Such respect for tradition, as embodied in the older generation, is an effective adaptation to agricultural techniques and ecology, a system in which land is controlled by the older generation (Collier 1990).

Our follow-up study two decades later (Greenfield 1999; Greenfield, Maynard, and Childs 1997) was based on the fact that by the 1990s, the ecocultural environment had changed. Men who formerly farmed were now in the transport business. They had become commercial entrepreneurs, running a van service back and forth to the former colonial city of San Cristóbal de las Casas. Other men were involved in trucking, and many either drove or owned their own trucks for transport businesses and commerce, the buying and selling of agricultural products.

In accord with our theory, a coordinated change had taken place in the apprenticeship process. The emphasis in weaving apprenticeship had shifted in many families (specifically, those families most involved in commerce) from scaffolding toward independent practice. A second shift was from teachers in the maternal generation to teenage teachers.

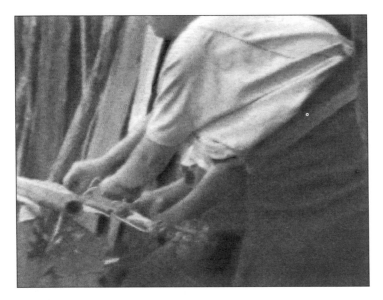

FIGURE 9.8.

Mother and daughter (Katal 1) work at the loom together. Four hands on one loom are often a part of Zinacantec weaving apprenticeship. Nabenchauk, Zinacantán, Chiapas, Mexico, 1970. Video by Patricia Greenfield.

As with chimpanzees (Matsuzawa and Yamakoshi 1996), innovation was concentrated in the younger generation. Compare a video frame of a girl learning to weave in 1970 (fig. 9.8) with a video frame of her daughter learning to weave at the same age (nine years old) in 1991 (fig. 9.9). In 1970, weaving instructors stayed close to their pupils, often resulting in four hands on a loom, with the teacher and the pupil working to keep the weaving going together (fig. 9.8). In the 1990s, the learner was often weaving more independently; weaving teachers were often away from their pupils and had to be called over for help (fig. 9.9) (Greenfield, Maynard, and Childs 1997). Collaborative activity still took place in the apprenticeship process, but, in comparison with a generation earlier, it was initiated more by the learner and less by the teacher. In figure 9.8, the learner is being taught by her mother. In figure 9.9, the mother is not teaching her own daughter but has assigned a teenage sibling to serve as teacher. Both independent practice and teenage teachers are better adapted to discovery learning and innovation, a value implicit in commercial entrepreneurship.

FIGURE 9.9.

A teenage sister stands near a girl (Loxa 1-201) who is learning to weave, but she does not offer assistance until she is summoned. These girls are the daughters of Katal 1, the learner in figures 9.2 and 9.8. Nabenchauk, Zinacantán, Chiapas, Mexico, 1991. Video by Patricia Greenfield.

Indeed, innovation was rampant in weaving and embroidery in 1991. Unlike the 1970 weavers (Childs and Greenfield 1980; Greenfield and Childs 1977), weavers of the 1990s were engaging in a constant process of pattern creation. No two pieces of clothing or other woven items were exactly alike. We saw both new motifs and new recombinations of old motifs. Both geometric designs and figurative representations had entered the scene. There had been no figurative representations in Zinacantán two decades earlier. Figure 9.10A depicts two brothers dressed almost identically in 1970. Examples of the variety of woven and embroidered patterns created in 1991 are depicted in figure 9.10B–D. As in chimpanzee culture (Matsuzawa and Yamakoshi 1996), innovation was concentrated in the younger generation (Greenfield 1999).

Here, then, as hypothesized, was a correlated historical change in ecocultural environment (from agriculture to commerce) and techniques of apprenticeship (from emphasis on scaffolding to emphasis on independent practice). Changes in the material culture of woven

FIGURE 9.10.

A: *Two Zinacantec brothers dressed almost identically, 1970. Photo courtesy of Sheldon Greenfield.* B *(above),* C, D *(next page): Three men's ponchos representing the variety and invention found in Zinacantec textiles in 1991. Photos courtesy of Lauren Greenfield.*

C

D

FIGURE 9.10 (CONT.).

C, D: *Close ups of men's ponchos from Zinacantán, 1991. Photos courtesy of Lauren Greenfield.*

artifacts to an innovative mode entailed change in the method by which weaving was taught and learned. The mode of cultural learning changed from one adapted to maintaining an unchanging stock of traditional artifacts to one adapted to creating cultural innovation and novelty. Thus, change in emphasis in cultural apprenticeship was interrelated with both general ecocultural change and further change in textile design.

Each mode of cultural learning contributes to maintaining a distinctive cultural environment. Hence, we find that certain methods of apprenticeship—such as closely guided participation by the older generation—are more culturally conservative. They tend to transmit the cultural status quo. In contrast, other methods—notably independent practice and teenage teachers—tend to lead to more rapid and radical cultural transformation.

As Pérusse and colleagues (1994) had hypothesized, Zinacantec mothers taught in a way that would maximize their children's fitness under current conditions. They did not simply repeat the way they had been taught weaving as children. Because the historical change was uneven (some families were still involved in agriculture and not commerce, some were involved in both, and some were involved in commerce alone), the change in apprenticeship was also uneven. One would expect that as the economic change became more complete, the techniques of apprenticeship would become more uniform again, as they had been in 1970.

Thus, Zinacantec weaving apprenticeship serves as an example of how cultural transitions may often begin as adaptive alternatives within a given group. If so, this would mirror phylogenetic evolution, in which "phenotypic transitions often begin as adaptive alternatives within species" (West-Eberhard 1988). Variability may be as important in processes of cultural change as it is in biological evolution. Insofar as there is a genetic basis for the more adaptive variants, they can be selected for over a long period of time. In this way cultural change has the potential to become biological in nature (Baldwin 1902).

CONCLUSION

Cultures provide models for the growth and survival of their members that are transmitted from one member to another. There are many aspects of cultural transmission that are shared between *Homo sapiens* and *Pan*. These include scaffolding, observational learning and imitation, experimentation with objects, independent practice, the use of both internal and external symbolic communication, and collaborative learning. In both species (albeit not with equal frequency), these techniques have been incorporated into deliberate teaching.

In terms of the evolution of cultural transmission, perhaps most impressive is the finding that a chimpanzee mother will spontaneously teach her child what she was taught and the way she was taught (by humans) during her own childhood. Although our example comes from a special captive situation, we emphasize that the potential of *Pan* is almost as important, in terms of preadaptations for becoming human, as what *Pan* actually does in the wild. While we can never know the special environmental circumstances that led humans to develop as they did, we can at least identify types of behavior in the wild and in captivity that enable us to speak about likely preadaptations.

In most instances, we have information on humans and on *Pan troglodytes*; in one instance (the use of external symbolic tools), we have information on humans and *Pan paniscus*. In only one instance (instruction about symbolic communication) do we have information for all three species. The evolutionary implications would be much clearer if all aspects of cultural transmission had been explored among all three sibling species. Cladistic methodology posits that when all sibling species share a trait, that trait is highly likely to have been possessed by the ancestor of those sibling species. When traits have been investigated in only two of three sibling species, as is often the case with mechanisms for cultural apprenticeship, the evidence is weaker. Nonetheless, our positive cross-species findings of shared features of cultural apprenticeship indicate that these features are strong candidates to be part of *Homo sapiens'* evolutionary heritage, going back five million years to the common ancestor of *Homo* and *Pan*. The more research that can be done in which the same behavioral features are investigated in all three species, *Homo sapiens, Pan troglodytes*, and

Pan paniscus, the surer the phylogenetic basis for these traits will become.

Unique in human cultural transmission is technical instruction by means of arbitrary symbols, such as language. But note that this unique feature results from a combining of elements that are shared with *Pan*: we have seen that technical instruction occurs in stone tool apprenticeship in the jungle of Ivory Coast (Boesch 1991, 1993) and that instruction incorporating arbitrary symbols occurs spontaneously in captive chimpanzees exposed to a human language environment (Fouts, Hirsch, and Fouts 1982; Ikeo, Jones, and Niio 1993; Tomlinson and Jones 1993).

In addition, of course, human technologies are much more complex than chimpanzee technologies, and the complexity of teaching must increase proportionately. Indeed, in industrial societies, human tool apprenticeship now requires a specialized institution, the school. Hence, the last five million years of human evolution have involved, among other things, the coordination of symbolic communication with collaborative learning and the elaboration of apprenticeship techniques to induct the young into a high level of technological complexity. After examining the evidence, however, we come to the unexpected conclusion that the last five million years of evolution of human tool apprenticeship seem not to have produced major differences in kind. Instead, they have involved the elaboration and combination of learning mechanisms that may go back at least as far as the split of *Pan* and *Homo* five million years ago.

We believe there is a co-evolutionary process in the development of technology, teaching, and cognition. As the complexity of technologies increases, the complexity of apprenticeship to use those technologies must also increase. The evolution of apprenticeship processes must also be coordinated with ontogeny and its evolution. That is, there must also be coordination between the skill development of the learner and the nature of the apprenticeship process across species. Current work on the development of teaching in young children indicates that children as young as six years are able to provide stage-sensitive instruction that is adapted to the developmental capacities of their pupils (Maynard 1999). This research also shows that apprenticeship itself has an ontogeny (Maynard 1999)—that is, skills in cultural teaching develop with age.

Processes of apprenticeship provide mechanisms for both the re-creation and the transformation of culture from one generation to the next. Given a pool of apprenticeship techniques, changes in emphasis on particular processes of learning and teaching provide ways of both responding to and creating cultural change or cultural continuity. While both human and chimpanzee cultures show constancy and change over time as a result of these apprenticeship modes, evidence for cultural accumulation is found only in the human species. Perhaps the combination of complex technology and symbolic instruction is responsible for the cumulative quality of human culture.

Human cultural change holds an implication for the evolution of ontogenies: it is not adult stages that evolve but rather ontogenies that must evolve in response to new tool systems or changing ecocultural conditions. Even across species, it must be *ontogenies* that evolve (Gould 1977; McKinney, this volume) and get transmitted genetically in response to new ecological niches. Given the absence of a fossil record of behavior and its development, the comparative study of human and nonhuman development and socialization is prerequisite to recon-structing the evolution of human behavioral development, including sequences and rates of development (Parker, this volume). A key con-tribution of the present chapter is that behavioral development must include the development of teaching processes as well as the develop-ment of learning processes.

Studying chimpanzees and humans in the transmission of tool sys-tems furthers our understanding of the biological and historical basis of culture, cultural transmission, and cultural change. We gain knowl-edge of which features of teaching and learning were most likely pres-ent in the common ancestor we share with *Pan*; we understand how we humans are both similar to and different from our closest primate rela-tives; and we gain insight into how cultures themselves are transmitted and transformed from one generation to the next. When we examine cultural apprenticeship across species, factors essential for cultural transmission and for cultural change become more evident, ultimately deepening our understanding of both.

Notes

We would like to acknowledge the School of American Research, the Wenner-Gren Foundation for Anthropological Research, the Fogarty International Center of the National Institutes of Health (Minority International Research Training Program to UCLA–El Colegio de la Frontera Sur, TW00061, Steven Lopez, principal investigator), and UCLA for supporting the preparation of this manuscript. We thank the Jane Goodall Research Center archives at the University of Southern California, the Spencer Foundation, the National Geographic Society, the UCLA Latin American Center, the UCLA Center for the Study of Women, the Center for Cognitive Studies (co-directed by Jerome Bruner), the Harvard Chiapas Project (directed by Evon Z. Vogt), the Radcliffe Institute, and the Milton Fund of Harvard Medical School for supporting the research on which this chapter is based. We also thank Carla Childs, Leslie Devereux, Hanna Carlson, and Lauren Greenfield for collaborating in the collection of the Zinacantec Maya data and Matthew Greenfield for managing the video database. Christopher Boehm thanks the H. F. Guggenheim Foundation for support of research that enabled him to assemble a substantial archive of videotaped footage of wild chimpanzee behavior, which is housed at the Jane Goodall Research Center at the University of Southern California. Ashley Maynard was supported by graduate fellowships from the National Science Foundation, the UCLA Center for the Study of Evolution and the Origin of Life (CSEOL), and the University of California Office of the President.

Portions of this chapter were presented as a poster on chimpanzee tool apprenticeship by Yut, Greenfield, and Boehm at the annual symposium of the Jean Piaget Society for the Study of Knowledge and Development, held in Berkeley, California, in June 1995 (Yut, Greenfield, and Boehm 1995).

1. Whiten (1999) concludes that lowland gorilla mothers (observed in a zoo) manifest in a basic form almost all of the features of scaffolding identified by Wood, Bruner, and Ross (1976). Boesch (1991) concludes that all of these features are present in chimpanzee apprenticeship for nut cracking with hammer and anvil.

10

Homo erectus Infancy and Childhood

The Turning Point in the Evolution of Behavioral Development in Hominids

Sue Taylor Parker

> In man, attachment is mediated by several different sorts of
> behaviour of which the most obvious are crying and calling,
> babbling and smiling, clinging, non-nutritional sucking, and
> locomotion as used in approach, following and seeking.
>
> —*John Bowlby,* Attachment

The evolution of hominid behavioral ontogeny can be reconstructed using two lines of evidence: first, comparative neontological
data on the behavior and development of living hominoid species
(humans and the great apes), and second, comparative paleontological and archaeological evidence associated with fossil hominids.
(Although behavior rarely fossilizes, it can leave significant traces.)[1]

In this chapter I focus on paleontological and neontological evidence relevant to modeling the evolution of the following hominid
adaptations: (1) bipedal locomotion and stance; (2) tool use and tool
making; (3) subsistence patterns; (4) growth and development and
other life history patterns; (5) childbirth; (6) childhood and child care;
and (7) cognition and cognitive development. In each case I present a
cladistic model for the origins of the characters in question.[2]

Specifically, I review pertinent data on the following widely recog-
nized hominid genera and species: *Australopithecus* species *(A. afarensis,
A. africanus*, and *A. robustus [Paranthropus robustus])*, early *Homo* species
(Australopithecus gahri, Homo habilis, and *Homo rudolfensis)*, and Middle
Pleistocene *Homo* species *(Homo erectus, Homo ergaster*, and others)*,
which I am calling *erectines*.

TABLE 10.1

Estimated Body Weights and Geological Ages of Fossil Hominids

Species	Geologic Age (MYA)	Male Weight (kg)	Female Weight (kg)
A. afarensis	4.0–2.9	44.6 ± 18.5	29.3 ± 15.7
A. africanus	3.0–2.4	40.8 ± 17.3	30.2 ± 19.5
A. robustus	1.8–1.6	40.2 ± 15.8	31.0 ± 21.5
A. boisei	2.0–1.3	48.6 ± 34.6	34.0 ± 13.7
H. habilis	2.4–1.6	51.6 ± 22.6	31.5 ± 22.5
African H. erectus	1.7–0.7	63.0	52.3

Source: McHenry 1994.

The australopithecines lived from about 4.5 million years ago (MYA) to about 1.5 MYA. They constituted one or more adaptive arrays of small-bodied species (weighing about 35–40 kg) with ape-sized brains. They differed in their cranial and dental features and in their body proportions. Their remains have been found in eastern and southern Africa.

Early *Homo* species *(Homo habilis* and *Homo rudolfensis)* lived from about 2.3 to about 1.6 MYA. They were the descendants of one or more *Australopithecus* species. They, too, were small-bodied creatures (weighing about 35–55 kg), but their brains were somewhat larger than those of the australopithecines. They produced simple worked stone tools. Their fossil remains are also restricted to Africa.

Erectines *(H. erectus* and *H. ergaster)* lived from about 1.8 million to about 300,000 years ago and perhaps longer in some areas. They were larger-bodied creatures (weighing about 57 kg) with larger brains who used more complex tools and technology than early *Homo* species. They were the first hominids to move out of Africa into the Old World. Archaic *Homo sapiens* first appeared about 200,000 to 300,000 years ago. They were larger-brained than the erectines. Modern *Homo sapiens* appeared about 80,000 years ago. They spread throughout the world. Table 10.1 summarizes data on geologic time and body weight for early hominid species (McHenry 1994; note the wide range of error).

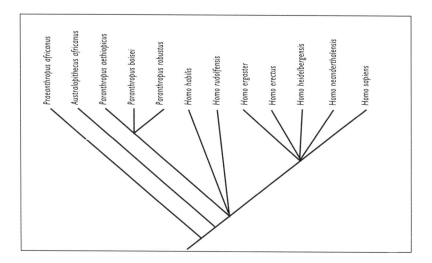

Figure 10.1.

Wood and Collard's hominid phylogeny. (Reproduced with permission from Wood and Collard 1999.)

The taxonomy and phylogeny of hominids continues to be debated and revised. Wood and Collard (1999) argue that early *Homo* species do not belong in the genus *Homo*. They would limit inclusion to species more closely related to *Homo sapiens* than to australopithecines—that is, species that are characterized by body masses, limb proportions, jaw and tooth morphologies, and life histories more similar to those of *Homo sapiens* than to those of australopithecines. Their meta-analysis of cladistic studies of *Homo* species (including that by Straight, Grine, and Moniz [1997]) suggests that *Homo habilis* and *Homo rudolfensis* do not unequivocally share a more recent common ancestor with erectines and sapiens than they do with australopithecines. Likewise, their review of the literature suggests that early *Homo* species are more similar in body mass, limb proportions, jaws and teeth, and life history to *Australopithecus* species than to erectines and sapiens.

In addition to disagreement about the number of valid taxa, widespread disagreement exists about which specimens should be assigned to which species and genera. The trend has been to recognize several synchronic species at any given time until after the emergence of modern *Homo sapiens* (Tattersall 1986). (Consensus is moving toward the conclusion that Neanderthals were a separate species that coexisted with modern *Homo sapiens* [Mercier et al. 1991], though some anthropologists disagree.)

Phylogenetic relationships among these species may never be known. All paleoanthropologists agree that one species of *Australopithecus* gave rise to early *Homo* species, that one early *Homo* species gave rise to Middle Pleistocene *Homo* species, and that one of these gave rise to modern *H. sapiens*. They disagree about which species gave rise to which (Conroy 1997). Figure 10.1 gives Wood and Collard's (1999) preferred phylogeny.

THE EVOLUTION OF BIPEDAL LOCOMOTION AND STANCE

The locomotor, postural, and manual behavior of fossil hominid species can be modeled from evidence regarding the size, shape, and proportions of their bones. This modeling is guided by comparative data on the behavior of living primates. Obviously, complete skeletons are most valuable for this kind of reconstruction and for the modeling of body size and sexual dimorphism. Unfortunately, nearly complete skeletons are rare. Indeed, postcranial remains are rare altogether.

Our closest living relatives—chimpanzees, bonobos, and gorillas—provide comparative models for reconstructing the locomotor behavior of fossil hominids. They display a variety of locomotor adaptations including quadrupedal knuckle walking and bimanual brachiation. Knuckle walking involves walking on the flexed knuckles of the hands, either in a cross-extension mode or in "crutch walking" by swinging both legs between the arms. Brachiation involves hand-over-hand suspensory locomotion under branches. Brachiation, a trait shared by all the apes, apparently first evolved in their common ancestor. Knuckle walking, a trait shared by the African apes, first evolved in their common ancestor or arose independently in the two genera. If knuckle walking arose in the common ancestor, it must have been a direct precursor to bipedal locomotion. A cladistic analysis of comparative locomotor behaviors in living apes suggests that bipedalism arose from terrestrial quadrupedalism like that of African apes (Gebo 1996). In the absence of relevant fossil evidence, the locomotor pattern of proto-hominids can only be surmised.

Bipedal locomotion is the defining characteristic of the family Hominidae (now demoted to the subfamily Homininae or tribe Hominini, according to recent revisions of taxonomic classification). It arose during or after the divergence of hominids from chimpanzees. Evidence that *Australopithecus afarensis* walked bipedally comes from fossilized footprints at Latoli, Tanzania (Leakey and Hay 1979), and from a few fossilized bones from the pelvis, leg, and foot; there are two *Australopithecus* pelves, AL 288-1 (Lucy) from the *A. afarensis* hypodigm and Sterkfontein (Sts) 14 from the *A. africanus* hypodigm (McHenry 1986). The new species *Ardipithecus ramidus*, dated about 4.3 million years ago (White, Suwa, and Asfaw 1994), and *Australopithecus amenensis*, dated about 4 MYA (Leakey et al. 1995), may have displayed an earlier, more primitive form of bipedalism (White, Suwa, and Asfaw 1994).

The two pelves seemed to indicate that *Australopithecus afarensis* (dated from about 3.4 MYA) and *A. africanus* (dated from about 2.5 MYA) had similar locomotor adaptations, though they differed in cranial, facial, and dental adaptations (McHenry 1986). Specifically, the *Australopithecus* pelvis is wider (longer from side to side) and shallower (shorter from front to back), or more platypeloid, than the human pelvis (McHenry 1986). It is also characterized by a relatively greater distance between the hip and sacral joints, a smaller sacral iliac joint surface, and larger pubic bones.

New postcranial remains of *A. africanus* from member 4 Sterkfontein, including a new partial skeleton, Stw 431, indicate that this species had a more primitive (i.e., *Pongid*-like) morphology of relatively large forelimb and small hind limb joints and a more adducted great toe than *A. afarensis* (McHenry and Berger 1998)

Abitbol's (1995) effort to model the posture of *A. afarensis*, specifically the curvature of the vertebral column, suggests that these creatures had not attained an upright orientation of the spine. In modern humans the lumbar-sacral articulation is virtually horizontal. The transition from the nearly vertical lumbar-sacral articulation that is typical of quadrupeds to the nearly horizontal orientation of modern humans had not occurred.

The pelvis of AL 288-1 is controversial. Paleoanthropologists differ over whether *A. afarensis* is a single, very sexually dimorphic species or two or more less dimorphic species. They also differ over whether AL 288-1 is the pelvis of a male or a female and whether the pelves of this species or these species are sexually dimorphic or not. Those who argue that *A. afarensis* is one species interpret the pelvis as female and argue that this otherwise dimorphic species lacked sexual dimorphism in the pelvis. They argue that the flattened shape of the pelvis was an adaptation for locomotion and support unmodified for birth of a large-brained offspring (Teague and Lovejoy 1986). In contrast, Hausler and Schmid (1995) argue on the basis of a comparison with the Sts 14 pelvis that AL 288-1 is the pelvis of a male and could not have supported delivery of an infant with an *Australopithecus*-size brain. They also argue on this basis that *A. afarensis* includes two species (Hausler and Schmid 1995). Recently, Teague and Lovejoy (1998) responded that both Sts 14 and AL 288-1 are female pelves and that AL 288-1 would have been obstetrically adequate. Sexing australopithecine pelves relies on identification of features that distinguish the pelves of modern human females from those of males. This procedure is questionable, however, because these features probably arose later in hominid evolution as a consequence of brain enlargement (Hager 1991).

The feet of australopithecines also display a unique complex of features, including longer, more curved toes (McHenry 1986; Susman, Stern, and Jungers 1984) and, in the case of *A. africanus*, a more adducted great toe (McHenry and Berger 1998). Like pelvis and foot, the forelimbs of all australopithecines show a unique complex of features (McHenry 1986), though this is more extreme in *A. africanus* than in *A. afarensis* (McHenry and Berger 1998). Their arms are longer relative to their legs as compared with those of modern humans. Their shoulder joints and humeral heads are narrower, and their shoulder joints face upward. Their wrist bones also display a unique pattern intermediate between those of modern humans and great apes. Their cone-shaped rib cages are similar to those of chimpanzees (Stanley 1992). Figure 10.2 provides a comparison of the torsos of humans, australopithecines, and chimpanzees (Hunt 1994).

McHenry (1986) concluded that the australopithecines displayed a locomotor adaptation unseen in living forms. Some anatomists have

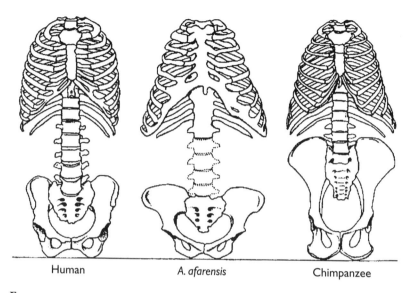

| Human | A. afarensis | Chimpanzee |

Figure 10.2.

A comparison of primate torsos. (Reproduced with permission from Hunt 1994.)

argued that their short lower limbs, long upper limbs, and curved toes reflect a continuing adaptation for tree climbing (Susman, Stern, and Jungers 1984), whereas others believe these are simply retentions of primitive features (Gebo 1996). The primitive australopithecine locomotor anatomy is consistent with evidence that these creatures lacked the inner ear configuration associated with bipedal balance in modern humans and *Homo erectus* (Spoor, Wood, and Zenneveld 1994). The evidence from inner ears is also consistent with the discovery that at least one early *Homo* species retained the primitive limb proportions of the australopithecines (Johanson et al. 1987).[3]

Consistent with this interpretation, Stanley (1992) argued that gracile australopithecines lived in small forest patches and maintained a semiarboreal mode of existence. Although they probably used tools, they depended on trees for defense against predators. He argued that immature young could still cling with their upper limbs. He argued that this mode of existence lasted for 1 or 2 million years, until a major climatic shift occurred at about 2.5 million years ago, resulting in the spread of grasslands. Sabater Pi and his colleagues have argued that early hominids not only climbed trees but also, like great apes, built and used nests in trees (Sabater Pi, Vea, and Serrallonga 1997). Accordingly, I call this pattern semiarboreal bipedalism in order to distinguish it from fully terrestrial bipedalism.

However we label it, bipedalism is the defining characteristic of hominids. Through the years, investigators have proposed a variety of selection pressures to explain the evolution of bipedalism. These include (1) tool use and missile throwing, (2) tool and food transport, (3) aquatic or semiaquatic foraging, (4) social-sexual displays, (5) efficient long-distance travel for hunting, (6) thermoregulation, and (7) terrestrial gathering from trees and bushes. Rose (1991), Tuttle, Webb, and Tuttle (1991), and Morgan (1993) have provided reviews.

Many anthropologists, when they realized that the earliest hominid sites yielded no evidence for hunting and/or worked stone tools, rejected tool use as an explanation for bipedalism. Given the evidence for tool use in chimpanzees, however, it seems likely that the Darwinian interpretation that bipedalism arose in conjunction with greater reliance on tool use and tool transport is correct. Selection for bipedalism might simply have involved a shift from seasonal to year-round tool use with concomitant demands on object carrying. Generally, explanations for bipedalism have been proposed as alternative rather than complementary hypotheses. It seems likely, however, that bipedalism conferred more than one advantage—for example, in male displays and missile throwing as well as tool use and transport (Tuttle 1992, 1994).

Whatever its adaptive significance, bipedal locomotion apparently evolved through at least two major stages: first, the semiarboreal bipedalism of the australopithecines and early *Homo*, and second, the fully terrestrial striding bipedalism of later *Homo*, beginning with *Homo erectus* (Tuttle 1994). This is indicated by the fact that *Homo erectus* was the first hominid to display essentially modern limb proportions, body size, and locomotor anatomy. Figure 10.3 gives a cladistic depiction of the evolution of locomotor behavior in apes and hominids.

THE EVOLUTION OF HOMINID SUBSISTENCE PATTERNS AND TOOL USE

Like Darwin (1871), most early students of human origins styled early hominids as hunters and tool users. As evidence accumulated that

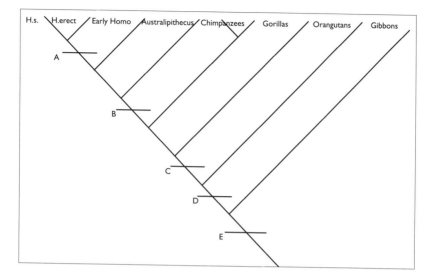

Figure 10.3.

Cladogram of putative hominoid locomotor patterns in a series of common ancestors. Key: A, fully terrestrial bipedalism; B, semiarboreal bipedalism; C, quadrupedal knuckle walking; D, fist walking; E, brachiation.

australopithecines did not manufacture stone tools, these interpretations were questioned and virtually discarded. Few anthropologists would deny that the earliest hominids used perishable tools, but many now discount the significance of tool use in early hominid evolution. More refined analysis of percussion marks on bones, however, may facilitate recognition of stone tool use prior to stone tool manufacture (Blumenschine and Selvaggio 1994).

The earliest evidence for percussion marks and the cracking open of animal bones is associated with the new species *A. garhi* (Asfaw et al. 1999), about 2.5 million years ago (Heinzelin et al. 1999). The earliest evidence for worked stone tools occurs at about 2.5 MYA in Gona, Ethiopia (Semaw et al. 1997). These early chopper tools, roughly flaked on one end, seem to be associated with early *Homo* species. *Homo habilis* shows new features of the hand, particularly broadening of the thumb, that support tool production. Indeed, these features inspired the name *Homo habilis*, or "handy man" (Leakey, Tobias, and Napier 1961). The coincidence of worked stone tools is also consistent with the increased brain size of this species. Susman (1988), however, has argued that *Parathropus* had hands that were equal to stone tool production.

Simple chopper tools persisted virtually unchanged for nearly a million years. Intriguingly, evidence from tool-flaking patterns suggests that the early tool makers were right-handed, like the majority of modern humans (Toth 1985). Simple chopping tools were superseded in most parts of the Old World by bifacial Acheulean tools by about 1.6 MYA, coincident with the appearance of *Homo erectus*. Bifacial hand axes dominate the Acheulean assemblage. Archaeological evidence from Lake Turkana suggests that the makers of hand axes had larger home ranges than the makers of cobble tools. It also suggests that they used a wider variety of materials to fashion their tools (Rogers, Harris, and Feibel 1994).

Acheulean tools persisted for another million years, until the appearance of the more diverse and sophisticated Middle Paleolithic (Mousterian) artifacts coincident with archaic *H. sapiens* at about 200,000 years ago. Upper Paleolithic tool cultures appeared sometime between 80,000 and 35,000 years ago (e.g., Conroy 1997; Klein 1989). Mousterian tools involved an early form of manufacture entailing production of many flakes from one prepared core. Upper Paleolithic tools encompassed many innovations, including micro tools, blades, hafting, and the use of bone. In contrast to all the earlier assemblages, Upper Paleolithic tool kits were regionally variable and show rapid cultural evolution. Accordingly, we distinguish the following major phases in hominid evolution: (1) modification of tools, characteristic of chimpanzees and australopithecines, as opposed to (2) manufacture of tools, characteristic of early *Homo*. Among manufactured tools we can distinguish Oldowan, Acheulean, Mousterian, and Upper Paleolithic assemblages.

The earliest evidence for cut marks on animal remains is associated with *A. gahri* at about 2.5 MYA and with *H. habilis* at about 2.0 MYA (Potts and Shipman 1981). The first clear evidence for systematic hunting of a single large species, however, occurs much later, at about 200,000 years ago, with archaic *Homo sapiens* (Klein 1984). In recent years, archaeologists have argued that the earliest manufactured tools were used for butchering animal carcasses and extracting marrow from animal bones rather than for capturing animals during hunting (Blumenschine and Selvaggio 1994). Many archaeologists now suggest that scavenging arose early in hominid evolution. Although archaeologists recognize that early hominids may have done some hunting, they discount the importance of hunting as an early hominid adaptation. This distinction is somewhat arbitrary, considering that most carnivores facultatively scavenge or hunt.

Tool use and hunting play important roles in chimpanzee subsistence, if only seasonally. Chimpanzees use tools primarily for extracting a variety of embedded foods, including termites, ants, hard-shelled nuts and fruits, honey, and roots during the dry season (Parker and Gibson 1977, 1979). Extractable foods are particularly significant during the dry season in many parts of Africa (e.g., Goodall 1986; McGrew 1992; Teleki 1975). The long apprenticeship required for efficient tool use has considerable significance for the life histories of chimpanzees and other great apes (Boesch et al. 1994; Greenfield and Maynard, this volume; Parker 1996a).

Likewise, studies of predation in wild chimpanzees suggest that hunting is a significant aspect of their subsistence. A recent study of hunting by red colobus monkeys reveals the nutritional importance of hunting for the predators and the impact of hunting on the prey (Stanford 1996). Like insect consumption, meat consumption is limited primarily to the dry season. Intriguingly, except for sponging out brain juices, chimpanzees rarely use tools in hunting and consuming animal prey. It is important to note that chimpanzees eat all the bones, skin, and hair of their prey, leaving no remains that could fossilize, except perhaps feces. Hunting not only is important nutritionally but also reverberates in the sexual and political lives of the hunters. Among western chimpanzees, cooperative hunting of red colobus monkeys is the dominant pattern (Boesch and Boesch 1989). Males increase their frequency of copulation and their political power through selective sharing of the meat they catch by trading food for sex (Teleki 1973).

Discounting the importance of hunting and tool use in early hominids was an overcorrection for an earlier tendency to equate the adaptations of early hominids with those of modern human gatherer-hunters. It also came from a dichotomous classification of both hominids and humans into hunters versus nonhunters. In opposition to this classification, I suggest the categories *omnivorous forager-hunters* (Teleki 1979) to describe early hominids, *gatherer-scavenger-hunters* to describe early *Homo*, and *true gatherer-hunters* to describe *Homo sapiens.*

The key distinctions here depend upon techniques of prey acquisition and prey consumption as well as prey size. Forager-hunters such as chimpanzees and early hominids hunt relatively small prey that they are able to consume without butchering. Scavenger-hunters such as early *Homo* scavenge and hunt relatively large prey that they can consume only by butchering. Specialized hunters hunt large prey cooperatively with the aid of spears and other sharp weapons wielded or thrown some distance from their own bodies. Of course, consumption of prey by chimpanzees involves the use of long canine teeth to rip open skin. The intermediate length of the canines of *A. afarensis* would barely have allowed them to rip open carcasses, and the reduced canines of *Homo* would have precluded this function.

The tendency to discount hunting in early hominids is based on the use of modern humans as a standard of reference. If we adopted chimpanzees as a standard of reference, we would certainly classify early hominids as extractive foraging tool users and probably as hunters. This policy seems more consistent with comparative approaches. Moreover, recent isotopic analysis of the dental enamel of *A. africanus* is consistent with the hypothesis that they were meat eaters (Sponheimer and Lee-Thorp 1999).

On the other hand, tool-mediated foraging for roots and tubers may have been a more reliable primary means of getting food beginning with *H. erectus* (O'Connell, Hawkes, and Blurton-Jones 1999). The shift to primary dependence on this food source probably occurred coincident with the long-term cooling trend at the base of the Pleistocene epoch. The exploitation of roots and tubers was made possible by the manufacture of tools suitable for making sharpened digging sticks and by the processing of these foods with fire.

Accordingly, I suggest the following reconstruction of the evolution of subsistence modes in apes and hominids: (1) year-round, tool-mediated extractive foraging and hunting of small prey by gracile australopithecines; (2) extension of extractive foraging to butchery of scavenged prey by habilines (and *A. garhi*); (3) extension of tool use to the production of sharpened digging sticks for excavating roots and

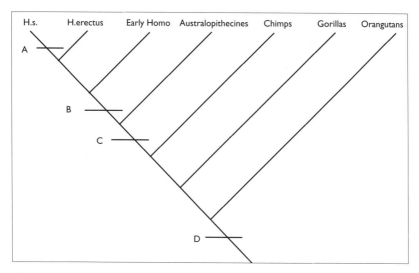

Figure 10.4.

Cladogram of putative hominoid subsistence modes in a series of common ancestors. Key: A, big game hunting with flaked tools; B, butchering of game with flaked tools; C, year-round extractive foraging with tools and small game hunting; D, seasonal extractive foraging with tools.

tubers, and use of fire for processing these foods, as well as production of shelters and clothings by erectines; and (4) extension of tool use to hunting through spearing and missile launching plus use of fire in food and tool processing by *Homo sapiens* (Parker and Gibson 1979). Figure 10.4 gives a cladistic interpretation of the evolution of subsistence.

THE EVOLUTION OF LIFE HISTORY, GROWTH, AND DEVELOPMENT IN HOMINIDS

Life history strategy theory views life cycles as products of selection operating within developmental constraints. From this perspective, the various subdivisions of the life cycle represent an optimization of energy expenditure in growth, maintenance, defense, and reproduction across the life span (Horn 1978). The length of gestation, infancy, immaturity, and life span, the number of offspring per reproductive effort, and the number and frequency of reproductive efforts are all parameters of life histories. Indeed, brain size seems to be a pace-maker in mammalian life histories (Gittleman et al., this volume). Both neonatal and adult brain sizes seem to correlate closely with or even determine birth weight, age of molar eruptions, and age at first breeding. Brain size also correlates significantly with gestation length and life span. There is a 99-percent correlation between brain size at birth and adult brain weight (Sacher 1959; Sacher and Staffeldt 1974; Smith 1989).

Each primate species has a peculiar life history that has been shaped by past selection as well as by common developmental constraints. These vary along a continuum from the r life history strategy of tiny mouse lemurs to the K life history strategy of great apes. Mouse lemurs weigh less than an ounce, mature within one year, and bear two offspring per year. Gorillas weigh more than 100 pounds at maturity, begin reproduction only after 10 or more years of life, and produce one offspring every 4 to 10 years. Among mammals, K life history strategies correlate with large body size and large brain size (Harvey, Martin, and Clutton-Brock 1986).

Molar eruption patterns are important indicators of primate life history. Eruption of the first molar correlates (about 93 percent) with weaning (Smith 1989, 1992). As Smith notes, this makes sense considering that infants must be able to feed themselves in order to survive without mother's milk. Eruption of the first molar also correlates with the achievement of 90 percent of brain mass (Portmann 1990). The eruption of the second molar correlates with the transition from middle childhood to adolescence. Completion of dentition correlates well (about 93 percent) with onset of reproduction (Smith 1989). Hence, the eruption dates of permanent molars correspond, respectively, to the end of infancy/early childhood and the onset of middle childhood (M1), the end of middle childhood and the onset of adolescence (M2), and the end of adolescence and the onset of reproductive life (M3) (Smith 1993).

Data on dental development in Old World monkeys, great apes, and humans provide a standard for comparing dental development in fossil hominids. The data reveal the following salient differences in developmental patterns: (1) permanent molar teeth develop in closer succession in apes than in humans; (2) molar roots develop about twice as fast in apes as in humans (though rates of crown development are similar); and (3) the canines of great apes develop over a longer period

TABLE 10.2

Age (in Years) of Tooth Eruption in Selected Anthropoid Primates

Species	Last Deciduous	First Molar	Second Molar	Third Molar
Macaca mulatta	0.44	1.35	3.2	6.0
Pan troglodytes	1.20	3.15	6.5	10.5
Homo erectus	Unknown	4.50	9.5	14.5
Homo sapiens	2.30	5.40	12.5	18.0

Source: Smith et al. 1994.

than those of modern humans (Bromage 1987). Specifically, M1 erupts at 3.15 years in great apes and at 5.4 years in humans, M2 erupts at 6.5 years in great apes and at 12 years in humans, and M3 erupts at 10.5 years in great apes and at 18–20 years in humans (Smith 1992, 1994; Smith, Crummett, and Brandt 1994).

In short, the developmental sequence in macaques begins about two years earlier than that in chimpanzees and occurs at approximately two-year intervals. The sequence in chimpanzees begins about two years later than that in macaques and occurs at approximately three-year intervals. The sequence in humans begins about two years later than that in chimpanzees and occurs at approximately six-year intervals.

Erectine molar development apparently began about one year later than that in chimpanzees and one year earlier than that in humans, and it occurred at approximately five-year intervals (Smith 1993). In other words, going from macaque to chimpanzee to human, the rate of molar development in each species shows later onset and offset and is uniformly decelerated (table 10.2). These comparative dental data suggest that the duration of various subdivisions of the life cycle has changed in a consistent direction during hominid evolution. The length of middle childhood has increased from three years in great apes to five years in modern humans. The length of adolescence has increased from four years in great apes to six years in modern humans. Data from fossil hominids confirm this.

Because teeth fossilize better than any other part of the body, there is a wealth of dental data on individuals of various hominid

TABLE 10.3

Estimated Life History Variables for Hominids

Species	Neonatal Brain Size (grams)	Gestation Period (months)	Age at Weaning (months)	Age at Puberty (years)	Age at 1st Breeding (years)	Life Span (years)
A. afarensis	162	7.6	28.7	9.3	11	42
A. africanus	166	7.6	29.2	9.4	11.2	43
A. robustus	175	7.6	30.1	9.7	11.4	43
A. boisei	185	7.7	31.2	10.0	11.8	44
H. habilis	173	7.6	29.8	9.7	11.4	43
H. erectus	270	8.2	39.6	12.5	14.2	50

Source: Adapted from McHenry 1994.

species of various ages. Analyses of these dental remains have revealed the following pattern: (1) the gracile australopithecines *(A. afarensis* and *A. africanus)* show a pattern of dental development similar to that of the great apes; (2) the robust australopithecines (*Paranthropus* species) show a unique pattern of dental development that is even more accelerated than that of the great apes; (3) habilines show a pattern similar to that of the gracile australopithecines; and (4) erectines show a pattern intermediate between that of great apes and humans (Bromage 1987; Bromage and Dean 1985; Conroy and Kuykendall 1995; Smith 1986). Table 10.3 summarizes life history parameters for great apes and hominids (McHenry 1994). These conclusions are based on analyses of the dentition of immature specimens of fossil hominids. Analyses include comparison of relative maturation of the molars and other teeth in a jaw (Bromage 1987; Smith 1986) and determination of the number of incremental growth lines on a single tooth (Bromage and Dean 1985).

Although jaw fragments and even single teeth provide important information, greater insight into growth and development comes from the analysis of dental and skeletal material from a single individual. Three fossilized skeletons have contributed significantly to the understanding of the evolution of human life history: (1) the skeleton of an adult female *Australopithecus afarensis,* Lucy (AL 288-1), from Hadar,

TABLE 10.4

Duration of Life History Stages Based on Molar Eruption Ages

Species	Infancy	Childhood (Juvenility)	Adolescence (Subadulthood)
Macaca mulatta	1.35	1.85	2.8
Pan troglodytes	3.15	3.35	4.0
Homo erectus	4.50	5.00	5.0
Homo sapiens	5.40	7.10	5.5

Ethiopia (Johanson et al. 1978); (2) the fragmentary skeleton of an adult female *Homo habilis* (OH 62) from Olduvai Gorge, Tanzania (Johanson et al. 1987); and (3) the skeleton of an adolescent male *Homo erectus* (or *Homo ergaster*), the Turkana boy (WT 15000), from Nariokotome in Kenya (Brown et al. 1985). Whereas OH 62 is fragmentary and Lucy's skeleton is about 40 percent complete, Turkana boy's skeleton is about 80 percent complete. He is also much larger, about 1.6 meter, or 5 feet 2 inches, projected to about 5 feet 8 inches in adult height.

The modeling of the life history of Turkana boy suggests that his developmental pattern was intermediate between that of modern humans and chimpanzees. Specifically, this reconstruction suggests that Turkana boy was 9–10 years of age at death. It suggests that his first molar erupted at 4.5 years and his second at 9 years. He would have lived 15 years longer than a chimpanzee (Smith 1993). The endocast of his brain has a volume of 880 cubic centimeters, about 97 percent of the 910 cc endocast of an adult *Homo erectus* brain (Begun and Walker 1993). Table 10.4 compares the duration of various life history stages of macaques, chimpanzees, modern humans, and erectines in terms of molar eruption patterns.

Turkana boy shows a mosaic of dental and skeletal development unlike that of either chimpanzees or humans. His skeletal age suggests that he was well into the adolescent growth spurt, whereas his dental age is younger than that of a human male in the midst of this growth spurt. This indicates that the growth spurt in *H. sapiens* typically occurs at a later stage in dental development than it did in *Homo erectus.* (In contrast to *Homo,* chimpanzees show a very small adolescent growth spurt.) Overall, humans achieve a higher percentage of their growth after adolescence than do chimpanzees. Specifically, humans achieve almost a third more growth in leg length during this period than do chimpanzees. Smith (1993) and Bogin (1997) argued that humans have added an additional developmental phase between weaning and sexual maturation. Specifically, they have extended the juvenile phase of development as compared with other primates. The life history pattern modeled for erectines contrasts both with that modeled for the gracile australopithecines and early *Homo* and with that of modern humans. Erectines seem to stand midway between early hominids and modern humans in their life history.

THE EVOLUTION OF CHILDBIRTH IN HOMINIDS

Neonatal mammals often display one of two general patterns described by Portmann (1990; see also Martin and MacLarnon 1985). Either they display the altricial pattern of immaturity and helplessness at birth (lacking hair, hearing, and vision as well as locomotoric capacities) or they display the precocial pattern of maturity at birth (having hair, hearing, vision, and well-developed locomotor capacities). Altricial young are generally born in large litters, whereas precocial young are generally born as singletons or twins. Precocial mammals also have larger brains than altricial mammals. This makes sense considering that precocial young go through a greater proportion of early development in the womb. Conversely, altricial young undergo a greater proportion of early development out of the womb, usually in a protected nest or den. Generally speaking, large-bodied K strategists give birth to precocial young, and small-bodied, rapidly developing r strategists give birth to altricial young (Portmann 1990).

Although most anthropoid primates are precocial, humans are secondarily altricial. Humans are born many months earlier than other primates as judged by their average degree of locomotor and social maturity (Portmann 1990). The human condition probably arose in response to the unusually large brain of human neonates. Recent work on molar eruption in australopithecines suggests that life history extension and hence modern birth patterns evolved relatively late in hominid evolution (Conroy and Kuykendall 1995).

As indicated in the earlier discussion of locomotion, the australopithecine pelvis is unique in its width as opposed to its depth (platypeloid). If AL 288-1 is female, then the *Australopithecus* pelvis differs from the human pelvis in its sexual dimorphism. If it is male, then we have no examples of females. According to Teague and Lovejoy (1986), the pelvis of the australopithecines is funnel-shaped, like that of modern human males but unlike the pelvis of modern human females. The angulation of the sacrum away from the pelvis is about 70 degrees in human females, as compared with 61 degrees in males and 63 degrees in australopithecines (Teague and Lovejoy 1986). According to Hausler and Schmid (1997), comparative data suggest that AL 288-1 is a male, which in turn suggests that it is part of a second, smaller species.

The birth position of hominid neonates has been a topic of some speculation among paleoanthropologists. Since the neonate's trajectory through the birth canal is determined by the shape of the mother's pelvis, species-typical features of the pelves could imply differences in mechanisms of birth. During birth, human fetuses undergo internal rotation as they move down through the pelvic outlet, and they are born with the nose facing the mother's sacrum (rotational position). Baboon and chimpanzee neonates are most often born with their noses facing the mother's pubis (Rosenberg and Trevathan 1996).

Teague and Lovejoy (1986) have argued on the basis of AL 288-1 that australopithecine neonates differed from both humans and chimpanzees in being born with their noses facing their mother's hip joint (nonrotational position), the orientation that allows them the greatest room. Figure 10.5 reproduces Teague and Lovejoy's (1986) comparison of neonatal head positions in the three species.

Hausler and Schmid (1995) argue that the *A. africanus* pelvis, Sts 14, is less platypeloid than Teague and Lovejoy suggest. They argue that the rotational position during delivery would have been easier than the nonrotational position in this species. Ruff (1995), however, claims that Teague and Lovejoy's own data do not support this interpretation. He therefore accepts the idea that *Australopithecus* infants were born in the nonrotational position.

The deeper front-to-back (anterior-posterior) dimensions of the human pelvis, especially the pelvic outlet, are apparently adaptations to

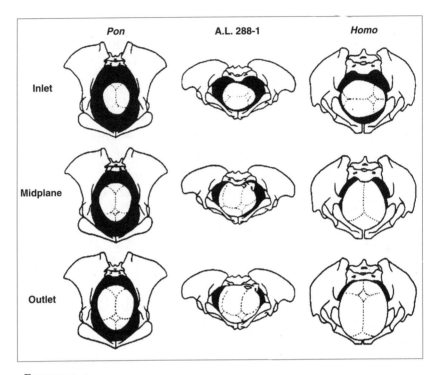

FIGURE 10.5.

Neonatal head position during birth in three hominoid pelves. (Reproduced with permission from Teague and Lovejoy 1986.)

the birth of large-headed neonates as opposed to adaptations for loco-motion (Ruff 1995). The small brains of adult australopithecines and the shape of their pelves indicate that these species faced no such obstetrical demands. This, in turn, indicates that the evolutionary shift from semiprecocial to secondarily altricial infancy occurred after the australopithecines (Teague and Lovejoy 1986).

Investigators who agree on the change in birth position disagree on when this shift might have occurred. Earlier scenarios suggested that australopithecines had already achieved a human life history pattern. Recent data on the shapes of the pelves and femora of hominids suggest that the change in the anterior-posterior dimensions of the pelvis and the pelvic outlet, and hence the birth orientation of fetuses, had not occurred in early *Homo* and probably did not occur until late in the evolution of erectines (Ruff 1995).

TABLE 10.5

Brain Size (in cc) in Neonatal and Adult Anthropoid Primates

Species	Neonate	Adult	% of Adult at Birth
Macaca mulatta	48	89	0.54
Pan troglodytes	176	382	0.46
Homo erectus	388	909	0.35
Homo sapiens	382	1250	0.25

The data on dental development, brain size, and other life history features of *Homo erectus* discussed earlier suggest that the trend toward secondary altriciality began with this species (Begun and Walker 1993). Brain size in early adult erectines was about 909 cc, which suggests a neonatal brain size of about 388 cc (Begun and Walker 1993). This compares with about 382 cc in human neonates and about 176 cc in chimpanzee neonates (Passingham 1982). These data suggest that *Homo erectus* females were selected to give birth to immature infants in order to get them through the birth canal (e.g., Begun and Walker 1993; Teague and Lovejoy 1986). This interpretation is also supported by the apparent sexual dimorphism in *Homo erectus* pelves (Begun and Walker 1993). Table 10.5 provides comparisons of adult and neonatal brain sizes in selected anthropoids, including hominid species.

In conclusion, australopithecines were probably semiarboreal bipeds that had life history patterns similar to those of chimpanzees. If so, it seems likely that their infants were capable of clinging at least with their upper limbs and that they continued to rely on nests in trees for defense, sleeping, and resting (Sabater Pi, Vea, and Serrallonga 1997; Stanley 1992). In addition, it seems likely that their maternal-infant behaviors, age at first reproduction, and birth intervals were similar to those of chimpanzees and other great apes.

LIFE HISTORY AND THE EVOLUTION OF MOTHER-INFANT INTERACTIONS AND CHILD CARE IN HOMINIDS

Life history strategy theory suggests that delayed maturation evolves when it increases lifetime reproductive success. Advantages conferred by large adult body size seem to be the major factor in the evolution of delayed maturation and the associated extension of the juvenile period. Also significant are advantages conferred by the extended opportunity for play and learning (Pagel and Harvey 1993). As we have seen, great apes have larger bodies and more extended maturation than monkeys, humans have larger bodies and more extended maturation than great apes, and erectines had body sizes and maturation schedules intermediate between those of great apes and modern humans. This suggests that erectines enjoyed an extended opportunity for play and learning and a concomitant increase in parental and/or other kin investment.

Great ape infants develop more slowly than macaque and other monkey infants do and therefore are weaned considerably later. Despite their longer infancy, great ape infants interact with their mothers primarily through the proximate tactile signals of clinging, sucking, and touching. Great ape mothers actively lick and groom the faces (and genitals) of their offspring. It is notable, however, that they fail to show the contingent, face-to-face interactions that Watson (1972) described in human mothers and infants (Parker 1993). In contrast to monkey mothers, great ape mothers engage in considerable play with their infants. They often dandle infants on their feet and tickle them. Their infants respond with chuckles and play faces (e.g., van Lawick-Goodall 1970). In contrast to human infants, great ape infants are usually silent except during play or separation from their mothers.

Unlike other anthropoid primates, modern human infants and mothers engage in a unique, gamelike, face-to-face interaction (Watson 1972). During this game infants seem to repeat their actions in order to elicit a contingent response in the mother. The game may involve either mutual imitation of facial expressions and vocalizations or other contingent responses. Human infants seem to recognize when they are being imitated and to prefer this mode of interaction to other kinds of contingent responses (Meltzoff 1990). These so-called circular reactions and mutual imitation games begin at about three months of age in human children (Piaget 1962).[4] They are crucial for the development of conversational turn taking (Stern 1977). They are also precursors to referential games in which mothers and infants achieve joint attention by responding to the direction of gaze and pointing (e.g., Bates et al. 1979; Bruner 1983).

These highly ritualized communicatory routines in modern humans are facilitated by visually salient distal displays: pink everted lips that highlight mouth movements and the white sclera surrounding the iris of the eyes that emphasizes the direction of gaze. Although the soft tissues involved in these behavioral displays have left no fossil record, various factors suggest that they evolved at the time of *Homo erectus*.[5]

First, secondarily altricial *Homo erectus* infants would have been unable to cling to their mothers. They may have been the first hominid infants who spent time off their mothers' bodies, lying on their backs. This postural shift is significant developmentally because it facilitates two trends: (1) canalization of hand-eye interactions to the dominant hand, which occurs in modern human infants (Gesell 1945), and (2) repetition of actions to test their effects on objects—so-called secondary circular reactions—which also occurs in modern babies. Similarly, delayed crawling and prolonged sitting, which likely evolved concomitant with slower locomotor development, facilitates two additional trends: (1) repetitive experimentation with objects relative to other objects, gravity, and friction—so-called tertiary circular reactions—and (2) creation and manipulation of object sets (e.g., Langer 1993), such as occurs in human infants.

Second, because they were on their mothers' bodies less often, *Home erectus* infants would have depended more than earlier hominid infants did on distal communicatory signals and attractiveness to their mothers and other caretakers. Therefore, it seems likely that some of the specialized care-eliciting signals that distinguish modern human babies from great ape infants evolved at this time. These include such morphological displays as fat cheeks, everted lips, and white sclera of the eyes. They evolved in conjunction with such behavioral displays as smiling, cooing (used to bond and express pleasure), and tears and wailing (used to convey anger, frustration, and discomfort). These distal mother-infant interactions would have facilitated mutual imitation and "contingency games" similar to those of modern humans. The evolution of new mother-infant signals probably accelerated (predisplaced) the development of social imitation from childhood to infancy.

The preceding review suggests that erectines were intermediate between chimpanzees and modern humans in their life history pattern and were no longer dependent on trees for refuge. If so, they must have displayed new adaptations for infant and child care. According to the preceding projections, *Homo erectus* offspring reached the end of infancy at 4.5 years, the end of childhood and the onset of adolescence at about 9.5 years, and adulthood at about 14.5 years of age (in the case of females). Slower development combined with loss of clinging and the addition of three years of juvenile dependency would have entailed additional parental and/or kin investment. These demands would have been exacerbated by the increased need for defense against predators and competitors attendant upon terrestrial life.

On the basis of archaeological remains and comparative data from other species, only a few potential solutions to these challenges suggest themselves. They include (1) use of natural and artificial shelters; (2) use of slings for carrying nonclinging infants; (3) use of weapons and fire for protection from predators; (4) more extensive food sharing; and (5) dependence on close kin and perhaps mates for aid in child care and perhaps even childbirth (Trevathan 1987). Recent reports suggest that early *Homo erectus* from Koobi Fora, Kenya, used fire for some purposes, apparently for warmth and defense. Whether they used fire for cooking or material processing is disputed (Bellomo 1994; O'Connell, Hawkes, and Blurton-Jones 1999). In any case, this evidence supports the traditional notion that the migration of this species out of Africa into colder regions of Europe and Asia depended on the use of fire. Various archaeological remains suggest that this species constructed shelters (Klein 1984), and it seems likely that they also made rude clothing and containers. If so, it is likely that they used slings to carry infants.

Within the primates, dependency on kin, especially older children, and on mates for help in child rearing is common only among modern humans and tamarins. It is also common in other mammalian and avian species (e.g., Goldizen 1990). Like hominids, tamarins experience constraints on neonatal development owing to the large size of the neonate head relative to the size of the mother's pelvis. In their case, this constraint (which results from dwarfing) may have favored twinning rather than secondary altriciality. The increased parental demands attendant on large litter birth weight versus mother's weight in tamarins has favored a variety of helping adaptations. In erectines, helping adaptations probably included allomothering by older siblings

(Weisner and Gallimore 1977; Whiting and Edwards 1988a) and provisioning of postweanlings by grandmothers (Gibbons 1997; Hawkes, O'Connell, and Blurton-Jones 1997; O'Connell, Hawkes, and Blurton-Jones 1999). These adaptations also favored flexible mating systems, which might have varied from monogamy to polyandry or polygyny. Whereas erectines responded to obstetrical constraints with secondary altriciality rather than twinning, they probably responded like tamarins and other vertebrates to increased demand for parental care with dependence on helpers.

In accord with these considerations, I propose the following two stages in the evolution of mother-infant interaction: (1) the great ape-like pattern of semiprecocial infant development involving semipassive infant transport with primary dependence on infant clinging and tactile communication during infancy, followed by a juvenile period of selective food sharing and apprenticeship in tool use and foraging until about eight years of age in gracile australopithecines and early *Homo*; and (2) a protohuman pattern of secondarily altricial infant development requiring active transport (perhaps in slings) and/or guarding of infants by caretakers using distal communicatory signals and vocal and facial imitation in infancy, followed by a juvenile period of continuing food provisioning and extended apprenticeship in a variety of subsistence skills, lasting until about 14 years. Figure 10.6 offers a cladistic interpretation of the evolution of symbolic communication.

APPRENTICESHIP AND THE EVOLUTION OF SYMBOLIC COMMUNICATION

Erectines, as compared with early *Homo*, faced additional demands for prolonged provisioning, socializing, and training of their juvenile offspring for subsistence activities. Substantial provisioning of difficult-to-process foods such as roots, tubers, and meat would have been necessary in the period of childhood between weaning and the child's achievement of semi-independence in foraging and food preparation. Food sharing probably extended beyond the mother-child relationship to grandmother-mother and perhaps mother-father relationships.

Extensive apprenticeship would have been required to learn to forage for deep roots and tubers, to process them, to make and use bifacial stone tools, to procure and butcher prey, to construct shelters and clothes, and to midwife and care for young. This training probably

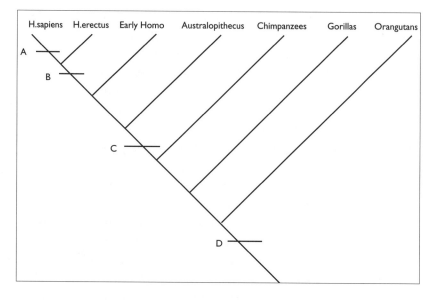

FIGURE 10.6.

Cladogram of putative mother-infant communication in a series of common ancestors. Key: A, full language acquisition; B, mother-infant imitation complex; C, childhood imitation-apprenticeship complex; D, incipient childhood imitation-apprenticeship complex.

took the form of observational learning and explicit teaching by demonstration. The latter form would have relied on the capacity to imitate complex motor sequences with and on objects and to play pretend games. Given the pattern of sexual dimorphism in subsistence activities among both chimpanzees and modern foraging peoples, it seems likely that apprenticeship differed for male and female roles in *Homo erectus*. Males likely relied on prolonged associations with male mentors, probably kin such as uncles, brothers, and fathers, to gain their training. This in turn suggests that infant and juvenile erectines played with gender-specific toys—that is, replicas of the material objects important for subsistence—and they may have played simple games with rules (Parker 1984). Extended juvenility and play probably correlated with prolonged brain development and hence increased brain size (see Fairbanks, this volume).

Because chimpanzees shared a common ancestor with humans approximately six million years ago (e.g., Cronin 1986), we can assume that the earliest hominids displayed symbolic abilities at least as great as those of chimpanzees and the other great apes. Since *Homo erectus* displayed a larger brain and more advanced technology than the earliest hominids, we can assume that it showed greater symbolic abilities than australopithecines and early *Homo* species.

Given the length of apprenticeship for subsistence activities in chimpanzees, we should expect an even more prolonged apprenticeship in a slow-developing, large-brained species such as *Homo erectus*. Western chimpanzees require at least eight years to become proficient in the use of hammers and anvils in nut cracking. Reports indicate that some mothers engage in demonstration teaching of techniques of tool use in nut cracking (Boesch 1991, 1993). Surely, competence in manufacturing Acheulean tools, in butchery, in shelter construction, and in other erectine technologies required a longer apprenticeship than those of earlier hominids. If the end of chimpanzee apprenticeship coincides roughly with third molar eruption, then we might expect erectine apprenticeship to have extended to 14 or 15 years, when third molar eruption occurred in this species.

Finally, we can infer that Acheulean technology required greater planning than did chimpanzee technology. It required location and identification of suitable raw materials for tools, planning for production and storage, and/or transportation of tools to appropriate sites for use. Production of bifacial hand axes implies greater planning than production of chopper tools. Appropriate use of Acheulean tools must have required considerable practice and likely apprenticeship. Some level of symbolic communication regarding the nature and location of raw materials and the use of these tools, if not their production, must have been necessary (Parker and Milbrath 1993).

By reason of homology, we would expect erectines to have engaged in symbolic communication at least as complex as that of great apes. Individuals of all four commonly recognized species of great apes have learned rudimentary sign language or some other form of symbolic communication through human tutelage. These individuals understand and use words to communicate meanings to others and to respond to communications by others (e.g., to ask and answer questions, to refer to and to act on objects) (e.g., Gardner, Gardner, and van Cantford 1989; Miles 1990; Patterson and Cohn

1994; Savage-Rumbaugh et al. 1993). They can even sign simple two-word utterances that carry semantic meanings comparable to those constructed by two-year-old children (e.g., Greenfield and Savage-Rumbaugh 1990). They recognize words as members of semantic categories (e.g., foods and tools or, within foods, fruits and nonfruits). More remarkably, a few individuals of two great apes species have learned symbols through observation and/or teaching by conspecifics (Savage-Rumbaugh et al. 1989).

In modern humans, imitation and tool use are closely associated with the emergence of symbolic communication in children. Likewise, pretend play and object combination are closely associated with the development of early grammar (e.g., Bates 1993; Bates et al. 1979; Greenfield and Savage-Rumbaugh 1990; Greenfield and Smith 1976). Table 10.6 summarizes some correlates of early language development in humans.

The argument by homology that leads us to expect symbolic communication in erectines is further supported by reports of rudimentary forms of imitation and symbolic play in great apes. Several investigators have reported imitation and pretend play in cross-fostered great apes (e.g., Hayes and Hayes 1952; Jensvold and Fouts 1994; Miles et al. 1996; Patterson and Linden 1981). More recently, two fieldworkers have reported them in wild chimpanzees (Matsuzawa 1995; Wrangham 1995a).[6]

I suggest the following stages of language evolution: (1) protosymbolic gestural communication among wild great apes and the earliest hominids, used primarily for locating resources and teaching tool use; (2) simple case grammar in gestures and sounds encoding semantic relations such as agent, action, patient, instrument, and location (typical of three-year-old children; see, e.g., Greenfield and Smith 1976) in erectines, used primarily for describing and locating resources and for describing simple activity routines; and (3) fully grammatical language in modern *Homo sapiens* (Parker 1985).

HETEROCHRONY AND THE EVOLUTION OF COGNITIVE ABILITIES IN HOMINIDS

Heterochrony is the study of processes by which the ontogeny of descendants is changed relative to that of their ancestors (Gould

TABLE 10.6

Relationships between Symbolic Play, Gesture, and Language Acquisition

Symbolic Play	Gesture	Language Acquisition
Contextualized use of objects and symbols	Child recognizes the appropriate use of an object by briefly carrying out an associated activity	Child uses a word as a procedure or part of a routine or game
Temporal decontextualization in symbolic play and symbol use	Child "pretends" to carry out his own familiar activities (e.g., sleeping) outside of their usual context	Child uses a word to anticipate or remember the scheme with which it is typically associated
Decontexualization through role reversal in symbolic play and symbol use	Child "pretends" to carry out his own familiar activities outside of their usual context	Child uses words to designate actions carried out by himself or others or to designate the agents or objects of such actions
Reference with decreased contextual support	Child carries out actions with objects that are inappropriate or are related quite abstractly to the original object	Child uses words to categorize new persons, objects, or events

Source: Bates et al. 1979:177–78.

1977; McKinney and McNamara 1991; Shea 1983a). Paedomorphosis (juvenilization, or arrest of development before the adult stage of ancestors) and peramorphosis (adultification, or growth beyond the adult stage of ancestors) are the two major categories of heterochrony. These patterns can occur through changes in the onset and/or offset of growth and through changes in rates of growth. Peramorphosis results from delayed offset of development (hypermorphosis), acceleration of development, and/or earlier onset of growth (predisplacement). Paedomorphosis can result from earlier offset of growth, reduced rate of development (neoteny), and/or delayed onset of growth (postdisplacement) (McKinney, this volume; McKinney and McNamara 1991).

Modeling of heterochronic processes depends upon life history data—that is, developmental data on the duration of ontogeny and the timing of growth (Shea 1983a). These data are necessary to distinguish changes in onset and offset from changes in developmental rates. Morphologically similar outcomes can result from changed offset and/or onset of growth (i.e., from time hypo- or hypermorphosis) or from changed rates of growth (i.e., from rate hyper- or hypomorphosis) (Shea 1983a). Data on the cognitive development of hominoids come from two sources: from comparisons of monkeys, great apes, and humans and from models of cognitive development in fossil hominids (see Shea, this volume).

How exactly can we compare the abilities of great apes and hominids? Piagetian stages of development provide one useful comparative framework for assessing species cognitive abilities. This framework is comprehensive across a range of cognitive domains (called "series" in the sensorimotor period) including physical causality (objects, space, time, causality) and logical-mathematical reasoning (number, classification) as well as some aspects of social cognition (imitation and moral judgment). The epigenetic sequences of Piagetian stages provide a built-in standard of complexity because within most series or domains, each succeeding stage is constructed on and depends upon the achievements of the previous stage (Parker and Gibson 1979).

Piaget described four periods of intellectual development in human children: (1) the sensorimotor period, from birth to 2 years; (2) the preoperations period, from 2 to 6 years; (3) the concrete operations period, from 6 to 12 years; and (4) the formal operations period, from 12 to 16 years. Each period has two subperiods: the early and late sensorimotor subperiods (comprising stages 1–4 and stages 5 and 6, respectively); the early and late preoperations subperiods (also known as the symbolic and intuitive subperiods); the early and late concrete operations subperiods; and finally the early and late formal operations subperiods. Children traverse these subperiods across the domains and subdomains of physical and logical cognition roughly in parallel (Piaget 1952, 1962; Piaget and Inhelder 1967a).[7] These periods and their subperiods can be diagnosed through a variety of so-called clinical methods that involve posing problems with various materials. Abilities characteristic of the stages of the various series (object concept, space, causality, time, circular reactions, and imitation) that develop during the sensorimotor period have been assessed in several anthropoid species, including cebus monkeys, macaques, and great apes (e.g., Parker 1990). Some studies have assessed preoperational abilities in great apes (Langer, this volume; Parker and McKinney 1999).

As in the case of symbolic abilities, I reason about the evolution of hominid cognition as follows. Because chimpanzees shared a common ancestor with humans approximately six million years ago (e.g., Cronin 1986), we can assume that the earliest hominids displayed cognitive abilities at least as great as those of chimpanzees. Since *Homo erectus* displays a larger brain and more advanced technology than the earliest hominids, we can assume that the erectines showed greater cognitive abilities than australopithecines and early *Homo* species. Because modern humans have, from their origins, displayed larger brains and more advanced technologies than those of *Homo erectus*, we can assume that our cognitive abilities are more advanced than theirs were.

Some general patterns emerge from comparative data on the cognitive abilities of monkeys, great apes, and human children. These are that (1) all species go through roughly the same developmental sequence within each domain; (2) humans go through more stages or subperiods than great apes do, and great apes go through more stages than monkeys do; (3) monkeys, great apes, and humans go through their developmental subperiods at different rates; and (4) within each subperiod, monkeys and great apes go through various sensorimotor series and preoperational domains at different rates (see Langer, this volume). In the sensorimotor period, for example, they complete the object concept series before completing the causality and imitation series (macaques and cebus monkeys fail to complete the imitation series, and macaques fail to complete the causality series).

Great apes complete all six stages of the sensorimotor period (though in an impoverished form) in all domains by three or four years of age, one or two years later than human infants. They traverse the first few stages of most series at about the same rate as human infants, but they traverse the later (fifth and sixth) stages at markedly slower rates in the causality series (Poti' and Spinozzi 1994) and the imitation series (Parker, personal observation; Russon 1996). Great apes progress partway through the first subperiod of the preoperations period—the symbolic subperiod—by adolescence. They show rudimentary symbolic capacities including the ability to draw, to engage in pretend play, and to learn symbols (e.g., Jensvold and Fouts 1994; Miles 1994; Patterson and Cohn 1994). Some individuals achieve the intuitive level of preoperations in logical-mathematical domains. None achieve any subsequent developmental milestones (for reviews, see, e.g., Antinucci 1989; Doré and Dumas 1987; Parker 1990; Parker and McKinney 1999; Tomasello and Call 1997).

Presenting the timing of stages of cognitive development in terms of years of age is one way to compare species. Another, more useful way is to present the timing of these stages in terms of dental development. If we look at cognitive development from this perspective, we see that the longer the period of molar development in a species, the greater the number of cognitive subperiods traversed.

Modern humans begin the molar eruption sequence at about 5 years of age and end at about 18 years; during this period they traverse 7 subperiods of cognitive development. Great apes begin molar eruption at 3 years and end at 10 years; during this time, they traverse 3 or 4 cognitive subperiods. Macaques begin molar development at less than 1.5 years and end at 5.5 years; during this time, they traverse only 1 cognitive subperiod. Preliminary analysis suggests that erectines began molar eruption at about 4.5 years and ended at 14.5 years and traversed 5 subperiods of cognitive development. Table 10.7 summarizes these putative relationships between molar eruption and cognitive development.

The developmental gap between the terminal levels of cognitive development in great apes and modern humans—from the middle of the preoperations period to the end of the formal operations period—is enormous. Cladistic analysis implies that over the course of hominid evolution this gap was filled by the sequential evolution of late preoperational, concrete operational, and formal operational abilities (Parker and Gibson 1979). (This assumes, of course, that all these developmental periods are based on evolutionary changes in the brain.)

Comparative data on the timing of development of the sensorimotor and early preoperations stages in great apes and humans suggest that predisplacement and acceleration in the *rate* of development of

Table 10.7

Putative Relationship between Molar Eruption and Completed Level of Cognitive Development in Selected Anthropoids

Species	First Molar	Second Molar	Third Molar
Pan troglodytes	2 subperiods (LSM)	3 subperiods (EPO)	4 subperiods (LPO)?
H. erectus	3 subperiods (EPO)	4 subperiods (LPO)	5 subperiods (ECO)
H. sapiens	4 subperiods (LPO)	6 subperiods (LCO)	7 subperiods (EFO)

Key to abbreviations:

LSM	*late sensorimotor period*
EPO	*early preoperations period*
LPO	*late preoperations period*
ECO	*early concrete operations period*
LCO	*late concrete operations period*
EFO	*early formal operations period*

sensorimotor and preoperational stages also occurred. These changes took place concomitant with the terminal addition of later stages of cognitive development (i.e., late preoperations, concrete operations, and formal operations). This explains why human infants complete the sensorimotor period by two years of age and the symbolic subperiod of preoperations by four years of age, whereas great apes complete the sensorimotor period by four years of age and the symbolic subperiod by six or eight years of age.

Comparative data on the displacement (or *decalage*) of series suggest that some series changed their developmental pace relative to one another during hominid evolution. There was apparently a "folding over" or developmental alignment of physical and logical-mathematical reasoning during human evolution. This process re-aligned two domains that are displaced or asynchronous relative to each other in nonhuman primates but synchronous in human development (Langer 1993, this volume). Likewise, there was apparently a developmental alignment of all the sensorimotor series during human evolution. This process realigned the fifth and sixth stages of the causality and imitation series relative to the object concept series, so that all the series develop synchronously. This occurred through acceleration and predisplacement of the causality and imitation series.

Evolutionary changes in rates of development were accompanied by changes in the richness and breadth of expression of abilities within series and domains. These changes included (1) expansion of imitation from the facial and gestural mode to the auditory and vocal modality; (2) increases in the number and kinds of object manipulation schemes and the number and kinds of circular reactions; and (3) increases in the kinds of sets and in the logical operations produced on these sets. These elaborations could have occurred only through differentiation and recombination of existing schemes. Evidence for these changes can be found in the greater richness and elaboration of sensorimotor and symbolic schemes of humans as compared with great apes.

Cladistic analysis provides clues to the nature of heterochronic changes in cognitive development that must have occurred between the origins of hominids and the appearance of modern humans. These changes occurred in the following heterochronic categories: (1) terminal addition of the late preoperations and the concrete and formal operations periods of cognitive development (time hypermorphosis); (2) predisplacement of the sensorimotor and preoperations periods of development; (3) acceleration of development of the sensorimotor and early preoperations periods (condensation, or rate hypermorphosis); (4) dissociated heterochrony, or realignment of particular series and domains relative to one another, resulting in synchronous development across series and domains in modern humans; and (5) elaboration of schemes and operations and their breadth and application within series and domains. In other words, the evolution of cognitive development in hominids seems to have occurred through peramorphosis rather than paedomorphosis (McKinney and McNamara 1991; McNamara 1997; Parker 1996b; Parker and Gibson 1979; Parker and McKinney 1999).

Exactly when each new level of ability arose and exactly when various domains evolved relative to one another within hominid evolution is unclear. Identification of the cognitive abilities of specific hominid species is prerequisite to making such a determination. Fortunately, just as they provide a framework for comparative studies of cognition, the Piagetian stages also provide a framework for assessing the cognitive abilities of fossil hominids from tools and other artifacts. Wynn

(1989:59) noted that except for the apparent absence of a discrete stage of development of parallel lines (an affine stage), "the phylogenetic sequence does resemble the ontogenetic sequence, indeed in a manner that I find rather striking."

Wynn (1989) used Piaget's stages of development of spatial cognition (see Piaget and Inhelder 1967b) to analyze Oldowan and Acheulean tools from three sites in East Africa spanning a period of 1.5 million years of hominid evolution. He found support for the following conclusions: (1) Oldowan stone knappers (beginning about 2 MYA) used topological notions of space that are typical of human children in the symbolic substage of the preoperations period. These notions include proximity, separation, and order and are reflected in a series of blows made in proximity to one another and in an ordered series. (2) Early Acheulean stone knappers (beginning about 1.2 MYA) added projective notions of interval and symmetry based on diameter and radius, typical of modern children in the intuitive stage of the preoperations period. These suggest an overall sense of design, though this notion remained internal to the object at hand. (3) Late Acheulean stone knappers (beginning about 300,000 years ago) added the projective notion of parallels and Euclidian notions of cross section. Cross sections are based on perspective taking—that is, an external frame of reference typical of modern children in the concrete operations period.[8]

Wynn's model pegged the highest cognitive attainments of late erectines at the level of concrete operations, at least in the realm of spatial cognition. If this was their highest attainment, we can conclude that they completed the development of this terminal stage by the time their third molar erupted—that is, at about 14 years of age. This compares with an age of about 12 years for the transition from concrete to formal operations in modern humans. Combining this projection with the previous projection regarding the onset of imitation and tertiary circular reactions during infancy, I predict the following parameters of cognitive development in *H. erectus*: (1) sensorimotor period development was completed before 3 years of age—about midway between the 2 years that human children require and the 4 years that chimpanzees seem to require; (2) the symbolic and intuitive subperiods were completed between about 3 and 8 years of age; and (3) concrete operations was completed by about 14 years of age, at the onset of reproductive life.

Symbolic capacities are important for practicing simple subsistence roles in pretend play, and intuitive and concrete capacities are important for apprenticeship in technology. Therefore, it seems likely that *H. erectus* children completed the symbolic subperiod earlier than chimpanzees, perhaps midway between the timing of chimpanzees and humans, by approximately six years. Likewise, they probably completed the intuitive subperiod and entered the concrete operations period by age nine, when their second molars erupted and they entered sex-specific apprenticeship for adult subsistence roles. Table 10.8 compares cognitive development in chimpanzees *(Pan)*, *H. erectus*, and *H. sapiens* based on these projections. (Note that the table gives a rough estimate based on the latest series or domain to develop within each period. It is based on the general intermediacy of development of erectines relative to humans and great apes.)

Various lines of evidence suggest that further stages of cognitive evolution occurred in the transformation of *H. erectus* into archaic *H. sapiens* and of archaic into modern *H. sapiens*. Specifically, archaeological evidence suggests that the highest stage of cognitive development, presumably the capacity for formal operations, evolved with the transition to modern *H. sapiens*. Dramatic increases in the complexity, regional specialization, and historical change of archaeological assemblages suggest that modern *H. sapiens* had the capacity for full-fledged declarative planning based on language and culture (Parker and Milbrath 1993).[9]

CONCLUSION

Comparative data on the life histories of hominids enable us to model a scenario for the social and cognitive development of erectine infants and children. Analyses of paleontological evidence from the skeleton of the Turkana boy suggest that early *Homo erectus* (or *H. ergaster*) marked a turning point in the evolution of hominid life history and behavioral development. This species marked the appearance of fully terrestrial bipedalism, modern limb proportions, increased body size, significantly enlarged brain size, and secondarily altricial neonatality. These changes apparently occurred in conjunction with brain enlargement and a greater dependence on technology, including bifacial tools, shelters, fire, and clothing.

TABLE 10.8

Estimated Age in Years at Completion of Subperiods of Cognitive Development in Selected Anthropoid Species

Subperiod	Macaca	Pan	H. erectus	H. sapiens
Early sensorimotor	0.5	1.0	1	1
Late sensorimotor	2.5–3	3.5	3	2
Symbolic subperiod of preoperations (ESO)	NA	8.0?	6	3.5 or 4
Intuitive subperiod of preoperations (LPO)	NA	10.0?	9	5
Early concrete operations	NA	NA	14	6 or 7
Late concrete operations	NA	NA	NA	12
Early formal operations	NA	NA	NA	16

Concomitant with their secondary altriciality and more extended immaturity, erectines required new adaptations for infant and child care by kindred, especially siblings and grandmothers. Data suggest that they depended on distal communicatory signals and associated displays between infants and caretakers associated with earlier onset of imitation and the extension of imitation to the vocal modality.

New technological modes of subsistence, including excavating and processing roots and tubers, entailed a longer, more complex apprenticeship than that of chimpanzees. This in turn required increased imitation and teaching skills based on more advanced cognition than that of their predecessors. It probably also involved symbolic communication based on simple case grammars of agent, action, instrument, patient, location, and so forth.

Comparative data on cognitive development in great apes and humans suggest that operational intelligence evolved after hominids diverged from great apes and before modern *H. sapiens* diverged from archaic *H. sapiens*. Analysis of spatial concepts involved in making Acheulean tools suggests that erectines were the first hominids to achieve the level of concrete operational intelligence. A model for the stages of cognitive development in erectines was derived by assuming that their development occurred at a pace midway between that of chimpanzees and humans. According to this model, erectines developed their highest stage of cognition, concrete operations, between the ages of 9 and 14 years. This model and other comparative data suggest that cognitive evolution in hominids occurred through peramorphosis (overdevelopment) rather than through paedomorphosis (underdevelopment).

Notes

1. Locomotor and digging behavior may leave traces in the fossil record in the form of footprints or tracking or burrows. Consumption of prey may leave characteristic traces in bone remains, including percussion and cut marks from predator teeth and/or stone tools. Hominid behavior may also leave traces in the archaeological record in the form of tools and other artifacts, built structures, and symbolic representations. If the present trend toward recognition of subtle indicators continues, other behavioral traces may be recognized in the future.

2. Cladistic analysis involves mapping the occurrence of shared derived characters (characters unique to a group of sister species) onto an existing phylogeny or cladogram to reveal the common ancestor that first displayed particular adaptations (e.g., Brooks and McClennan 1993; Wiley 1991). This mapping is done on a cladogram of monkeys, apes, and fossil hominids. A cladogram differs from a phylogeny or family tree in two respects: (1) it represents the dimension of time indirectly and nonquantitatively through the sequence of branching points or speciation events, and (2) it places each species at the end of a branch independently of whether it is living or extinct.

3. One of the major events in the recent history of paleoanthropology was the 1986 announcement of the discovery of a new partial skeleton of *Homo habilis* (OH 62) from Olduvai Gorge in Tanzania (Johanson et al. 1987). The diminutive size and primitive limb proportions of this skeleton are similar to those of the small (gracile) australopithecines, suggesting that early *Homo* species were more apelike than previously imagined. This discovery has stimulated some taxonomists to suggest that there were two species of early *Homo*, *H. habilis* and *H. rudolfensis* (Wood 1992).

4. Piaget (1962) used the term *circular reactions* to characterize repetitive patterns of behavior that develop during infancy. He contrasted the following three categories of circular reactions: (1) primary circular reactions, involving the infant's repeated coordination of actions on his own body, such as repeated thumb sucking; (2) secondary circular reactions, involving the infant's repeated coordination of actions on objects in order to create interesting effects, such as

repeatedly shaking the crib in order to see a mobile move; and (3) tertiary circular reactions, involving the infant's repeated actions with objects in order to see the effects of variations in intensity of actions, such as repeatedly dropping objects from different heights.

5. Pink everted lips may represent an evolutionary ritualization of a behavioral display by young chimpanzees that involves extending and everting the lips while pouting and whimpering. Pink is a highly salient color that is associated with immaturity and with sex. Pink everted lips create a permanent display of infantlike vulnerability. Similar ritualizations have been postulated for many communicatory signals in other species (e.g., Eibl-Ebesfeldt 1975).

6. Pretend play has been reported in several cross-fostered great apes. These reports have been dismissed by some as artifacts of "enculturation" by human caretakers. Therefore, reports of pretend play in wild great apes should be more compelling to skeptics. There is at least one such report: "I watched a lonely boy chimpanzee, eight-year-old Kakama, playing for four hours with a log. He carried it on his back, on his belly, in his groin, on his shoulders. He took it with him up four trees, and down again. He lay in his nest and held it above him like a mother with her baby. And he made a special nest that he didn't use himself, except to put the log in" (Wrangham 1995a:5). According to Piaget (1962), pretend play reveals symbolic levels of cognition. It is also necessary for demonstration teaching (Parker 1996a).

7. Each succeeding period is marked by more complex and powerful intellectual adaptations. Sensorimotor intelligence is limited to a practical understanding of relations among objects. Preoperational intelligence involves manipulation of symbols—for example, in pretend play, drawing, and language. Operational intelligence involves true concepts such as hierarchical classification and conservation of quantities under transformations. It is based on reversibility of mental operations. Formal operational intelligence involves hypothetical-deductive reasoning. It is based on the ability to systematically test hypotheses by holding variables constant.

8. As he noted, Wynn's (1989) interpretation that late Acheulean tool knapping involved a significant shift toward an understanding of perspective and frames of reference goes against the widespread idea of stasis in Acheulean tools. His interpretation of a cognitive advance by late erectines is consistent with evidence for increased brain size in this group (Begun and Walker 1993).

9. Wynn (1989:62) concluded that "projective and Euclidian concepts would appear to be sufficient to account for all of the stone tools archaeologists

know of or, indeed, can imagine. Stone tools of later times are different and, arguably, more specialized in function. But their spatial prerequisites are no more complex than those we can recognize by 300,000 years ago." Whereas concrete operational intelligence may be sufficient to produce Upper Paleolithic stone tools—as Wynn argued—higher levels of cognition may be implied by other kinds of technologies and artifacts.

References

Aarsleff, H.

1976　　An outline of language-origins theory since the Renaissance. In S. Harnad, H. Steklis, and J. Lancaster, eds., *Origins and evolution of language and speech,* 4–13. New York: New York Academy of Sciences.

Abitbol, M. M.

1995　　Lateral view of *Australopithecus afarensis:* Primitive aspects of bipedal positional behavior in the earliest hominids. *Journal of Human Evolution* 28:211–29.

Ahl, V. A.

1993　　Cognitive development in infants prenatally exposed to cocaine. Ph.D. diss., University of California, Berkeley.

Alberch, P., Gould, S. J., Oster, G. F., and Wake, D. B.

1979　　Size and shape in ontogeny and phylogeny. *Paleobiology* 5:296–317.

Altmann, G. T. M., and Steedman, M.

1988　　Interaction with context during human sentence processing. *Cognition* 30:191–238.

Annett, M.

1985　　*Left, right, hand and brain: The right shift theory.* Hillsdale, NJ: Erlbaum.

Antinucci, F., ed.

1989　　*Cognitive structure and development in nonhuman primates.* Hillsdale, NJ: Erlbaum.

Aram, D. M., Ekelman, B., Rose, D., and Whitaker, H.

1985   Verbal and cognitive sequelae following unilateral lesions acquired in early childhood. *Journal of Clinical and Experimental Neuropsychology* 7:55–78.

Asfaw, B., White, T., Lovejoy, O., Latmer, B., Simpson, S., and Suwa, G.

1999   *Australopithecus garhi:* A new species of early hominid from Ethiopia. *Science* 284:629–35.

Balaban, E., Teillet, M. A., and Le Douarin, N.

1988   Application of the quail-chick chimera system to the study of brain development and behavior. *Science* 241:1339–42.

Baldwin, J. D.

1969   The ontogeny of social behaviour of squirrel monkeys *(Saimiri sciureus)* in a seminatural environment. *Folia Primatologica* 11:35–79.

Baldwin, J. D., and Baldwin, J. I.

1974   Exploration and social play in squirrel monkeys *(Saimiri). American Zoologist* 14:303–15.

1976   Effects of food ecology on social play: A laboratory simulation. *Zeitschrift für Tierpsychologie.* 40:1–14.

Baldwin, J. M.

1896   A new factor in evolution. *American Naturalist* 30:354, 442–51, 536–53.

1902   *Development and evolution.* New York: Macmillan.

1915   *Genetic theory of reality.* New York: Putnam.

Bandura, A.

1977   *Social learning theory.* Englewood Cliffs, NJ: Prentice-Hall.

Bard, K. A.

1990   "Social tool use" by free-ranging orangutans: A Piagetian and developmental perspective on the manipulation of an animate object. In S. T. Parker and K. R. Gibson, eds., *"Language" and intelligence in monkeys and apes: Comparative and developmental perspectives,* 356–78. New York: Cambridge University Press.

Bates, E.

1979   *The emergence of symbols.* New York: Academic Press.

1993   Comprehension and production in early language development. *Language comprehension in ape and child* 58:22–41. Chicago: University of Chicago Press.

Bates, E., ed.

1991   Special issue: Cross-linguistic studies of aphasia. *Brain and Language* 41(2).

Bates, E., Benigni, L., Bretherton, I., Camaioni, L., and Volterra, V.

1979   *The emergence of symbols: Cognition and communication in infancy.* New York: Academic Press.

Bates, E., Devescovi, A., Hernandez, A., and Pizzamiglio, L.
1996   Gender priming in Italian. *Perception and Psychophysics* 58(7):992–1004.

Bates, E., Friederici, A., and Wulfeck, B.
1987   Comprehension in aphasia: A crosslinguistic study. *Brain and Language* 32:19–67.

Bates, E., and Goodman, J.
1997   On the inseparability of grammar and the lexicon: Evidence from acquisition, aphasia, and real-time processing. *Language and Cognitive Processes* 12(5–6):507–84.

Bates, E., Thal, D., Aram, D., Eisele, J., Nass, R., and Trauner, D.
1997   From first words to grammar in children with focal brain injury. *Developmental Neuropsychology* 13(3):275–343.

Bates, E., Thal, D., Finlay, B. L., and Clancy, B.
n.d.   Early language development and its neural correlates. In I. Rapin and S. Segalowitz, eds., *Handbook of neuropsychology, vol. 7: Child neurology,* 2d ed., 69–110. Amsterdam: Elsevier. In press.

Bates, E., Thal, D., and Marchman, V.
1991   Symbols and syntax: A Darwinian approach to language development. In N. Krasnegor et al., eds., *Biological and behavioral determinants of language development,* 29–65. Hillsdale, NJ: Erlbaum.

Bateson, P.
1982   Behavioural development and evolutionary processes. In King's College Sociobiology Group, ed., *Current problems in sociobiology,* 133–51. Cambridge: Cambridge University Press.

Beach, F.
1965   The snark is a boojum. In T. E. McGill, ed., *Readings in animal behavior,* 3–14. New York: Holt, Rinehart and Winston.

Begun, D., and Walker, A.
1993   The endocast. In A. Walker and R. Leakey, eds., *The Nariokotome* Homo erectus *skeleton,* 326–58. Cambridge, MA: Harvard University Press.

Bekoff, M.
1974.   Social play in mammals. *American Zoologist* 14:265–436.
1975   The communication of play intention: Are play signals functional? *Semiotica* 15:231–39.

Bellomo, R. V.
1994   Methods of determining early hominid behavioral activities associated with controlled use of fire at FxJj 20 IV at Koobi Fora, Kenya. *Journal of Human Evolution* 27:173–95.

Berrigan, D., Charnov, E. L., Purvis, A., and Harvey, P. H.
1993   Phylogenetic contrasts and the evolution of mammalian life histories. *Evolutionary Ecology* 7:270–78.

Biben, M.

1986   Individual and sex-related strategies of wrestling play in captive squirrel monkeys. *Ethology* 71:229–41.

1989   Effects of social environment on play in squirrel monkeys: Resolving Harlequin's dilemma. *Ethology* 81:72–82.

Blackstone, N. W.

1987   Allometry and relative growth: Pattern and process in evolutionary studies. *Systematic Zoology* 35:76–78.

Blackwell, A., and Bates, E.

1995   Inducing agrammatic profiles in normals: Evidence for the selective vulnerability of morphology under cognitive resource limitation. *Journal of Cognitive Neuroscience* 7(2):228–57.

Blinkov, S., and Glezer, I.

1968   *The human brain in figures and tables.* New York: Plenum.

Bloch, M. N.

1989   Young boys' and girls' play at home and in the community: A cultural-ecological framework. In M. N. Bloch and A. D. Pellegrini, eds., *The ecological context of children's play,* 120–54. Norwood, NJ: Ablex.

Bloom, L., Lightbown, P., and Hood, L.

1974   Imitation in language development: If, when and why? *Cognitive Psychology* 6:380–420.

Blumenschine, R. J., and Selvaggio, M. M.

1994   Percussion marks on bone surfaces as a new diagnostic of hominid behavior. *Nature* 333:763–65.

Boesch, C.

1991   Teaching among wild chimpanzees. *Animal Behavior* 41:530–32.

1993   Aspects of transmission of tool-use in wild chimpanzees. In K. R. Gibson and T. Ingold, eds., *Tools, language and cognition in human evolution,* 171–83. Cambridge: Cambridge University Press.

1996   Three approaches for assessing chimpanzee culture. In A. E. Russon, K. A. Bard, and S. T. Parker, eds., *Reaching into thought: The mind of the great apes,* 404–29. Cambridge: Cambridge University Press.

Boesch, C., and Boesch, H.

1989   Hunting behavior of wild chimpanzees in the Tai National Park. *American Journal of Physical Anthropology* 78:547–73.

Boesch, C., Marchesi, P., Marchesi, N., Fruth, B., and Joulian, F.

1994   Is nut cracking in wild chimpanzees a cultural behavior? *Journal of Human Evolution* 26:54–77.

Bogin, B.

1997   Evolutionary hypotheses for human childhood. *Yearbook of Physical Anthropology* 40:1–27.

Bolk, L.

1926   On the problem of anthropogenesis. *Proceedings of the Sciences Section of the Koninklijk Akademie Wetenschappen* 29:465–75. Amsterdam.

Bonner, J. T.

1980   *The evolution of culture in animals.* Princeton, NJ: Princeton University Press.

Borer, H., and Wexler, K.

1987   The maturation of syntax. In T. Roeper and E. Williams, eds., *Parameter setting*, 123–72. Dordrecht, Netherlands: Reidel.

Bornstein, M. H., and O'Reilly, A. W., eds.

1993   *The role of play in the development of thought.* New Directions for Child Development 59. San Francisco: Jossey-Bass.

Bourgeois, J.-P., Goldman-Rakic, P. S., and Rakic, P.

1994   Synaptogenesis in the prefrontal cortex of rhesus monkeys. *Cerebral Cortex* 4:78–96.

Bourgeois, J.-P., and Rakic, P.

1993   Changes in synaptic density in the primary visual cortex of the macaque monkey from fetal to adult stage. *Journal of Neuroscience* 13:2801–20.

1996   Synaptogenesis in the occipital cortex of macaque monkey devoid of retinal input from early embryonic stages. *European Journal of Neuroscience* 8:942–50.

Bowlby, J.

1969   *Attachment*, vol. 1. New York: Basic Books.

Boyd, R., and Richerson, P. J.

1985   *Culture and the evolutionary process.* Chicago: University of Chicago Press.

1996   Why culture is common, but cultural evolution is rare. *Evolution of social behaviour patterns in primates and man: Proceedings of the British Academy* 88:77–93.

Bozdogan, H.

1988   ICOMP: A new model-selection criterion. In H. H. Bock, ed., *Classification and related methods of data analysis: Proceedings of the First Conference of the International Classification Societies*, 599–608. Amsterdam: North-Holland.

1990   On the information-based measure of covariance complexity and its application to the evaluation of multivariate linear models. *Communications in Statistics: Theory and Methods* 19:221–78.

Braine, M. D. S.

1990   The "natural logic" approach to reasoning. In W. F. Overton, ed., *Reasoning, necessity, and logic: Developmental perspectives*, 135–58. Hillsdale, NJ: Erlbaum.

Bramblett, C. A.

1978   Sex differences in the acquisition of play in juvenile vervet monkeys. In E. O. Smith, ed., *Social play in primates*, 33–48. New York: Academic Press.

Bromage, T.

1987   The biological and chronological maturation of early hominids. *Journal of Human Evolution* 16:257–72.

Bromage, T., and Dean, C.

1985   Reevaluation of the age at death of immature fossil hominids. *Nature* 317:525–27.

Brooks, D., and McLennan, D.

1991   *Phylogeny, ecology, and behavior.* Chicago: University of Chicago Press.

Brown, F., Harris, J., Leakey, R., and Walker, A.

1985   Early *Homo erectus* skeleton from West Lake Turkana, Kenya. *Nature* 316:788–92.

Bruner, J. S.

1972   Nature and uses of immaturity. *American Psychologist* 27:687–708.

1983   *Child's talk.* New York: Norton.

Bub, D. N.

1994   Language activation and positron emission topography: The relative contribution of task-based and stimulus-based effects. In D. C. Gajdusek, G. M. McKhann, and C. L. Bolis, eds., *Evolution and neurology of language*, 124–30. *Discussions in Neuroscience* 10(1–2).

Burghardt, G. M.

1984   On the origins of play. In P. K. Smith, ed., *Play in animals and humans*, 5–41. Oxford: Blackwell.

Burke, A. C.

1989   Development of the turtle carapace: Implications for the evolution of a novel bauplan. *Journal of Morphology* 199:363–78.

Burnham, K. P., and Anderson, D. R.

1992   Data-based selection of an appropriate biological model: The key to modern data analysis. In D. R. McCullough and R. H. Barrett, eds., *Wildlife 2001: Populations*, 16–30. London: Elsevier.

Butler, A. B., and Hodos, W.

1996   *Comparative vertebrate neuroanatomy.* New York: Wiley-Liss.

Byers, J. A., and Walker, C.

1995   Refining the motor training hypothesis for the evolution of play. *American Naturalist* 146:25–40.

Byrne, G., and Soumi, S. J.

1995   Development of activity patterns, social interactions, and exploratory behavior in tufted capuchins *(Cebus apella). American Journal of Primatology* 35:255–70.

Caccone, A., and Powell, J. R.

1989   DNA divergence among hominoids. *Evolution* 43:925–42.

Caine, N., and Mitchell, G.

1979   A review of play in the genus *Macaca*: Social correlates. *Primates* 20:535–46.

Calder, W. A. III.

1984   *Size, function, and life history*. Cambridge, MA: Harvard University Press.

Caplan, D.

1987   *Neurolinguistics and linguistic aphasiology: An introduction*. Cambridge: Cambridge University Press.

Caramazza, A., and Berndt, R.

1985   A multicomponent view of agrammatic Broca's aphasia. In M.-L. Kean, ed., *Agrammatism*, 27–63. Orlando, FL: Academic Press.

Caramazza, A., Berndt, R. S., Basili, A. G., and Koller, J. J.

1981   Syntactic processing deficits in aphasia. *Cortex* 17:333–48.

Caramazza, A., and Zurif, E.

1976   Dissociation of algorithmic and heuristic processes in language comprehension: Evidence from aphasia. *Brain and Language* 3:572–82.

Carnap, R.

1960   *The logical syntax of language*. Paterson, NJ: Littlefield and Adams.

Caro, T. M.

1988   Adaptive significance of play: Are we getting closer? *Trends in Ecology and Evolution* 3:50–54.

1994   *Cheetahs of the Serengeti Plains: Group living in an asocial species*. Chicago: University of Chicago Press.

1995   Short-term costs and correlates of play in cheetahs. *Animal Behavior* 49:333–45.

Caro, T. M., and Hauser, M. D.

1992   Is there teaching in nonhuman animals? *Quarterly Review of Biology* 67(2):151–74.

Case, R.

1992   The role of the frontal lobes in the regulation of cognitive development. *Brain and Cognition* 20:51–73.

Cassirer, E.

1953   *Substance and function*. New York: Dover.

Cates, S. E., and Gittleman, J. L.

1997   Reading between the lines: Is allometric scaling useful? *TREE* 12(9):338–39.

Chalmers, N. R.

1984   Social play in monkeys: Theories and data. In P. K. Smith, ed., *Play in animals and humans*, 119–41. Oxford: Blackwell.

Chalmers, N. R., and Locke-Haydon, J.

1984   Correlations among measures of playfulness and skillfulness in captive common marmosets *(Callithrix jacchus jacchus). Developmental Psychobiology* 17:191–208.

Changeux, J. P., Courrège, P., and Danchin, A.

1973   A theory of the epigenesis of neural networks by selective stabilization of synapses. *Proceedings of the National Academy of Sciences* 70:2974–78.

Changeux, J. P., and Danchin, A.

1976   Selective stabilization of developing synapses as a mechanism for the specification of neuronal networks. *Nature* 264:705–12.

Charnov, E. L.

1991   Evolution of life history variation among female mammals. *Proceedings of the National Academy of Sciences* 88:1134–37.

1993   *Life history invariants: Some explorations of symmetry in evolutionary ecology.* Oxford: Oxford University Press.

Cheney, D. L., and Seyfarth, R. M.

1990   *How monkeys see the world.* Chicago: University of Chicago Press.

Cheverud, J. M., Dow, M. M., and Leutenegger, W.

1985   The quantitative assessment of phylogenetic constraints in comparative analyses: Sexual dimorphism in body weight among primates. *Evolution* 39:1335–51.

Childs, C. P., and Greenfield, P. M.

1980   Informal modes of learning and teaching: The case of Zinacanteco weaving. In N. Warren, ed., *Studies in cross-cultural psychology,* 2:269–316. London: Academic Press.

Chism, J.

1991   Ontogeny of behavior in humans and nonhuman primates: The search for common ground. In J. D. Loy and C. B. Peters, eds., *Understanding behavior,* 90–120. New York: Oxford University Press.

Chomsky, N.

1957   *Syntactic structures.* The Hague: Mouton.

1965   *Aspects of the theory of syntax.* Cambridge, MA: MIT Press.

1980   On cognitive structures and their development: A reply to Piaget. In M. Piatelli-Palmarini, ed., *Language and learning: The debate between Jean Piaget and Noam Chomsky,* 35–54. Cambridge, MA: Harvard University Press.

Chugani, H. T., Hovda, D. A., Villablanca, J. R., Phelps, M. R., and Xu, W.-F.

1991   Metabolic maturation of the brain: A study of local cerebral glucose utilization in the developing cat. *Journal of Cerebral Blood Flow Metabolism* 11:35–47.

Churchland, P. S., and Sejnowski, T. J.

1992   *The computational brain.* Cambridge, MA: MIT Press.

Cliff, A. D., and Ord, J. K.

1981   *Spatial processes: Models and applications.* London: Pion Press.

Cole, L.

1954   The population consequences of life history phenomena. *Quarterly Review of Biology* 29(2):103–37.

Collier, G. A.

1990   *Seeking food and seeking money: Changing productive relations in a highland Mexican community.* Geneva: United Nations Research Institute for Social Development.

Conroy, G. C.

1997   *Reconstructing human origins.* New York: Norton.

Conroy, G. C., and Kuykendall, K.

1995   Paleopediatrics: Or when did human infants really become human? *American Journal of Physical Anthropology* 98:121–34.

Coppinger, R., and Schneider, R.

1995   Evolution of working dogs. In J. Serpell, ed., *The domestic dog: Its evolution, behavior, and interactions with people,* 21–47. Cambridge: Cambridge University Press.

Corballis, M. C., and Morgan, M. J.

1978   On the biological basis of human laterality: I. Evidence for a maturational left-right gradient. *Behavioral and Brain Sciences* 1:261–69.

Cosmides, L., and Toobey, J.

1991   Cognitive adaptations for social exchange. In J. Barkow, L. Cosmides, and J. Toobey, eds., *The adapted mind,* 163–228. New York: Oxford University Press.

Count, E. W.

1947   Brain and body weight in man: Their antecedents in growth and evolution. *Annals of the New York Academy of Sciences* 46:993–1122.

Courchesne, E., Townsend, J., and Chase, C.

1995   Neurodevelopmental principles guide research on developmental psychopathologies. In D. Cicchetti and D. Cohen, eds., *A manual of developmental psychopathology,* vol. 2, 195–226. New York: Wiley.

Crain, S.

1991   Language acquisition in the absence of experience. *Behavioral and Brain Sciences* 14:597–611.

Crick, F.

1994   *The astonishing hypothesis: The scientific search for the soul.* New York: Scribner.

Cronin, J.

1986   Molecular insights into the nature and timing of ancient speciation events: Correlates with paleoclimate and paleogeography. *South African Journal of Science* 82:83–85.

Custance, D. M., Whiten, A., and Bard, K. A.

1995   Can young chimpanzees *(Pan troglodytes)* imitate arbitrary actions? Hayes and Hayes (1952) revisited. *Behavior* 132:837–59.

Damasio, A. R.

1994a   *Descartes' error.* New York: G. P. Putnam.

1994b   Descartes' error and the future of human life. *Scientific American* 271(4):144.

Damasio, H., Grabowski, T. J., Tranel, D., Hichwa, R. D., et al.

1996   A neural basis for lexical retrieval. *Nature* 381(6585):810.

Darwin, C.

1859   *The origin of species.* London: John Murray.

1871   *The descent of man and selection with respect to sex.* New York: Modern Library.

Davidoff, J., and De Bleser, R.

1994   Impaired picture recognition with preserved object naming and reading. *Brain and Cognition* 24(1):1–23.

Deacon, T. W.

1988   Human brain evolution: II. Embryology and brain allometry. In H. Jerison and I. Jerison, eds., *Intelligence and evolutionary biology,* 383–415. Berlin: Springer-Verlag.

1990a   Problems of ontogeny and phylogeny in brain size evolution. *International Journal of Primatology* 11:237–82.

1990b   Rethinking mammalian brain evolution. *American Zoologist* 30:629–705.

1990c   Fallacies of progression in theories of brain-size evolution. *International Journal of Primatology* 11:193–236.

1995   On telling growth from parcellation in brain evolution. In E. Alleva et al., eds., *Behavioral brain research in naturalistic and seminaturalistic settings.* NATO ASI Series. Dordrecht, Netherlands: Kluwer Academic Press.

1997   *The symbolic species: The coevolution of language and the brain.* New York: Norton.

Deacon, T., Dinsmore J., Costantini, L., Ratliff, J., and Isacson, O.

1998   Blastula-stage stem cells can differentiate into dopaminergic and serotonergic neurons after transplantation. *Experimental Neurology* 149:28–41.

Deacon, T., Pakzaban, P., Burns, L., Dinsmore, J., and Isacson, O.

1994   Cytoarchitectonic development, axon-glia relationships, and long-distance axon growth of porcine striatal xenografts in rats. *Experimental Neurology* 130:151–67.

de Beer, G.

1930   *Embryology and evolution.* Oxford: Clarendon Press.

De Bleser, R., and Luzzatti, C.

1994   Morphological processing in Italian agrammatic speakers' syntactic implementation of inflectional morphology. *Brain and Language* 46(1):21–40.

Desmond, A.

1982   *Archetypes and ancestors.* Chicago: University of Chicago Press.

D'Esposito, M., Detre, J., Alsop, D., Shin, R., Atlas, S., and Grossman, M.

1995   The neural basis of the central executive system of working memory. *Science* 378:279–81.

Dettlaff, T. A., Ignatieva, G. M., and Vassetsky, S. G.

1987   The problem of time in developmental biology: Its study by the use of relative characteristics of developmental duration. *Soviet Scientific Research in Physiology and General Biology* 1:1–88.

Dickerson, J. W. T., and Dobbing, J.

1967   Prenatal and postnatal growth and development of the central nervous system of the pig. *Proceedings of the Royal Society B* 166:384–95.

Dobbing, J., and Sands, J.

1973   Quantitative growth and development of human brain. *Archives of Disease in Childhood* 48:757–67.

Dolle, P., Dierich, A., LeMeur, M., Schimmang, T., Schuhbaur, B., Chambon, P., and Duboule, D.

1993   Disruption of the Hoxd-13 gene induces localized heterochrony leading to mice with neotenic limbs. *Cell* 75:431–41.

Donald, M.

1991   *Origins of the modern mind.* Cambridge, MA: Harvard University Press.

Doré, F. Y., and Dumas, C.

1987   Psychology of animal cognition: Piagetian studies. *Psychological Bulletin* 102(2):219–33.

Doré, F. Y., and Goulet, S.

1998   The comparative analysis of object knowledge. In J. Langer and M. Killen, eds., *Piaget, evolution and development,* 55–72. Mahwah, NJ: Erlbaum.

Edelman, G. M.

1987   *Neural Darwinism: The theory of neuronal group selection.* New York: Basic Books.

Eggleton, P., and Vane-Wright, R. I.

1994   Some principles of phylogenetics and their implications for comparative biology. In P. Eggleton and R. I. Vane-Wright, eds., *Phylogenetics and ecology,* 345–66. London: Academic Press.

Eibl-Ebesfeldt, I.

1975   *Ethology: The biology of behavior.* 2d ed. New York: Holt, Rinehart and Winston.

Eisele, J., and Aram, D.

1995   Lexical and grammatical development in children with early hemisphere damage: A cross-sectional view from birth to adolescence. In P. Fletcher and B. MacWhinney, eds., *The handbook of child language*, 664–89. Oxford: Blackwell.

Ekstig, B.

1994   Condensation of developmental stages and evolution. *Bioscience* 44:158–64.

Elman, J. L.

1990   Finding structure in time. *Cognitive Science* 14:179–211.

1991   Distributed representations, simple recurrent networks, and grammatical structure. *Machine Learning* 7:195–225.

1993   Learning and development in neural networks: The importance of starting small. *Cognition* 48:71–99.

Elman, J. L., Bates, E., Johnson, M., Karmiloff-Smith, A., Parisi, D., and Plunkett, K.

1996   *Rethinking innateness: A connectionist perspective on development.* Cambridge, MA: MIT Press.

Erhard, P., Kato, T., Strick, P. L., and Ugurbil, K.

1996   Functional MRI activation pattern of motor and language tasks in Broca's area. Abstract. *Society for Neuroscience* 22:260.4.

Fagen R.

1981   *Animal play behaviour.* New York: Oxford University Press.

1993   Primate juveniles and primate play. In M. E. Pereira and L. A. Fairbanks, eds., *Juvenile primates*, 182–96. New York: Oxford University Press.

Fairbanks, L. A.

1990   Reciprocal benefits of allomothering for female vervet monkeys. *Animal Behavior* 40:553–62.

1993a   Risk taking in juvenile vervet monkeys. *Behaviour* 124:57–72.

1993b   Juvenile vervet monkeys: Establishing relationships and practicing skills for the future. In M. E. Pereira and L. A. Fairbanks, eds., *Juvenile Primates*, 211–27. New York: Oxford University Press.

1996   Individual differences in maternal style: Causes and consequences for mothers and offspring. *Advances in the Study of Behavior* 25:579–611.

Fairbanks, L. A., and McGuire, M. T.

1985   Relationships of vervet mothers with sons and daughters from one through three years of age. *Animal Behavior* 33:40–50.

1988   Long-term effects of early mothering behavior on responsiveness to the environment in vervet monkeys. *Developmental Psychobiology* 21:711–74.

Falmagne, R. J.

1990   Language and the acquisition of logical knowledge. In W. F. Overton, ed., *Reasoning, necessity, and logic: Developmental respectives*, 111–34. Hillsdale, NJ: Erlbaum.

Farah, M. J.

1991   Patterns of co-occurrence among the associative agnosias: Implications for the nature of visual object representation. *Cognitive Neuropsychology* 8(1):1–19.

Federmeier, K., and Bates, E.

1997   Contexts that pack a punch: Lexical class priming of picture naming. *Center for Research in Language Newsletter* 11(2). La Jolla: University of California, San Diego.

Fedigan, L.

1972   Social and solitary play in a colony of vervet monkeys *(Cercopithecus aethiops)*. *Primates* 13:347–64.

Feldman, H., Holland, A., Kemp, S., and Janosky, J.

1992   Language development after unilateral brain injury. *Brain and Language* 42:89–102.

Finlay, B. L., and Darlington, R. B.

1995   Linked regularities in the development and evolution of mammalian brains. *Science* 268:1578–84.

Fiske, A. P., Kitayama, S., Markus, H., and Nisbett, R. E.

1998   The cultural matrix of social psychology. In D. Gilbert, S. Fiske, and G. Lindzey, eds., *Handbook of social psychology* 4:915–81. New York: McGraw Hill.

Fodor, J. A.

1983   *The modularity of mind: An essay on faculty psychology.* Cambridge, MA: MIT Press.

Fodor, J. A., Bever, T. G., and Garrett, M. F.

1974   *The psychology of language: An introduction to psycholinguistics and generative grammar.* New York: McGraw-Hill.

Foley, R.

1987   *Another unique species: Patterns in human evolutionary ecology.* New York: Longman.

Fontaine, R. P.

1994   Play as physical flexibility training in five ceboid primates. *Journal of Comparative Psychology* 108:203–12.

Fouts, R. S., Hirsch, A. D., and Fouts, D. H.

1982   Cultural transmission of a human language in a chimpanzee mother-infant relationship. In H. E. Fitzgerald, J. A. Mullins, and P. Gage, eds., *Psychological perspectives,* 159–93. Child Nurturance Series, vol. 3. New York: Plenum.

Fragaszy, D. M., and Adams-Curtis, L. E.

1991   Generative aspects of manipulation in tufted capuchin monkeys. *Journal of Comparative Psychology* 105:387–97.

Frantz, G. D., Weimann, J. M., Levin, M. E., and McConnell, S. K.
1994   Otx1 and Otx2 define layers and regions in developing cerebral cortex and cerebellum. *Journal of Neuroscience* 14:5725–40.

Freud, A.
1953   *On aphasia: A critical study.* New York: International Universities Press.
[1891]

Friederici, A. D., and Schriefers, H.
1994   The nature of semantic and morphosyntactic context effects on word recognition in young healthy and aphasic adults. *Linguistische Berichte* 6:9–32.

Friedlander, M. J., Martin, K. A. C., and Wassenhove-McCarthy, D.
1991   Effects of monocular visual deprivation on geniculocortical innervation of area 18 in cat. *Journal of Neuroscience* 11:3268–88.

Frost, D. O.
1982   Anomalous visual connections to somatosensory and auditory systems following brain lesions in early life. *Brain Research* 255(4):627–35.
1990   Sensory processing by novel, experimentally induced cross-modal circuits. *Annals of the New York Academy of Sciences* 608:92–109; discussion 109–12.

Futuyma, D. J.
1986   *Evolutionary biology.* 2d ed. Sunderland, MA: Sinnauer Associates.

Gardner, B. T., and Gardner, R. A.
1969   Teaching sign language to a chimpanzee. *Science* 165:664–72.

Gardner, R. A., Gardner, B. T., and van Cantford, T.
1989   *Teaching sign language to chimpanzees.* Albany: State University of New York Press.

Garrett, M.
1980   Levels of processing in sentence production. In B. Butterworth, ed., *Language production, vol. 1: Speech and talk.* New York: Academic Press.
1992   Disorders of lexical selection. Cognition 42:143–80.

Gebo, D. L.
1996   Climbing, brachiation, and terrestrial quadrupedalism: Historical precursors of hominid bipedalism. *American Journal of Physical Anthropology* 101:55–92.

German, R. Z., and Myers, L. L.
1989   The role of time and size in ontogenetic allometry: 1. Review. *Growth, Development, and Aging* 53:101–6.

Gesell, A.
1945   *The embryology of behavior.* Westport, CT: Greenwood Publishers.

Gibbons, A.
1997   Why life after menopause? *Science* 276:536.

Gibson, K. R.

1990   New perspectives on instincts and intelligence: Brain size and the emergence of hierarchical mental construction skills. In S. T. Parker and K. R. Gibson, eds., *"Language" and intelligence in monkeys and apes*, 97–128. New York: Cambridge University Press.

1991   Myelination and brain development: A comparative perspective on questions of neoteny, altriciality, and intelligence. In K. R. Gibson and A. C. Petersen, eds., *Brain maturation and cognitive development: Comparative and cross-cultural perspectives*, 29–64. New York: Aldine de Gruyter.

1993   Tool use, language and social behavior in relationship to information processing capacities. In K. R. Gibson and T. Ingold, eds., *Tools, language, and cognition in human evolution*, 251–270. Cambridge: Cambridge University Press.

Gibson, K. R., and Ingold, T., eds.

1993   *Tools, language and cognition in human evolution*. Cambridge: Cambridge University Press.

Gittleman, J. L.

1993   Carnivore life histories: A re-analysis in the light of new models. In N. Dunstone and M. L. Gorman, eds., *Mammals as predators*, 65–86. Oxford: Oxford University Press.

1994a   Are pandas successful specialists or evolutionary failures? *BioScience* 44:456–64.

1994b   Female brain size and parental care in carnivores. *Proceedings of the National Academy of Sciences* 91:5495–97.

Gittleman, J. L., Anderson, C. G., Kot, M., and Luh, H.-K.

1996a   Phylogenetic lability and rates of evolution: Comparison of behavioral, morphological and life history traits. In E. P. Martins, ed., *Phylogenies and the comparative method in animal behavior*, 166–205. New York: Oxford University Press.

1996b   Comparative tests of evolutionary lability and rates using molecular phylogenies. In P. H. Harvey et al., eds., *New uses for new phylogenies*, 289–307. Oxford: Oxford University Press.

Gittleman, J. L., and Kot, M.

1991   Adaptation: Statistics and a null model for estimating phylogenetic effects. *Systematic Zoology* 39:227–41.

Gittleman, J. L., and Luh, H.-K.

1992   Comparing comparative methods. *Annual Review of Ecology and Systematics* 23:383–404.

Glickman, S. E., and Sroges, R. W.

1966   Curiosity in zoo animals. *Behaviour* 26:151–88.

Godfrey, L. R., King, S. J., and Sutherland, M. R.

1998   Heterochronic approaches to the study of locomotion. In E. Strasser, J. Fleagle, A. Rosenberger, and H. M. McHenry, eds., *Primate Locomotion*, 277–307. New York: Plenum.

Godfrey, L. R., and Sutherland, M. R.

1994   What's growth got to do with it? Process and product in the evolution of ontogeny. *Journal of Human Evolution* 29:405–31.

1995a   Flawed inference: Why size-based tests of heterochronic processes do not work. *Journal of Theoretical Biology* 172:43–61.

1995b   What's growth got to do with it? Process and product in the evolution of ontogeny. *Journal of Human Evolution* 29:405–31.

1996   Paradox of peramorphic paedomorphosis: Heterochrony and human evolution. *American Journal of Physical Anthropology* 99:17–42.

Gold, E. M.

1967   Language identification in the limit. *Information and Control* 16:447–74.

Goldizen, A. W.

1990   A comparative perspective on the evolution of tamarin and marmoset social systems. *International Journal of Primatology* 11:63–83.

Goldowitz, D.

1989   Cell allocation in mammalian CNS formation: Evidence from murine interspecies aggregation chimeras. *Neuron* 3:705–13.

Goodall, J.

1986   *The chimpanzees of Gombe: Patterns of behavior.* Cambridge, MA: Harvard University Press.

1995   The acquisition of culture in young chimpanzees. Lecture presented at the University of Southern California.

Goodenough, W.

1981   *Language, culture, and society.* 2d ed. Menlo Park: Benjamin Cummings.

Goodglass, H.

1993   *Understanding aphasia.* San Diego: Academic Press.

Goodwin, B. C.

1984   Changing from an evolutionary to a generative paradigm in biology. In J. W. Pollard, ed., *Evolutionary theory: Paths into the future,* 99–120. Chichester, UK: Wiley.

Goodwin, D., Bradshaw, J., and Wickens, S.

1997   Paedomorphosis affects agnostic visual signals of domestic dogs. *Animal Behavior* 53:297–304.

Gopnik, M., and Crago, M. B.

1991   Familial aggregation of a developmental language disorder. *Cognition* 39:1–50.

Gottlieb, G.

1992   *Individual development and evolution.* New York: Oxford University Press.

Gould, L.

1990   The social development of free-ranging infant *Lemur catta* at Berenty Reserve, Madagascar. *International Journal of Primatology* 11:297–318.

Gould, S. J.

1966   Allometry and size in ontogeny and phylogeny. *Biological Reviews* 41:587–640.

1968   Ontogeny and the explanation of form: An allometric analysis. *Paleontological Society Memoirs* 2:81–98.

1969   An evolutionary microcosm: Pleistocene and recent history of the land snail *Poecilozonites* in Bermuda. *Bulletin of the Museum of Comparative Zoology* 138:407–532.

1975   Allometry in primates, with emphasis on scaling and evolution of the brain. In F. S. Szalay, ed., *Approaches to primate paleobiology*, 244–92. *Contributions to Primatology* 5. Basel, Switzerland: Karger.

1977   *Ontogeny and phylogeny*. Cambridge, MA: Harvard University Press.

1981   *The mismeasure of man*. New York: Norton.

1984   Relationship of individual and group care. *Human Development* 27:233–39.

1996a   Creating the creators. *Discover* 17(10):43–54.

1996b   *Full house*. New York: Harmony Books.

Govindarajulu, P., Hunte, W., Vermeer, L. A., and Horrocks, J. A.

1993   The ontogeny of social play in a feral troop of vervet monkeys (*Cercopithecus aethiops sabaeus*): The function of early play. *International Journal of Primatology* 14:701–19.

Greenfield, P. M.

1984   A theory of the teacher in the learning activities of everyday life. In B. Rogoff and J. Lave, eds., *Everyday cognition: Its development in social context*, 117–38. Cambridge, MA: Harvard University Press.

1991   Language, tools and brain: The ontogeny and phylogeny of hierarchically organized sequential behavior. *Behavior and Brain Science* 14:531–95.

1999   Cultural change and human development. In E. Turiel, ed., *Development, evolution, and culture*, 37–59. San Francisco: Jossey-Bass.

2000   Culture and universals integrating social and cognitive development. In L. Nucci, G. Saxe, and E. Turiel, eds., *Culture, thought, and development*. Mahwah, NJ: Erlbaum.

Greenfield, P. M., Brazelton, T. B., and Childs, C.

1989   From birth to maturity in Zinacantan: Ontogenesis in cultural context. In V. Bricker and G. Gossen, eds., *Ethnographic encounters in southern Mesoamerica: Celebratory essays in honor of Evon Z. Vogt*, 177–216. Albany: Institute of Mesoamerican Studies, State University of New York.

Greenfield, P. M., and Childs, C. P.

1977   Weaving, color terms, and pattern representation: Cultural influences and cognitive development among the Zinacantecos of southern Mexico. *Inter-American Journal of Psychology* 11:23–48.

1991   Developmental continuity in biocultural context. In R. Cohen and A. W. Siegel, eds., *Context and development*, 135–59. Hillsdale, NJ: Erlbaum.

Greenfield, P. M., and Lave, J.

1982   Cognitive aspects of informal education. In D. Wagner and H. Stevenson, eds., *Cultural perspectives on child development*, 181–207. San Francisco: Freeman.

Greenfield, P. M., Maynard, A. E., and Childs, C. P.

1997   History, culture, learning, and development. Paper presented at the annual meeting of the Society for Research in Child Development, Washington, D.C.

Greenfield, P. M., and Savage-Rumbaugh, E. S.

1990   Grammatical combination in *Pan paniscus:* Processes of learning and invention in the evolution and development of language. In S. T. Parker and K. R. Gibson, eds., *"Language" and intelligence in monkeys and apes*, 540–48. New York: Cambridge University Press.

Greenfield, P. M., and Smith, J. H.

1976   *The structure of communication in early language development.* New York: Academic Press.

Greenough, W. T., Black, J. E., and Wallace, C. S.

1993   Experience and brain development. In M. Johnson, ed., *Brain development and cognition: A reader*, 290–322. Oxford: Blackwell.

Greenough, W. T., McDonald, J. W., Parnisari, R. M., and Camel, J. E.

1986   Environmental conditions modulate degeneration and new dendrite growth in cerebellum of senescent rats. *Brain Research* 380:136–43.

Grodzinsky, Y.

1990   *Theoretical perspectives on language deficits.* Cambridge, MA: MIT Press.

Grosjean, F.

1980   Spoken word recognition processes and the gating paradigm. *Perception and Psychophysics* 28:267–83.

Grosjean, F., Dommergues, J-Y., Cornu, E., Guillelmon, D., and Besson, C.

1994   The gender-marking effect in spoken word recognition. *Perception and Psychophysics* 56(5):590–98.

Gurjanov, M., Lukatela, G., Moskovljevic, J., and Turvey, M. T.

1985   Grammatical priming of inflected nouns by inflected adjectives. *Cognition* 19:55–71.

Haeckel, E.

1874   *The evolution of man*, vol. 1. Akron, OH: Werner Company.

Hager, L.

1991   The evidence for sex differences in the hominid fossil record. In D. Walde and N. Willows, eds., *The archaeology of gender*, 46–49. Calgary, Alberta: University of Calgary Archaeological Association.

Haining, R.

1994   Diagnostics for regression modeling in spatial econometrics. *Journal of Regional Science* 34:325–41.

Hall, B. K.

1992   *Evolutionary developmental biology.* London: Chapman and Hall.

Hall, B. K., and Miyake, T.

1995   How do embryos measure time? In K. J. McNamara, ed., *Evolutionary change and heterochrony,* 3–20. Chichester, UK: Wiley.

Hall, G. S.

1897   A study of fears. *American Journal of Psychology* 8:147–249.

1904   *Adolescence: Its psychology and its relations to physiology, anthropology, sociology, sex, crime, religion and education,* vols. 1 and 2. New York: Appleton.

Hannah, A. C., and McGrew, W. C.

1987   Chimpanzees using stones to crack open oil palm nuts in Liberia. *Primates* 28:31–46.

Harnad, S., Steklis, H., and Lancaster, J., eds.

1976   *Origins and evolution of language and speech.* New York: New York Academy of Sciences.

Hartwig-Scherer, S., and Martin, R. D.

1992   Allometry and prediction in hominoids: A solution to the problem of intervening variables. *American Journal of Physical Anthropology* 88:37–57.

Harvey, P. H.

1990   Life-history variation: Size and mortality patterns. In C. J. D. Rousseau, ed., *Primate life history and evolution,* 81–88. New York: Wiley-Liss.

Harvey, P. H., Brown, A. J. L., Smith, J. M., and Nee, S., eds.

1996   *New uses for new phylogenetics.* Oxford: Oxford University Press.

Harvey, P. H., and Clutton-Brock, T. H.

1985   Life history variation in primates. *Evolution* 39:559–81.

Harvey, P. H., and Krebs, J. R.

1990   Comparing brains. *Science* 249:140–46.

Harvey, P. H., Martin, L., and Clutton-Brock, T.

1986   Life histories in comparative perspective. In B. B. Smuts et al., eds., *Primate societies,* 181–96. Chicago: University of Chicago Press.

Harvey, P. H., and Pagel, M. D.

1991   *The comparative method in evolutionary biology.* Oxford: Oxford University Press.

Harvey, P. H., Read, A. F., and Promislow, D. E. L.

1989   Life history variation in placental mammals: Unifying the data with theory. *Oxford Surveys in Evolutionary Biology* 6:13–31.

Harwerth, R. S., Smith, E. L. III, Duncan, G. C., Crawford, M. L. J., and von Noorden, G. K.

1986   Multiple sensitive periods in the development of the primate visual system. *Science* 232:235–37.

Hauser, M., and Fairbanks, L. A.

1988   Mother-offspring conflict in vervet monkeys: Variation in response to ecological conditions. *Animal Behavior* 36:802–13.

Hausler, M., and Schmidt, P.

1995   Comparison of the pelves of Sts 14 and AL 288-1: Implications for birth and sexual dimorphism in australopithecines. *Journal of Human Evolution* 29:363–83.

1997   Assessing the pelvis of AL 288-1: A reply to Wood and Quinney. *Journal of Human Evolution* 32:99–102.

Hawkes, K., O'Connell, J., and Blurton-Jones, N.

1997   Hadza women's time allocation: Offspring, provisioning, and the evolution of long postmenopausal life spans. *Current Anthropology* 18(4):551–78.

Hayes, K., and Hayes, C.

1952   Imitation in a home-raised chimpanzee. *Journal of Comparative and Physiological Psychology* 45:450–59.

Head, H.

1926   *Aphasia and kindred disorders of speech.* Cambridge: Cambridge University Press.

Heilman, K. M., and Scholes, R. J.

1976   The nature of comprehension errors in Broca's conduction and Wernicke's aphasics. *Cortex* 12:258–65.

Heinzelin, J., Clark, J. D., White, T., Hart, W., Renne, P., WoldeGabriel, G., Beyene, Y., and Vrba, E.

1999   Environment and behavior of 2.5-million-year-old Bouri hominids. *Science* 284:625–29.

Helluy, S., Sandeman, R., Beltz, B., and Sandeman, D.

1993   Comparative brain ontogeny of the crayfish and clawed lobster: Implications of direct and larval development. *Journal of Comparative Neurology* 335:343–54.

Hennig, W.

1979   *Phylogenetic systematics.* Urbana: University of Illinois Press.

Hernandez, A. E., and Bates, E.

1994   Interactive activation in normal and brain-damaged individuals: Can context penetrate the lexical 'module'? *Linguistische Berichte* 6:145–67.

Hernandez, A. E., Fennema-Notestine, C., and Udell, C.

1995   Selective or exhaustive access with unambiguous words? A chronometric study of lexical and sentential context effects. Unpublished manuscript.

Herron, D., and Bates, E.

1997   Sentential and acoustic factors in the recognition of open- and closed-class words. *Journal of Memory and Language* 37(2):217–39.

Hickok, G., Kritchevsky, M., Bellugi, U., and Klima, E. S.

1995   The role of the left frontal operculum in sign language aphasia: Clues to the function of Broca's area. Abstract. *Brain and Language* 51(1):179–81.

Hillert, D., and Bates, E.

1996   *Morphological constraints on lexical access: Gender priming in German.* Center for Research in Language, Technical Report no. 9601. La Jolla: University of California, San Diego.

Hofman, M. A.

1983   Energy metabolism, brain size and longevity in mammals. *Quarterly Review of Biology* 58:495–512.

Holland, P., Ingham, P., and Krauss, S.

1992   Development and evolution: Mice and flies head to head. *Nature* 358:627–28.

Holt, A. B., Cheek, D. B., Mellits, E. D., and Hill, D. E.

1975   Brain size and the relation of the primate to the nonprimate. In D. B. Cheek, ed., *Fetal and postnatal cellular growth: Hormones and nutrition.* New York: Wiley.

Horn, H.

1978   Optimal tactics of reproduction and life history. In J. R. Krebs and N. B. Davies, eds., *Behavioral ecology: An evolutionary approach,* 411–29. Oxford: Blackwell.

Horrocks, J. A., and Hunte, W.

1983   Maternal rank and offspring rank in vervet monkeys: An appraisal of the mechanisms of rank acquisition. *Animal Behavior* 31:772–82.

Howes, C., and Matheson, C. C.

1992   Sequences in the development of competent play with peers: Social and social pretend play. *Developmental Psychology* 28:961–74.

Hughes, F. P.

1991   *Children, play, and development.* Boston: Allyn and Bacon.

Humphreys, A. P., and Smith, P. K.

1984   Rough and tumble in preschool and playground. In P. K. Smith, ed., *Play in animals and humans,* 241–66. Oxford: Blackwell.

Hunt, K. D.

1994   The evolution of human bipedality: Ecology and functional morphology. *Journal of Human Evolution* 26:183–202.

Huttenlocher, P. R.

1979   Synaptic density in human frontal cortex: Developmental changes and effects of aging. *Brain Research* 163:195–205.

1990   Morphometric study of human cerebral cortex development. *Neuropsychologia* 28:517–27.

Huxley, J. S.

1932   *Problems of relative growth.* London: MacVeagh.

Ikeo, M. (producer), Jones, J. (director), and Niio, G. (director).

1993   *Can chimps talk?* Film. Boston, MA: NOVA.

Inhelder, B., and Piaget, J.

1958   *The growth of logical thinking from childhood to adolescence.* New York: Basic Books.

Isacson, O., and Deacon, T.

1996   Specific axon guidance factors persist in the adult brain as demonstrated by pig neuroblasts transplanted to the rat. *Neuroscience* 75:827–37.

1997   Neural transplantation studies reveal the brain's capacity for continuous reconstruction. *Trends in Neuroscience* 20:477–82.

Itani, J., and Nishimura, A.

1973   The study of infrahuman culture in Japan. In E. Menzel, ed., *Symposia of the Fourth International Congress of Primatology* 1:26–50. Basel: Karger.

Jacobs, B., Chugani, H. T., Allada, V., Chen, S., Phelps, M. E., Pollack, D. B., and Raleigh, M. J.

1995   Developmental changes in brain metabolism in sedated rhesus macaques and vervet monkeys revealed by positron emission tomography. *Cerebral Cortex* 3:222–33.

Jacobsen, T. A.

1984   The construction and regulation of early structures of logic: A cross-cultural study of infant cognitive development. Ph.D. diss., University of California, Berkeley.

Jaeger, J. J., Lockwood, A. H., Kemmerer, D. L., van Valin, R. D., Murphy, B. W., and Khalak, H. G.

1996   A positron emission tomographic study of regular and irregular verb morphology in English. *Language* 72(3):451–97.

Janson, C. H., and van Schaik, C. P.

1993   Ecological risk aversion in juvenile primates: Slow and steady wins the race. In M. E. Pereira and L. A. Fairbanks, eds., *Juvenile primates,* 57–74. New York: Oxford University Press.

Jarema, G., and Friederici, A.

1994   Processing articles and pronouns in agrammatic aphasia: Evidence from French. *Brain and Language* 46(4):683–94.

Jensvold, M. L. A., and Fouts, R. S.

1994   Imaginary play in chimpanzees. *Human Evolution* 8:217–27.

Jerison, H. J.

1973   *Evolution of the brain and intelligence.* New York: Academic Press.

Johanson, D., Masao, F. T., Eck, G., White, T. D., Walter, R. C., Kimble, W. H., Asfaw, B., Manega, P., Ndessokia, P., and Suwa, G.

1987   A new partial skeleton of *Homo habilis* from Olduvai Gorge, Tanzania. *Nature* 327:205–9.

Johanson, D., White, T., et al.

1978   A new species of the genus *Australopithecus* (Primates: Hominidae) from the Pliocene of eastern Africa. *Kirtlandia* 28:1–14.

Johnson, J. S., and Newport, M.

1989   Critical period effects in second language learning: The influence of maturational state on the acquisition of English as second language. *Cognitive Psychology* 21:60–99.

Johnson, M. H.

1997   *Developmental cognitive neuroscience: An introduction.* Cambridge, MA: Blackwell.

Johnson, M. H., ed.

1993   *Brain development and cognition: A reader.* Oxford: Blackwell.

Jones, D. S.

1988   Sclerochronology and the size vs. age problem. In M. L. McKinney, ed., *Heterochrony in evolution: A multidisciplinary approach,* 93–108. New York: Plenum.

Jordan, M. I.

1986   *Serial order: A parallel distributed processing approach.* ICS Technical Report no. 8604. La Jolla: University of California, San Diego.

Jungers, W. L., and Susman, R. L.

1984   Body size and skeletal allometry in African apes. In R. L. Susman, ed., *The pygmy chimpanzee: Evolutionary biology and behavior,* 131–78. New York: Plenum.

Just, M., Carpenter, P. A., Keller, T., and Thulborn, K. R.

1996   Movies of the brain: Imaging a sequence of cognitive processes. In A. W. Toga, R. S. J. Frackowiak, and J. C. Mazziotta, eds., *Abstracts of the Second International Conference on Functional Mapping of the Human Brain* 3(3):S250. San Diego: Academic Press.

Kano, T.

1980   The social group of pygmy chimpanzees (*Pan paniscus*) of Wamba. *Primates* 23:171–88.

Karmiloff-Smith, A.

1992   *Beyond modularity: A developmental perspective on cognitive science.* Cambridge, MA: MIT Press.

Katz, M. J.

1987   Is evolution random? In R. Raff and E. Raff, eds., *Development as an evolutionary process,* 285–315. New York: Liss.

Kawai, M.

1965   Newly acquired pre-cultural behavior of the natural troop of Japanese macaques. In S. Altmann, ed., *Japanese monkeys,* 1–30. Calgary: University of Alberta Press.

Kawamura, S.

1959   The process of sub-cultural propagation among Japanese macaques. *Primates* 2:43–60.

Kaye, K.

1977   Infants' effects upon their mothers' teaching strategies. In J. Glidewell, ed., *The social context of learning and development,* 173–206. New York: Gardner Press.

Kean, M.-L., ed.

1985   *Agrammatism.* Orlando, FL: Academic Press.

Keene, R.

1991   Heterochrony. *Yearbook of Physical Anthropology* 34:251–82.

Kilborn, K.

1991   Selective impairment of grammatical morphology due to induced stress in normal listeners: Implications for aphasia. *Brain and Language* 41:275–88.

Killackey, H. P.

1990   Neocortical expansion: An attempt toward relating phylogeny and ontogeny. *Journal of Cognitive Neuroscience* 2:1–17.

Killackey, H. P., Chiaia, N. L., Bennett-Clarke, C. A., Eck, M., and Rhoades, R.

1994   Peripheral influences on the size and organization of somatotopic representations in the fetal rat cortex. *Journal of Neuroscience* 14:1496–1506.

Kimbel, W. H., and Martin, L. B., eds.

1993   *Species, species concepts, and primate evolution.* New York: Plenum.

Kinsbourne, M., and Hiscock, M.

1983   The normal and deviant development of functional lateralization of the brain. In M. Haith and J. Campos, eds., *Handbook of child psychology,* vol. 2 (4th ed.), 157–280. New York: Wiley.

Kintsch, W., and Mross, E. F.

1985   Context effects in word identification. *Journal of Memory and Language* 24(3):336–49.

Klein, R.

1984   Mammalian extinctions and Stone Age people in Africa. In P. S. Martin and R. Klein, eds., *Quarternary extinctions: A prehistoric revolution,* 553–73. Tucson: University of Arizona Press.

1989   *The human career: Human biological and cultural origins.* Chicago: University of Chicago Press.

Klima, E. S., Kritchevsky, M., and Hickok, G.

1993   The neural substrate for sign language. Symposium at the 31st annual meeting of the Academy of Aphasia, Tucson, AZ.

Klingenberg, C. P.

1988   Heterochrony and allometry: The analysis of evolutionary change in ontogeny. *Biological Review* 73:79–123.

Kluender, R., and Kutas, M.

1993   Bridging the gap: Evidence from ERPs on the processing of unbounded dependencies. *Journal of Cognitive Neuroscience* 5(2):196–214.

n.d.   Interaction of lexical and syntactic effects in the processing of unbounded dependencies. *Language and Cognitive Processes.* In press.

Kohler, W.

1917   *The mentality of apes.* New York: Liveright.

Konner, M.

1991   Universals of behavioral development in relation to brain myelination. In K. R. Gibson and A. C. Petersen, eds., *Brain maturation and cognitive development: Comparative and cross-cultural perspectives,* 181–224. New York: Aldine de Gruyter.

Krashen, S.D.

1973   Lateralization, language learning, and the critical period: Some new evidence. *Language Learning* 23:63–74.

Krasnegor, N., Rumbaugh, D., Schiefelbush, R., and Studdert-Kennedy, M., eds.

1991   *Biological and behavioral determinants of language development.* Hillsdale, NJ: Erlbaum.

Kuhn, D., Langer, J., Kohlberg, L., and Haan, N. S.

1977   The development of formal operations in logical and moral judgment. *Genetic Psychology Monographs* 95:97–188.

Kummer, H.

1971   *Primate societies.* New York: Aldine.

Kuroda, S.

1979   Grouping of the pygmy chimpanzees. *Primates* 20:161–83.

Kutas, M., and King, J. W.

1996   The potentials for basic sentence processing: Differentiating integrative processes. In I. Ikeda and J. L. McClelland, eds., *Attention and performance,* vol. 16, 501–46. Cambridge, MA: MIT Press.

Laitman, J. T., and Crelin, E. S.

1976   Postnatal development of the basicranium and vocal tract region in man. In J. F. Bosma, ed., *Symposium of Development of the Basicranium,* 206–19. Washington, DC: U.S. Government Printing Office.

1980   Developmental change in the upper respiratory system of human infants. *Perinatology-Neonatology* 4:15–22.

Laitman, J. T., and Heimbuch, R. C.

1982   The basicranium of Plio-Pleistocene hominids as an indicator of their upper respiratory systems. *American Journal of Physical Anthropology* 59:323–43.

Laitman, J. T., Heimbuch, R. C., and Crelin, E. S.

1978   Developmental change in a basicranial line and its relationship to the upper respiratory system in living primates. *American Journal of Anatomy* 152:467–82.

1979   The basicranium of fossil hominids as an indicator of their upper respiratory systems. *American Journal of Physical Anthropology* 51:15–34.

Langer, J.

1969   *Theories of development.* New York: Holt, Rinehart and Winston.

1980   *The origins of logic: Six to twelve months.* New York: Academic Press.

1982a   From prerepresentational to representational cognition. In G. Forman, ed., *Action and thought,* 37–64. New York: Academic Press.

1982b   Dialectics of development. In T. G. R. Bever, ed., *Regression in development,* 233–66. Hillsdale, NJ: Erlbaum.

1983   Concept and symbol formation by infants. In S. Wapner and B. Kaplan, eds., *Toward a holistic developmental psychology,* 221–34. Hillsdale, NJ: Erlbaum.

1985   Necessity and possibility during infancy. *Archives de Psychologie* 53:61–75.

1986   *The origins of logic: One to two years.* New York: Academic Press.

1990   Early cognitive development: Basic functions. In C. A. Hauert, ed., *Developmental psychology: Cognitive, perceptuo-motor, and neuropsychological perspectives,* 19–42. Amsterdam: North Holland.

1993   Comparative cognitive development. In K. R. Gibson and T. Ingold, eds., *Tools, language and cognition in human evolution,* 300–13. Cambridge: Cambridge University Press.

1994a   From acting to understanding: The comparative development of meaning. In W. F. Overton and D. Palermo, eds., *The nature and ontogenesis of meaning,* 191–214. Hillsdale, NJ: Erlbaum.

1994b   Logic. In V. S. Ramachandran, ed., *Encyclopedia of human behavior,* vol. 3, 83–91. San Diego: Academic Press.

1996   Heterochrony and the evolution of primate cognitive development. In A. E. Russon, K. A. Bard, and S. T. Parker, eds., *Reaching into thought: The mind of the great apes,* 257–77. Cambridge: Cambridge University Press.

1998   Phylogenetic and ontogenetic origins of logic: Classification. In J. Langer and M. Killen, eds., *Piaget, evolution and development,* 33–54. Mahweh, NJ: Erlbaum.

n.d.*a*   The mosaic evolution of cognitive and linguistic ontogeny. In M. Bowerman and S. Levinson, eds., *Conceptual development and language acquisition.* Cambridge: Cambridge University Press. In press.

n.d.*b*   The descent of cognitive development. *Developmental Science.* In press.

n.d.*c*   The origins and early development of cognition in comparative perspective. Manuscript in preparation.

Langer, J., Schlesinger, M., Spinozzi, G., and Natale, F.

1998   Developing classification in action: 1. Human infants. *Human Evolution* 13:107–24.

Laurenson, M. K., Wielebnowski, N., and Caro, T. M.

1995   Extrinsic factors and juvenile mortality on cheetahs. *Conservation Biology* 9:1329–31.

Lave, J., and Wenger, E.

1991   *Situated learning: Legitimate peripheral participation.* New York: Cambridge University Press.

Leakey, L. S. B., Tobias, P. V., and Napier, J. R.

1961   A new species of the genus *Homo* from Olduvai Gorge. *Nature* 202:7–9.

Leakey, M. G., Feibel, C. S., McDougall, L., and Walker, A.

1995   New four-million-year-old hominid species from Kanapoi and Allia Bay, Kenya. *Nature* 376:565–71.

Leakey, M. G., and Hay, R.

1979   Pliocene footprints in Laetoli Beds at Laetoli, northern Tanzania. *Nature* 278:383–85.

Le Douarin, N. M.

1993   Embryonic neural chimaeras in the study of brain development. *Trends in Neuroscience* 16:64–72.

Lee, P. C.

1984   Ecological constraints on the social development of vervet monkeys. *Behaviour* 91:245–62.

Lenneberg, E. H.

1967   *Biological foundations of language.* New York: Wiley.

Leslie, A. M.

1994a   Pretending and believing: Issues in the theory of ToMM. *Cognition* 50(1–3):211–38.

1994b   ToMM, ToBy, and Agency: Core architecture and domain specificity. In L. A. Hirschfeld and S. A. Gelman, eds., *Mapping the mind: Domain specificity in cognition and culture,* 119–48. Cambridge: Cambridge University Press.

Levy, J.

1978   *Play behavior.* New York: Wiley.

Lieberman, P.

1984   *The biology and evolution of language.* Cambridge, MA: Harvard University Press.

Lieberman, P., Kako, E., Friedman, J., Tajchman, G., Feldman, L., and Jiminez, E.

1992   Speech production, syntax comprehension, and cognitive deficits in Parkinson's disease. *Brain and Language* 43:169–89.

Lightfoot, D.

1991   The child's trigger experience: Degree-0 learnability. *Behavioral and Brain Sciences* 14:364.

Linebarger, M., Schwartz, M., and Saffran, E.

1983   Sensitivity to grammatical structure in so-called agrammatic aphasics. *Cognition* 13:361–92.

Liu, H.

1996   Lexical access and processing dissociations in nouns and verbs in a second language. Ph.D. diss., University of California, San Diego.

Lock, A. J.

1980   *The guided reinvention of language.* London: Academic Press.

Locke, J. L.

1993   Learning to speak. *Journal of Phonetics* 21(2):141–46.

Loy, J.
1970   Behavioral responses in free-ranging rhesus monkeys to food shortages. *American Journal of Physical Anthropology* 33:263–72.

Loy, J., Loy, K., Patterson, D., Keifer, G., and Conaway, C.
1978   The behavior of gonadectomized rhesus monkeys: I. Play. In E. O. Smith, ed., *Social play in primates,* 49–78. New York: Academic Press.

Luh, H.-K., Gittleman, J. L., Bozdogan, H., and Anderson, C. G.
n.d.   An information-based approach for multivariate correlated trait evolution. Manuscript in preparation.

Lukatela, G., Kostic, A., Feldman, L., and Turvey, M.
1983   Grammatical priming of inflected nouns. *Memory and Cognition* 11:59–63.

Lumsden, C., and Wilson, E. O.
1981   *Genes, mind, and culture.* Cambridge, MA: Harvard University Press.

MacArthur, R. H., and Wilson, E. O.
1967   *Theory of island biogeography.* Princeton, NJ: Princeton University Press.

MacLean, P. D.
1990   *The triune brain in evolution.* New York: Plenum.

Marchman, V.
1993   Constraints on plasticity in a connectionist model of the English past tense. *Journal of Cognitive Neuroscience* 5(2):215–34.

Markus, N., and Croft, D. B.
1995   Play behaviour and its effects on social development of common chimpanzees *(Pan troglodytes). Primates* 36:213–25.

Marslen-Wilson, W., and Tyler, L.
1980   The temporal structure of spoken language understanding. *Cognition* 8:1–71.
1987   Against modularity. In J. L. Garfield, ed., *Modularity in knowledge representation and natural-language understanding,* 37–62. Cambridge, MA: MIT Press.

Martin, A., Wiggs, C. L., Ungerleider, L. G., and Haxby, J. V.
1996   Neural correlates of category-specific knowledge. *Nature* 379(6566):649–52.

Martin, P., and Caro, T. M.
1985   On the function of play and its role in behavioral development. In J. S. Rosenblatt et al., eds., *Advances in the study of behavior,* 15:59–103. New York: Academic Press.

Martin, R. D., and Harvey, P. H.
1985   Brain size allometry: Ontogeny and phylogeny. In W. Jungers, ed., *Size and scaling in primate biology.* New York: Plenum.

Martin, R. D., and MacLarnon, A. M.
1985   Gestation period, neonatal size and maternal investment in placental mammals. *Nature* 313:220–23.

Matsuzawa, T.

1994   Field experiments on use of stone tools by chimpanzees in the wild. In R. W. Wrangham, W. C. McGrew, F. B. M. de Waal, and P. G. Heltne, eds., *Chimpanzee cultures*, 351–70. Cambridge, MA: Harvard University Press

1995   *Chimpanzee being.* (Excerpt translated by Tetsuro Matsuzawa.) Tokyo: Iwanami-shoten.

Matsuzawa, T., and Yamakoshi, G.

1996   Comparison of chimpanzee material culture between Bossou and Nimba, West Africa. In A. E. Russon, K. A. Bard, and S. T. Parker, eds., *Reaching into thought: The mind of the great apes*, 211–32. Cambridge: Cambridge University Press.

Maynard, A. E.

1999   Cultural teaching: The social organization and development of teaching in Zinacantec Maya sibling interactions. Ph.D. diss., University of California, Los Angeles.

Maynard, A. E., Greenfield, P. M., and Childs, C. P.

1999   Culture, history, biology, and body: Native and non-native acquisition of technological skill. *Ethos* 27:379–402.

Mayr, E.

1963   *Animal species and evolution.* Cambridge, MA: Harvard University.

Mazoyer, B. M., Tzourio, N., Frak, V., Syrota, A., Murayama, N., Levrier, O., Salamon, G., Dehaene, S., Cohen, L., and Mehler, J.

1993   The cortical representation of speech. *Journal of Cognitive Neuroscience* 5(4):467–79.

McConnell, S. K.

1988   Fates of visual cortical neurons in the ferret after isochronic and heterochronic transplantation. *Journal of Neuroscience* 8:945–74.

McCune, L.

1995   A normative study of representational play at the transition to language. *Developmental Psychology* 31:198–206.

McGinnis, W., and Kuziora, M.

1994   The molecular architects of body design. *Scientific American* 270:58–66.

McGrew, W. C.

1972   *An ethological study of children's behavior.* New York: Academic Press.

1992   *Chimpanzee material culture.* New York: Cambridge University Press.

McHenry, H. M.

1986   The first bipeds: A comparison of the *A. afarensis* and *A. africanus* post-cranium and implications for the evolution of bipedalism. *Journal of Human Evolution* 15:177–91.

1994   Behavioral ecological implications of early hominid body size. *Journal of Human Evolution* 27:77–87.

McHenry, H. M., and Berger, L. R.

1998   Body proportions in *Australopithecus afarensis* and *A. africanus* and the origin of the genus *Homo. Journal of Human Evolution* 35:1–22.

McKinney, M. L.

1998   The juvenilized ape myth: Our "overdeveloped" brain. *Bioscience* 48:109–23.

McKinney, M. L., ed.

1988a   *Heterochrony in evolution: A multidisciplinary approach.* New York: Plenum.

1988b   Classifying heterochrony: Allometry, size, and time. In M. L. McKinney, ed., *Heterochrony in evoution: A multidisciplinary approach,* 17–34. New York: Plenum.

McKinney, M. L., and Gittleman, J. L.

1995   Ontogeny and phylogeny: Tinkering with covariation in life history, morphology and behaviour. In K. J. McNamara, ed., *Evolutionary change and heterochrony,* 21–47. London: Wiley.

McKinney, M. L., and McNamara, K. J.

1991   *Heterochrony: The evolution of ontogeny.* New York: Plenum.

MacNamara, K. J.

1986   *A border dispute: The place of logic in psychology.* Cambridge, MA: MIT Press.

1997   *The shapes of time.* Baltimore, MD: Johns Hopkins University Press.

Meltzoff, A.

1990   Foundations for developing a concept of self: The role of imitation in relating self to others and the value of social mirroring, social modeling, and self practice in infancy. In D. Cicchetti and M. Beeghly, eds., *The self in transition: Infancy to childhood,* 139–64. Chicago: University of Chicago Press.

Mendoza-Granados, D., and Sommer, V.

1995   Play in chimpanzees of the Arnhem Zoo: Self-serving compromises. *Primates* 36:57–68.

Menn, L., and Obler, L. K., eds.

1990   *Agrammatic aphasia: Cross-language narrative sourcebook.* Amsterdam: John Benjamins.

Mercier, N., Valladas, H., Joron, J., Reyss, J., Leveque, F., and Vandermeersch, B.

1991   Thermoluminescence dating of the late Neanderthal remains from Saint-Cesaire. *Nature* 351:737–39.

Merzenich, M. M., Recanzone, G., Jenkins, W. M., Allard, T. T., and Nudo, R. J.

1988   Cortical representational plasticity. In P. Rakic and W. Singer, eds., *Neurobiology of neocortex,* 41–67. New York: Wiley.

Miceli, C., and Mazzucchi, A.

1990   The nature of speech production deficits in so-called agrammatic aphasia: Evidence from two Italian patients. In L. Menn and L. K. Obler, eds., *Agrammatic aphasia: Cross-language narrative sourcebook.* Amsterdam: John Benjamins.

Miceli, C., Mazzucchi, A., Menn, L., and Goodglass, H.

1983   Contrasting cases of Italian agrammatic aphasia without comprehension disorder. *Brain and Language* 19:65–97.

Miles, H. L.

1990   The cognitive foundations for reference in a signing orangutan. In S. T. Parker and K. R. Gibson, eds., *"Language" and intelligence in monkeys and apes*, 511–39. New York: Cambridge University Press.

1994   Me Chantek: The development of self-awareness in a signing gorilla. In S. T. Parker, R. W. Mitchell, and M. L. Boccia, eds., *Self-awareness in animals and humans*, 254–72. New York: Cambridge University Press.

Miles, H. L., Mitchell, R. W., and Harper, S.

1996   Simon says: The development of imitation in an enculturated orangutan. In A. E. Russon, K. A. Bard, and S. T. Parker, eds., *Reaching into thought: The mind of the great apes*, 278–98. Cambridge: Cambridge University Press.

Mills, D., Coffey-Corina, S., DiIulio, L., and Neville, H.

1995   *The development of cerebral specialization for different lexical items in normal infants and infants with focal brain lesions.* Center for Research in Language, Technical Report CND-9507. La Jolla: University of California, San Diego.

Mithen, S.

1996   *The prehistory of the mind.* London: Thames and Hudson.

Miyake, A., Carpenter, P. A., and Just, M. A.

1995   Reduced resources and specific impairments in normal and aphasic sentence comprehension. *Cognitive Neuropsychology* 12(6):651–79.

Molnár, Z., and Blakemore, C.

1991   Lack of regional specificity for connections formed between thalamus and cortex in coculture. *Nature* 351(6326):475–77.

Montagu, A.

1981   *Growing young.* New York: McGraw-Hill.

Moore, A. H., Cherry, S. R., Pollack, D. B., Hovda, D. A., and Phelps, M. E.

1999   Application of positron emission tomography to determine glucose utilization in conscious infant monkeys. *Journal of Neuroscience Methods* 88:123–33.

Morbeck, M. E., Galloway, A., and Zihlman, A. L., eds.

1997   *The evolving female.* Princeton: Princeton University Press.

Morgan, E.

1993   Bipedalism. *Nutrition and Health* 9:193–203.

Morris, D.

1962   *The biology of art.* New York: Alfred A. Knopf.

Morss, J.

1990   *The biologizing of childhood.* Hillsdale, NJ: Erlebaum.

349

Nagell, K., Olguin, R. S., and Tomasello, M.

1993   Processes of social learning in the tool use of chimpanzees *(Pan troglodytes)* and human children *(Homo sapiens). Journal of Comparative Psychology* 7:174–86.

Nash, L. T.

1993   Juveniles in nongregarious primates. In M. E. Pereira and L. A. Fairbanks, eds., *Juvenile primates,* 119–37. New York: Oxford University Press.

Natale, F.

1989   Stage 5 object-concept. In F. Antinucci, ed., *Cognitive structure and development of nonhuman primates,* 89–95. Hillsdale, NJ: Erlbaum.

Nelson, K.

1973   Some evidence for the primacy of categorization and its functional basis. *Merrill-Palmer Quarterly* 19:21–39.

Neville, H. J., and Lawson, D.

1987   Attention to central and peripheral visual space in a movement detection task: An event-related potential and behavioral study. II. Congenitally deaf adults. *Brain Research* 45:268–83.

Neville, H. J., Nicol, J., Barss, A., Forster, K., and Garrett, M.

1991   Syntactically based sentence-processing classes: Evidence from event-related brain potentials. *Journal of Cognitive Neuroscience* 3:155–70.

Newmeyer, F. J., ed.

1988   *Linguistics: The Cambridge Survey.* Cambridge: Cambridge University Press.

Nishikawa, K. C.

1997   Emergence of novel functions during brain evolution. *BioScience* 47:341–54.

Nobre, A. C., and Plunkett K.

1997   The neural system of language: Structure and development. *Current Opinion in Neurobiology* 7(2):262–68.

O'Connell, J. F., Hawkes, K., and Blurton-Jones, N. G.

1999   Grandmothering and the evolution of *Homo erectus. Journal of Human Evolution* 36:461–85.

O'Leary, D. D.

1993   Do cortical areas emerge from a protocortex? In M. Johnson, ed., *Brain development and cognition: A reader,* 323–37. Oxford: Blackwell.

O'Leary, D. D., and Stanfield, B. B.

1989   Selective elimination of axons extended by developing cortical neurons is dependent on regional locale: Experiments utilizing fetal cortical transplants. *Journal of Neuroscience* 9(7):2230–46.

Olmo E.

1983 Nucleotype and cell size in vertebrates: A review. *Basic and Applied Histochemistry* 27:227–56.

Onifer, W., and Swinney, D.

1981 Accessing lexical ambiguities during sentence comprehension: Effects of frequency of meaning and contextual bias. *Memory and Cognition* 9:225–36.

Osterhout, L., and Holcomb, P. J.

1993 Event-related potentials and syntactic anomaly: Evidence of anomaly detection during the perception of continuous speech. *Language and Cognitive Processes* 8(4):413–37.

Owens, N. W.

1975 Social play behaviour in free-living baboons, *Papio anubis. Animal Behavior* 23:387–408.

Oyama, S.

1992 The problem of change. In M. Johnson, ed., *Brain development and cognition: A reader,* 19–30. Oxford: Blackwell.

Pagel, M. D., and Harvey, P. H.

1988 How do mammals produce large-brained offspring? *Evolution* 42:948–57.

1990 Diversity in the brain sizes of newborn mammals. *BioScience* 40:116–22.

1993 Evolution of the juvenile period in mammals. In L. Fairbanks and M. Pereira, eds., *Juvenile primates: Life history, development and behavior,* 28–37. New York: Oxford University Press.

Pallas, S. L., and Sur, M.

1993 Visual projections induced into the auditory pathway of ferrets: II. Corticocortical connections of primary auditory cortex. *Journal of Comparative Neurology* 337(2):317–33.

Paraskevopoulous, J., and Hunt, J. McV.

1971 Object construction and imitation under different conditions of rearing. *Journal of Genetic Psychology* 119:301–21.

Parker, S. T.

1977 Piaget's sensorimotor series in an infant macaque. In S. Chevalier-Skolnikoff and F. E. Poirier, eds., *Primate biosocial development: Biological, social, and ecological determinants,* 43–112. New York: Garland.

1984 Playing for keeps: An evolutionary perspective on human games. In P. K. Smith, ed., *Play in animals and humans,* 271–93. Oxford: Blackwell.

1985 A social technological model for the evolution of language. *Current Anthropology* 26(5):617–39.

1990 The origins of comparative developmental evolutionary studies of primate mental abilities. In S. T. Parker and K. R. Gibson, eds., *"Language" and intelligence in monkeys and apes,* 3–64. New York: Cambridge University Press.

1993   Imitation and circular reactions as evolved mechanisms for cognitive construction. *Human Development* 36:309–23.

1996a   Apprenticeship in tool-mediated extractive foraging: The origins of imitation, teaching and self-awareness in great apes. In A. E. Russon, K. A. Bard, and S. T. Parker, eds., *Reaching into thought: The mind of the great apes*, 348–70. Cambridge: Cambridge University Press.

1996b   Using cladistic analysis of comparative data to reconstruct the evolution of cognitive development in hominids. In E. Martins, ed., *Phylogenies and the comparative method in animal behavior*, 361–98. New York: Oxford University Press.

Parker, S. T., and Gibson, K. R.

1977   Object manipulation, tool use, and sensorimotor intelligence as feeding adaptations in cebus monkeys and great apes. *Journal of Human Evolution* 6:623–41.

1979   A developmental model for the evolution of language and intelligence in early hominids. *Behavioral and Brain Sciences* 2:367–408.

Parker, S. T., and Gibson, K. R., eds.

1990   *"Language" and intelligence in monkeys and apes.* New York: Cambridge University Press.

Parker, S. T., and McKinney, M. L.

1999   *Origins of intelligence: The evolution of cognitive development in monkeys, apes, and humans.* Baltimore, MD: Johns Hopkins University Press.

Parker, S. T., and Milbrath, C.

1993   Higher intelligence, propositional language, and culture as adaptations for planning. In K. R. Gibson and T. Ingold, eds., *Tools, language, and cognition in human evolution*, 314–33. Cambridge: Cambridge University Press.

Parker, S. T., and Poti', P.

1990   The role of innate motor patterns in ontogenetic and experiential development of intelligent use of sticks in cebus monkeys. In S. T. Parker and K. R. Gibson, eds., *"Language" and intelligence in monkeys and apes: Comparative and developmental perspectives*, 219–43. New York: Cambridge University Press.

Parker, S. T., and Russon, A. E.

1996   On the wild side of culture and cognition in the great apes. In A. E. Russon, K. A. Bard, and S. T. Parker, eds., *Reaching into thought: The mind of the great apes*, 430–50. Cambridge: Cambridge University Press.

Parmelee, A. H., and Sigman, M. D.

1983   Perinatal brain development and behavior. In M. M. Haith and J. Campos, eds., *Infancy and the biology of development, vol. 2: Handbook of child psychology.* New York: Wiley.

Passingham, R.

1982   *The human primate.* New York: Freeman.

Patterson, C.

1983   How does phylogeny differ from ontogeny? In B. C. Goodwin, N. Holder, and C. C. Wylie, eds., *Development and Evolution*, 1–31. Cambridge: Cambridge University Press.

Patterson, F., and Cohn, R.

1994   Self-recognition and self-awareness in lowland gorillas. In S. T. Parker, R. W. Mitchell, and M. L. Boccia, eds., *Self-awareness in animals and humans*, 273–90. New York: Cambridge University Press.

Patterson, F., and Linden, E.

1981   *The education of Koko*. New York: Holt, Rinehart and Winston.

Paus, T., Zijdenbos, A., Worsley, K., Collins, D. L., Blumenthal, J., Giedd, J. N., Rapoport, J. L., and Evans. A. C.

1999   Structural maturation of neural pathways in children and adolescents: In vivo study. *Science* 283:1908–11.

Pellegrini, A. D., and Smith, P. K.

1998   Physical activity play: The nature and function of a neglected aspect of play. *Child Development* 69:577–98.

Pennisi, E., and Roush, W.

1997   Developing a new view of evolution. *Science* 277:34–37.

Penny, D., Hendy, M. D., and Steel, M. A.

1992   Progress with methods for constructing evolutionary trees. *Trends in Ecology and Evolution* 7:73–79.

Pereira, M. E., and Altmann, J.

1985   Development of social behavior in free-living nonhuman primates. In E. S. Watts, ed., *Nonhuman primate models for growth and development*, 217–309. New York: Liss.

Pereira, M. E., and Fairbanks, L., eds.

1994   *Juvenile primates*. New York: Oxford University Press.

Pérusse, D., Neale, M. C., Heath, A. C., and Eaves, L. J.

1994   Human parental behavior: Evidence for genetic influence and potential implication for gene-culture transmission. *Behavior Genetics* 24:327–35.

Piaget, J.

1952   *The origins of intelligence in children*. New York: Norton.

1954   *The construction of reality in the child*. New York: Basic Books.

1962   *Play, dreams, and imitation in childhood*. New York: Norton.

1966   *Psychology of intelligence*. New York: Littlefield Adams.

1970   *Genetic epistemology*. New York: Columbia University Press.

1972   *Essai de logique opératoire*. Paris: Dunod.

1977   *Researches sur l'abstraction reflechisante*. Paris: PUF.

1987a   *Possibility and necessity, vol. 1: The role of possibility in cognitive development*. Minneapolis: University of Minnesota Press.

1987b   *Possibility and necessity, vol. 2: The role of necessity in cognitive development*. Minneapolis: University of Minnesota Press.

Piaget, J., and Inhelder, B.

1967a   *The psychology of intelligence.* New York: Norton.

1967b   *The child's conception of space.* New York: Norton.

1969   *The psychology of the child.* New York: Basic Books.

Piatelli-Palmarini, M.

1989   Evolution, selection, and cognition: From "learning" to parameter setting in biology and the study of language. *Cognition* 31(1):1–44.

Pilbeam, D., and Gould, S. J.

1974   Size and scaling in human evolution. *Science* 186:892–901.

Pinker, S.

1991   Rules of language. *Science* 253:530–35.

1994a   *The language instinct: How the mind creates language.* New York: William Morrow.

1994b   On language. *Journal of Cognitive Neuroscience* 6(1):92–97.

Pinker, S., and Bloom, P.

1990   Natural language and natural selection. *Behavioral and Brain Sciences* 13:707–84.

Plooij, F. X.

1978   Some basic traits of language in wild chimpanzees. In A. Lock, ed., *Action, gesture, and symbol: The emergence of language,* 111–31. New York: Academic Press.

Poeppel, D.

1996   A critical review of PET studies of phonological processing. *Brain and Language* 55(3):317–51.

Poizner, H., Klima, E., and Bellugi, U.

1987   *What the hands reveal about the brain.* Cambridge, MA: MIT Press.

Pons, T. P., Garraghty, P. E., Ommaya, A. K., Kaas, J. H., Taub, E., and Mishkin, M.

1991   Massive cortical reorganization after sensory deafferentation in adult macaques [see comments]. *Science* 252(5014):1857–60.

Portmann, A.

1990   *A zoologist looks at humankind.* New York: Columbia University Press.

Poti', P.

1989   Early sensorimotor development in macaques. In F. Antinucci, ed., *Cognitive structure and development of nonhuman primates,* 39–54. Hillsdale, NJ: Erlbaum.

1996   Spatial aspects of spontaneous object grouping by young chimpanzees *(Pan troglodytes). International Journal of Primatology* 17:101–16.

1997   Logical structure in young chimpanzees' spontaneous object grouping. *International Journal of Primatology* 18:33–59.

Poti', P., and Antinucci, F.

1989   Logical operations. In F. Antinucci, ed., *Cognitive structure and development of nonhuman primates,* 189–228. Hillsdale, NJ: Erlbaum.

Poti', P., Langer, J., Savage-Rumbaugh, E. S., and Brakke, K. E.

1999   Spontaneous logicomathematical constructions by chimpanzees *(Pan troglodytes, Pan paniscus). Animal Cognition* 2:147–56.

Poti', P., and Spinozzi, G.

1994   Early sensorimotor development in chimpanzee *(Pan troglodytes). Journal of Comparative Psychology* 108(1):93–103.

Potts, R., and Shipman, P.

1981   Cut marks made by stone tools on bones from Olduvai Gorge, Tanzania. *Nature* 291:577–80.

Promislow, D. E. L., and Harvey, P. H.

1990   Living fast and dying young: A comparative analysis of life history variation among mammals. *Journal of Zoology* 220:417–37.

Purugganan, M. D.

1996   Evolution of development: Molecules, mechanisms and phylogenetics. *TREE* 111:5–7.

Purves, D.

1988   *Body and brain: A trophic theory of neural connections.* Cambridge, MA: Harvard University Press.

Purves, D., and Lichtman, J. W.

1985   Elimination of synapses in the developing nervous system. *Science* 210:153–57.

Purvis, A., and Harvey, P. H.

1995   Mammal life history evolution: A comparative test of Charnov's model. *Journal of Zoology* 237:259–83.

Pusey, A. E.

1990   Behavioural changes at adolescence in chimpanzees. *Behaviour* 115:203–46.

Pysh, J. J., and Weiss, G. M.

1979   Exercise during development induces an increase in Purkinje cell dendritic tree size. *Science* 206:230–31.

Raff, R. A.

1996   *The shape of life: Genes, development, and the evolution of animal form.* Chicago: University of Chicago Press.

Raff, R. A., and Wray, G. A.

1989   Heterochrony: Developmental mechanisms and evolutionary results. *Journal of Evolutionary Biology* 2:409–34.

Rakic, P., Bourgeois, J.-P., Eckenhoff, M. F., Zecevic, N., and Goldman-Rakic, P. S.

1986   Concurrent overproduction of synapses in diverse regions of the primate cerebral cortex. *Science* 232:232–35.

Ramachandran, V. S.

1993   Behavioral and magnetoencephalographic correlates of plasticity in the adult human brain. *Proceedings of the National Academy of Sciences* 90:10413–20.

Ratcliff, R., and McKoon, G.

1994   Retrieving information from memory: Spreading-activation theories versus compound-cue theories. *Psychological Review* 101(1):177–84.

Reilly, J., Bates, E., and Marchman, V.

1998   Narrative discourse in children with early focal brain injury. *Brain and Language* 61:334–75.

Reiss, J. O.

1989   The meaning of developmental time: A metric for comparative embryology. *American Naturalist* 134:170–89.

Reyes, I.

1995   Interaction of sentential and gender context in bilingual and monolingual Spanish speakers. Unpublished manuscript, University of California, San Diego.

Reynolds, P.

1993   The complementation theory of language and tool use. In K. R. Gibson and T. Ingold, eds., *Tools, language, and cognition in human evolution,* 407–28. Cambridge: Cambridge University Press.

Riccuiti, H. N.

1965   Object grouping and selective ordering behavior in infants 12 to 24 months. *Merrill-Palmer Quarterly* 11:129–48.

Rice, S. H.

1997   The analysis of ontogenetic trajectories: When a change in size or shape is not heterochrony. *Proceedings of the National Academy of Science* 94:907–12.

Richtsmeier, J. T., and Lele, S.

1993   A coordinate-free approach to the analysis of growth patterns: Models and theoretical considerations. *Biological Review* 68:381–411.

Riska, B., and Atchley, W. R.

1985   Genetics of growth predict patterns of brain-size evolution. *Science* 229:668–71.

Roe, A. W., Pallas, S. L., Hahm, J. O., and Sur, M.

1990   A map of visual space induced in primary auditory cortex. *Science* 250(4982):818–20.

Roff, D. A.

1992   *The evolution of life histories: Theory and analysis.* New York: Chapman and Hall.

Rogers, M. J., Harris, J. W. K., and Feibel, C. S.

1994   Changing patterns of land use by Plio-Pleistocene hominids in the Lake Turkana basin. *Journal of Human Evolution* 27:139–58.

Rogoff, B.

1990   *Apprenticeship in thinking.* Cambridge: Cambridge University Press.

Rose, M. D.

1991   The process of bipedalization in hominids. In Y. Coppens and B. Senut, eds., *Origin(s) de la bipedie chez les hominides,* 37–48. Paris: Cahiers de Paleoanthropologie, Editions du CNRS.

Rosenberg, K., and Trevathan, W.

1996   Bipedalism and human birth: The obstetrical dilemma revisited. *Evolutionary Anthropology* 4(5):161–68.

Roth, G., Blanke, J., and Wake, D. B.

1994   Cell size predicts morphological complexity in the brains of frogs and salamanders. *Proceedings of the National Academy of Sciences* 91:4796–80.

Roth G., Nishikawa, K. C., Naujoks-Manteuffel, C., Schmidt, A., and Wake, D. B.

1993   Paedomorphosis and simplification in the nervous system of salamanders. *Brain Behavior and Evolution* 42:137–70.

Rowell, T. E., and Chism, J.

1986   The ontogeny of sex differences in the behavior of patas monkeys. *International Journal of Primatology* 7:83–107.

Roy, R. R., Baldwin, K. M., and Edgerton, V. R.

1991   The plasticity of skeletal muscle: Effects of neuromuscular activity. *Exercise and Sport Sciences Reviews* 19:269–312.

Ruff, C. B.

1995   Biomechanisms of the hip and birth in early *Homo. American Journal of Physical Anthropology* 98:527–74.

Rumelhart, D. E., Hinton, G., and Williams, R.

1986   Learning internal representations by error propagation. In D. E. Rumelhart and J. L. McClelland, eds., *Parallel distributed processing: Explorations in the microstructure of cognition. Vol. 1, Foundations,* 318–62. Cambridge, MA: MIT Press.

Rumelhart, D. E., and McClelland J. L., eds.

1986   *Parallel distributed processing: Explorations in the microstructure of cognition.* Cambridge, MA: MIT Press.

Russon, A. E.

1990   The development of peer social interaction in infant chimpanzees: Comparative social, Piagetian, and brain perspective. In S. T. Parker and K. R. Gibson, eds., *"Language" and intelligence in monkeys and apes: Comparative and developmental perspectives,* 379–419. New York: Cambridge University Press.

1996   Imitation in everyday use: Matching and rehearsal in the spontaneous imitation of rehabilitant orangutans *(Pongo pygmaeus).* In A. E. Russon, K. A. Bard, and S. T. Parker, eds., *Reaching into thought: The mind of the great apes,* 152–76. Cambridge: Cambridge University Press.

Russon, A. E., and Galdikas, B. M. F.

1995   Constraints on great ape imitation: Model and action selectivity in rehabilitant orangutan *(Pongo pygmaeus)* imitation. *Journal of Comparative Psychology* 109(1):5–17.

Ruvkun, G., Wrightman, B., Burglin, T., and Arusa, P.

1991   Dominant gain-of-function mutations that lead to misregulation of the *C. elegans* heterochronic gene *lin-14,* and the evolutionary implications of dominant mutations in pattern-formation genes. *Development* (supplement) 1:47–54.

Sabater Pi, J., Vea, J. J., and Serrallonga, J.

1997   Did the first hominids build nests? *Current Anthropology* 38(5):914–16.

Sacher, G. A.

1959   Relation of lifespan to brain weight and body weight in mammals. In G. Wolstenholme and M. O'Connor, eds., *CIBA Foundation colloquia on aging, vol. 5: The lifespan of animals,* 115–33. London: Churchill.

1982   Relation of lifespan to brain weight and body weight in mammals. In E. Armstrong and D. Falk, eds., *Primate brain evolution: Methods and concepts.* New York: Plenum.

Sacher, G. A., and Staffeldt, E. F.

1974   Relation of gestation time to brain weight for placental mammals: Implications for the theory of vertebrate growth. *American Naturalist* 108:593–615.

Sadato, N., Pascualleone, A., Grafman, J., Ibanez, V., et al.

1996   Activation of the primary visual cortex by Braille reading in blind. *Nature* 380(6574):526–28.

Saffran, E. M., and Schwartz, M. F.

1988   "Agrammatic" comprehension it's not: Alternatives and implications. *Aphasiology* 2:389–94.

Sagan, C.

1977   *The dragons of eden.* New York: Random House.

Satoh, N.

1982   Timing mechanisms in early embryonic development. *Differentiation* 22:156–63.

Sattler, R.

1994   Homology, homeosis, and process morphology in plants. In B. K. Hall, ed., *Homology: The hierarchical basis of comparative biology,* 423–75. New York: Academic Press.

Savage-Rumbaugh, E. S., Murphy, J., et al.

1993   *Language comprehension in ape and child.* Chicago: University of Chicago Press.

Savage-Rumbaugh, E. S., Romski, M. A., Hopkins, W. D., and Sevcik, R.

1989   Symbol acquisition and use by *Pan troglodytes, Pan paniscus,* and *Homo sapiens.* In P. G. Heltne and L. A. Marquardt, eds., *Understanding chimpanzees,* 266–95. Cambridge, MA: Harvard University Press.

Savage-Rumbaugh, E. S., Williams, S. L., Furuichi, T., and Kano, T.

1996   Language perceived: *Paniscus* branches out. In W. C. McGrew, L. T. Marchant, and T. Nishida, eds., *Great ape societies,* 173–84. Cambridge: Cambridge University Press.

Saxe, G. B.

1991   *Culture and cognitive development: Studies in mathematical understanding.* Hillsdale, NJ: Erlbaum.

Schmidt-Nielsen, K.

1984   *Scaling: Why is animal size so important?* Cambridge: Cambridge University Press.

Semaw, S., Renne, P., Harris, J. W. K., Feibel, C. S., Bernor, R. L., Fesseha, N., and Mowbray, K.

1997   2.5-million-year-old stone tools from Gona, Ethiopia. *Nature* 385:333–36.

Shankweiler, D., Crain, S., Gorrell, P., and Tuller, B.

1989   Reception of language in Broca's aphasia. *Language and Cognitive Processes* 4(1):1–33.

Shea, B. T.

1981   Relative growth of the limbs and trunk in the African apes. *American Journal of Physical Anthropology* 56:179–202.

1982   Growth and size allometry in the African Pongidae: Cranial and post-cranial analyses. Ph.D. diss., Duke University, Durham, North Carolina.

1983a   Allometry and heterochrony in the African apes. *American Journal of Physical Anthropology* 62:275–89.

1983b   Paedomorphosis and neoteny in the pygmy chimpanzee. *Science* 222:521–22.

1983c   Size and diet in the evolution of African ape craniodental form. *Folia Primatologica* 40:32–68.

1983d   Phyletic size change and brain/body allometry: A consideration based on the African pongids and other primates. *International Journal of Primatology* 4:33–62.

1984   An allometric perspective on the morphological and evolutionary relationships between pygmy *(Pan paniscus)* and common *(Pan troglodytes)* chimpanzees. In R. L. Susman, ed., *The pygmy chimpanzee: Evolutionary biology and behavior,* 89–130. New York: Plenum.

1985   Ontogenetic allometry and scaling: A discussion based on growth and form of the skull in African apes. In W. L. Jungers, ed., *Size and scaling in primate biology,* 175–206. New York: Plenum.

1988   Heterochrony in primates. In M. L. McKinney, ed., *Heterochrony in evolution: A multidisciplinary approach,* 237–66. New York: Plenum.

1989   Heterochrony in human evolution: The case for neoteny reconsidered. *Yearbook of Physical Anthropology* 32:69–101.

1990   Dynamic morphology: Growth, life history, and ecology in primate evolution. In C. J. DeRousseau, ed., *Primate Life History and Evolution*, 325–52. New York: Wiley-Liss.

1992a   Ontogenetic scaling of skeletal proportions in the talapoin monkey. *Journal of Human Evolution* 23:283–307.

1992b   Developmental perspective on size change and allometry in evolution. *Evolutionary Anthropology* 1:125–34.

1993   Review of M. L. McKinney and K. J. McNamara, eds., *Heterochrony: The evolution of ontogeny*. *International Journal of Primatology* 14:805–8.

1996   Ontogenetic scaling and size correction in the comparative study of primate adaptations. *Anthropologie* 33:1–16.

n.d.   Are some heterochronic transformations more likely than others? In K. J. McNamara and N. Minugh-Purvis, eds., *Human evolution through developmental change*. New York: Plenum. In press.

Shea, B. T., and Bailey, R. C.

1996   Allometry and adaptation of body proportions and stature in African pygmies. *American Journal of Physical Anthropology* 100:311–40.

Shea, B. T., and Gomez, A.

1988   Tooth scaling and evolutionary dwarfism: An investigation of allometry in human pygmies. *American Journal of Physical Anthropology* 77:117–32.

Simeone, A., Acampora, D., Gulisano, M., Stornaiuolo, A., and Boncinelli, E.

1992   Nested expression domains of four homeobox genes in developing rostral brain. *Nature* 358:687–90.

Simonds, R. J., and Scheibel, A. B.

1989   The postnatal development of the motor speech area: A preliminary study. *Brain and Language* 37:42–58.

Simpson, G. B., and Kang, H. W.

1994   Inhibitory processes in the recognition of homograph meanings. In D. Dagenbach and T. H. Carr, eds., *Inhibitory processes in attention, memory, and language*, 359–81. San Diego: Academic Press.

Skelton, R. R., McHenry, H. M., and Drawhorn, G.

1986   Phylogenetic analysis of early hominids. *Current Anthropology* 271:21–43.

Slobin, D. I.

1979   *Psycholinguistics*. 2d ed. Glenview, IL: Scott, Foresman.

Smith, B. H.

1986   Dental development in *Australopithecus* and early *Homo*. *Nature* 323:327–30.

1989   Dental development as a measure of life history in primates. *Evolution* 43:683–88.

1992   Life history and the evolution of human maturation. *Evolutionary Anthropology* 1(4):134–42.

1993 The physiological age of KNM-WT 15000. In A. Walker and R. Leakey, eds., *The Nariokotome Homo erectus skeleton,* 195–220. Cambridge, MA: Harvard University Press.

1994 Patterns of dental development in *Homo, Australopithecus, Pan,* and *Gorilla. American Journal of Physical Anthropology* 94:307–25.

Smith, B. H., Crummett, T. L., and Brandt, K. L.

1994 Ages of eruption of primate teeth: A compendium for aging individuals and comparing life histories. *Yearbook of Physical Anthropology* 37:177–231.

Smith, E. O., ed.

1978 *Social play in primates.* New York: Academic Press.

Smith, P. K., ed.

1984 *Play in animals and humans.* Oxford: Blackwell.

Smith, P. K., and Connolly, K.

1980 *The ecology of preschool behaviour.* Cambridge: Cambridge University Press.

Smith, P. K., and Simon, T.

1984 Object play, problem-solving and creativity in children. In P. K. Smith, ed., *Play in animals and humans,* 199–216. Oxford: Blackwell.

Smith, R. J., Gannon, P. J., and Smith, H.

1995 Ontogeny of australopithecines and early *Homo. Journal of Human Evolution* 29:155–68.

Sommer, R. J., and Sternberg, P. W.

1996 Evolution of nematode vulval fate patterning. *Developmental Biology* 173:396–407.

Sommer, V., and Mendoza-Granados, D.

1995 Play as an indicator of habitat quality: A field study of langur monkeys *(Presbytis entellus). Ethology* 99:177–92.

Spelke, E. S., Breinlinger, K., Macomber, J., and Jacobson, K.

1992 Origins of knowledge. *Psychological Review* 99(4):605–32.

Spinozzi, G.

1989 Early sensorimotor development in Cebus. In F. Antinucci, ed., *Cognitive structure and development of nonhuman primates,* 55–66. Hillsdale, NJ: Erlbaum.

1993 The development of spontaneous classificatory behavior in chimpanzees *(Pan troglodytes). Journal of Comparative Psychology* 107:193–200.

Spinozzi, G., and Langer, J.

1999 Spontaneous classification in action by a human-enculturated and language-reared bonobo *(Pan paniscus)* and common chimpanzees *(Pan troglodytes). Journal of Comparative Psychology* 113:286–96.

Spinozzi, G., and Natale, F.

1989 Classification. In F. Antinucci, ed., *Cognitive structure and development of nonhuman primates,* 163–88. Hillsdale, NJ: Erlbaum.

Spinozzi, G., Natale, F., Langer, J., and Brakke, K. E.
1999   Spontaneous class grouping behavior by bonobos *(Pan paniscus)* and common chimpanzees *(Pan troglodytes)*. *Animal Cognition* 2:157–70.

Spinozzi, G., Natale, F., Langer, J., and Schlesinger, M.
1998   Developing classification in action: 2. Young chimpanzees *(Pan troglodytes)*. *Human Evolution* 13:125–39.

Spinozzi, G., and Poti', P.
1993   Piagetian Stage 5 in two infant chimpanzees *(Pan troglodytes)*: The development of permanence of objects and the spatialization of causality. *International Journal of Primatology* 14:905–17.

Sponheimer, M., and Lee-Thorp, J.
1999   Isotopic evidence for the diet of an early hominid, *Australopithecus africanus*. *Science* 283:368–69.

Spoor, F., Wood, B., and Zenneveld, F.
1994   Implications of early hominid labyrinthine morphology for evolution of human bipedal locomotion. *Nature* 369:645–48.

Stanford, C.
1996   The hunting ecology of wild chimpanzees: Implications for the behavioral ecology of Pliocene hominids. *American Anthropologist* 98:1–18.

Stanley, S. M.
1992   An ecological theory for the origins of *Homo*. *Paleobiology* 18:237–59.
1996   *Children of the Ice Age*. New York: Basic Books.

Starkey, D.
1981   The origins of concept formation: Object sorting and object preference in early infancy. *Child Development* 52:489–97.

Stearns, S. C.
1992   *The evolution of life histories*. Oxford: Oxford University Press.

Stephan, H., Frahm, H., and Baron, G.
1981   New and revised data on volumes of brain structures in insectivores and primates. *Folia Primatologica* 35:1–29.

Stern, D.
1977   *The first relationship*. Cambridge, MA: Harvard University Press.

Stevenson, M., and Poole, T. B.
1982   Playful interactions in family groups of the common marmoset *(Callithrix jacchus jacchus)*. *Animal Behavior* 30:886–900.

Stornetta, W. S., Hogg, T., and Huberman, B. A.
1988   A dynamical approach to temporal pattern processing. In D. Z. Anderson, ed., *Neural information processing systems, Denver 1987,* 750–59. New York: American Institute of Physics.

Straight, D. S., Grine, F. E., and Moniz, M. A.
1997   A reappraisal of early hominid phylogeny. *Journal of Human Evolution* 32:17–82.

Struhsaker, T. T.
1967   Ecology of vervet monkeys *(Cercopithecus aethiops)* in the Masai-Amboseli Game Reserve, Kenya. *Ecology* 48:891–904.

Sugarman, S.
1983   *Children's early thought: Developments in classification.* New York: Cambridge University Press.

Sur, M., Garraghty, P. E., and Roe, A. W.
1988   Experimentally induced visual projections into auditory thalamus and cortex. *Science* 242:1437–41.

Sur, M., Pallas, S. L., and Roe, A. W.
1990   Cross-modal plasticity in cortical development: Differentiation and specification of sensory neocortex. *TINS* 13:227–33.

Susman, R. L.
1988   Hand of *Paranthropus robustus* from Member 1, Swartkrans: Fossil evidence for tool behavior. *Science* 240:781–84.

Susman, R. L., Stern, J. T., and Jungers, W. L.
1984   Arboreality and bipedality in the Hadar hominids. *Folia Primatologica* 43:113–56.

Symons, D.
1978   *Play and aggression: A study of rhesus monkeys.* New York: Columbia University Press.

Tabossi, P., and Zardon, F.
1993   Processing ambiguous words in context. *Journal of Memory and Language* 32(3):359–72.

Tank, D. W., and Hopfield, J. J.
1987   Neural computation by time compression. *Proceedings of the National Academy of Sciences* 84:1896–1900.

Tattersall, I.
1986   Species recognition in human paleontology. *Journal of Human Evolution* 15:165–75.

Teague, R. G., and Lovejoy, C. O.
1986   The obstetric pelvis of AL 288-1 Lucy. *Journal of Human Evolution* 15:237–55.

1998   AL 288-1 Lucy or Lucifer: Gender confusion in the Pliocene. *Journal of Human Evolution* 35:75–94.

Teleki, G.
1973   *The predatory behavior of wild chimpanzees.* Lewisburg, PA: Bucknell University Press.

1974   Chimpanzee subsistence technology: Materials and skills. *Journal of Human Evolution* 4:125–84.

1975   Primate subsistence patterns: Collector-predators and gatherer hunters. *Journal of Human Evolution* 4:125–84.

Thompson, D.

1917   *On growth and form.* Cambridge: Cambridge University Press.

Thomson, K. S.

1988   *Morphogenesis and evolution.* New York: Oxford University Press.

Toga, A. W., Frackowiak, R. S. J., and Mazziotta, J. C., eds.

1996   *Neuroimage, a Journal of Brain Function: Second international conference on functional mapping of the human brain,* vol. 3, no. 3, pt. 2. San Diego: Academic Press.

Tomasello, M.

1989   Chimpanzee culture? *Newsletter of the Society for Research in Child Development,* Winter, 1–3.

1994   The question of chimpanzee culture. In R. W. Wrangham, W. C. McGrew, F. B. M. de Waal, and P. G. Heltne, eds., *Chimpanzee cultures,* 301–17. Cambridge, MA: Harvard University Press.

Tomasello, M., and Call, J.

1997   *Primate cognition.* New York: Oxford University Press.

Tomasello, M., Davis-Dasilva, M., Camak, L., and Bard, K.

1987   Observational learning of tool use by young chimpanzees. *Journal of Human Evolution* 2:175–83.

Tomasello, M., Kruger, A. C., and Ratner, H. H.

1993   Cultural learning. *Behavioral and Brain Sciences* 16:495–552.

Tomasello, M., Savage-Rumbaugh, S., and Kruger, A. C.

1993   Imitative learning of actions on objects by children, chimpanzees, and enculturated chimpanzees. *Child Development* 64:1688–1705.

Tomlinson, M. (producer), and Jones, J. (director)

1993   *Horizon: Chimp talk.* Film. London: Orlando Productions.

Toth, N.

1985   Archaeological evidence for preferential right-handedness in the Lower and Middle Pleistocene and its possible implications. *Journal of Human Evolution* 14:607–14.

Trevarthan, C.

1973   Behavioral embryology. In E. C. Carterette and M. P. Friedman, eds., *Handbook of perception,* vol. 3, 89–117.

1980   The foundations of intersubjectivity: Development of interpersonal and cooperative understanding in infants. In D. R. Olson, ed., *The social foundations of language and thought,* 316–42. New York: Wiley.

Trevathan, W.

1987   *Human birth: An evolutionary perspective.* Hawthorne, NY: Aldine de Gruyter.

Tuttle, R.

1992   Hands from Newt to Napier. In S. Matano, R. Tuttle, H. Ishida, and M. Goodman, eds., *Evolutionary biology, reproductive endocrinology and virology,* vol. 3, 3–20. Tokyo: University of Tokyo Press.

1994   Up from electromyography: Primate energetics and the evolution of human bipedalism. In R. Corruccini and R. Ciochion, eds., *Integrative paths to the past: Paleoanthropological advances in honor of F. Clark Howell, vol. 2*, 269–84. Englewood Cliffs, NJ: Prentice Hall.

Tuttle, R., Webb, D. M., and Tuttle, N., eds.

1991   *Laetoli footprint trails and the evolution of hominid bipedalism.* Paris: Editions du CNRS.

Ullman, M.

n.d.   Evidence that lexical memory is part of the temporal lobe declarative memory, and that grammatical rules are processed by the frontal/ basal-ganglia procedural system. Unpublished manuscript.

Ullman, M., Corkin, S., Coppola, M., Hickok, G., Growdon, J. H., Koroshetz, W. J., and Pinker, S.

1997   A neural dissociation within language: Evidence that the mental dictionary is part of declarative memory, and that grammatical rules are processed by the procedural system. *Journal of Cognitive Neuroscience* 9(2):266–76.

Upton, G. J., and Fingleton, G.

1985   *Spatial data analysis by example, vol. 1: Point pattern and quantitative data.* New York: Wiley.

van Berkum, J. J. A.

1996   The psycholinguistics of grammatical gender: Studies in language comprehension and production. Doctoral diss., Max Planck Institute for Psycholinguistics, Nijmegen, Netherlands.

van Lawick-Goodall, J.

1970   Mother chimpanzees play with their infants. In J. S. Bruner, A. Jolly, and K. Sylva, eds., *Play*, 262–67. New York: Basic Books.

van Petten, C., and Kutas, M.

1991   Electrophysiological evidence for the flexibility of lexical processing. In G. B. Simpson, ed., *Understanding word and sentence*, 129–74. Amsterdam: Elsevier.

van Petten, C., Rubin, S., Plante, E., and Parks, M.

1999   Timecourse of word identification and semantic integration in spoken language. *Journal of Experimental Psychology: Learning, Memory, and Cognition* 25(2):394–417.

Vargha-Khadem, F., and Polkey, C. E.

1992   A review of cognitive outcome after hemi-decortication in humans. In F. D. Rose and D. A. Johnson, eds., *Recovery from brain damage: Reflections and directions*, 137–51. Advances in Experimental Medicine and Biology, 325: New York: Plenum.

Visalberghi, E.

1988   Responsiveness to objects in two social groups of tufted capuchin monkeys (*Cebus apella*). *American Journal of Primatology* 15:349–60.

Visalberghi, E., and Fragaszy, D. M.

1996   Pedagogy and imitation in monkeys: Yes, no, or maybe? In D. R. Olson and N. Torrance, eds., *The handbook of education and human development,* 277–301. Cambridge, MA: Blackwell.

Visalberghi, E., Fragaszy, D. M., and Savage-Rumbaugh, S. S.

1995   Performance in a tool-using task by common chimpanzees *(Pan troglodytes)*, bonobos *(Pan paniscus)*, an orangutan *(Pongo pygmaeus)*, and capuchin monkeys *(Cebus apella). Journal of Comparative Psychology* 109:52–60.

Visalberghi, E., and Limongelli, L.

1994   Lack of comprehension of cause-effect relations in tool-using capuchin monkeys *(Cebus apella). Journal of Comparative Psychology* 198:15–22.

Vygotsky, L. S.

1962   *Thought and language.* Cambridge, MA: MIT Press.

1978   *Mind in society: The development of higher psychological processes.* Cambridge, MA: Harvard University Press.

Waibel, A., Hanazawa, T., Hinton, G., Shikano, K., and Lang, K.

1989   Phoneme recognition using time-delay neural networks. *IEEE Transactions on Acoustics, Speech, and Signal Processing, ASSP-37,* 328–39.

Walters, J. R.

1987   Transition to adulthood. In B. B. Smuts, D. L. Cheney, R. M. Seyfarth, R. W. Wrangham, and T. T. Struhsaker, eds., *Primate societies,* 358–69. Chicago: University of Chicago Press.

Washburn, S. L.

1951   The new physical anthropology. *Transactions of the New York Academy of Sciences,* series 2, 13:298–304.

Watrous, R. L., and Shastri, L.

1987   Learning phonetic features using connectionist networks. *Proceedings of the 10th International Joint Conference on Artificial Intelligence,* 851–54. Milan, Italy.

Watson, J. S.

1972   Smiling, cooing and "the game." *Merrill-Palmer Quarterly* 18:323–39.

Watts, D. P., and Pusey, A. E.

1993   Behavior of juvenile and adolescent great apes. In M. E. Pereira and L. A. Fairbanks, eds., *Juvenile primates,* 148–67. New York: Oxford University Press.

Watts, E. S.

1990   Evolutionary trends in primate growth and development. In C. J. DeRousseau, ed., *Primate life history and evolution,* 89–104. New York: Wiley-Liss.

Wayne, R. K.
1986 Cranial morphology of domestic and wild canids: The influence of development on morphological change. *Evolution* 40:243–61.

Webster, M. J., Bachevalier, J., and Ungerleider, L. G.
1995 Development and plasticity of visual memory circuits. In B. Julesz and I. Kovacs, eds., *Maturational windows and adult cortical plasticity in human development: Is there reason for an optimistic view?* Reading, MA: Addison-Wesley.

Weisner, T. S., and Gallimore, R.
1977 My brother's keeper: Child and sibling caretaking. *Current Anthropology* 18:169–90.

Wernicke, C.
1977 *The aphasia symptom complex: A psychological study on an anatomic basis.*
[1874] N. Geschwind, trans. In G. H. Eggert, ed., *Wernicke's works on aphasia: A sourcebook and review.* The Hague: Mouton.

West, G. B., Bronw, J. H., and Enquist, B. J.
1997 A general model for the origin of allometric scaling laws in biology. *Science* 276:122–26.

West-Eberhard, M. J.
1988 Phenotypic plasticity and "genetic" theories of insect sociality. In G. Greenberg and E. Tobach, eds., *Evolution of social behavior and integrative levels: The T. C. Schneirla Conference Series* 3:123–33. Hillsdale, NJ: Erlbaum.

White, F. J.
1996 Comparative socio-ecology of *Pan paniscus.* In W. C. McGrew, L. F. Marchant, and T. Nishida, eds., *Great Ape Societies,* 29–41. Cambridge: Cambridge University Press.

White, L. A.
1959 *The evolution of culture.* New York: McGraw Hill.

White, T. D., Suwa, G., and Asfaw, B.
1994 *Australopithecus ramidus:* A new species of early hominid from Aramis, Ethiopia. *Nature* 371:306–12.

Whiten, A.
1998 Imitation of the sequential structure of actions in chimpanzees *(Pan troglodytes). Journal of Comparative Psychology* 112:270–81.
1999 Parental encouragement in *Gorilla* in comparative perspective: Implications for social cognition. In S. T. Parker, R. W. Mitchell, and H. L. Miles, eds., *The mentality of gorillas and orangutans,* 342–66. Cambridge: Cambridge University Press.
n.d. Primate culture and social learning. In M. Tomasello, ed., *Cognitive Science: Special Issue on Primate Cognition.* In press.

Whiten, A., Goodall, J., McGrew, W. C., Nishida, T., Reynolds, V., Sugiyama, Y., Tutin, C. E. G., Wrangham, R. W., and Boesch, C.

1999   Cultures and chimpanzees. *Nature* 399:682–85.

Whiting, B. B., and Edwards, C. P.

1988a   *Children of different worlds: The formation of social behavior.* Cambridge, MA.: Harvard University Press.

1988b   A cross-cultural analysis of sex differences in the behavior of children aged three through eleven. *Journal of Social Psychology* 91:171–88.

Wictorin, K., and Björklund, A.

1992   Axon outgrowth from grafts of human embryonic spinal cord in the lesioned adult rat spinal cord. *Neuroreport* 3:1045–48.

Wictorin, K., Brundin, P., Gustavii, B., Lindval, O., and Björklund, A.

1990   Reformation of long axon pathways in adult rat central nervous system by human forebrain neuroblasts. *Nature* 347:556–58.

Wictorin, K., Brundin, P., Sauer, H., Lindvall, O., and Björklund, A.

1992   Long distance directed axonal growth from human dopaminergic mesencephalic neuroblasts implanted along the nigrostriatal pathway in 6-hydroxydopamine lesioned adult rats. *Journal of Comparative Neurology* 323:475–94.

Widdowson, E. M.

1981   Growth of creatures great and small. *Symposium of the Zoological Society of London* 46:5–17.

Wiley, E. O.

1981   *Phylogenetics: The theory and practice of phylogenetic systematics.* New York: Wiley.

Williams, R. J., and Zipser, D.

1989   A learning algorithm for continually running fully recurrent neural networks. *Neural Computation* 1:270–80.

Wolpert, L.

1991   *The triumph of the embryo.* Oxford: Oxford University Press.

Wood, B.

1992   Origin and evolution of the genus *Homo*. *Nature* 355:783–90.

Wood, B., and Collard, M.

1999   The human genus. *Science* 284:65–71.

Wood, D., Bruner, J., and Ross, G.

1976   The role of tutoring in problem solving. *Journal of Child Psychology and Psychiatry* 17:89–100.

Wrangham, R.

1995a   Ape cultures and missing links. *Symbols,* 2–20. Cambridge, MA: Peabody Museum, Harvard University.

1995b   Presentation given at a meeting of the Southern California Primate Society, University of Southern California.

Wrangham, R., Goodall, J., and Uehara, S.

1983   Local differences in plant-feeding habits of chimpanzees between the Mahali Mountains and Gombe National Park, Tanzania. *Journal of Human Evolution* 12:467–80.

Wrangham, R., McGrew, W., de Waal, F., and Heltne, P., eds.

1994   *Chimpanzee cultures.* Cambridge, MA: Harvard University Press.

Wright, S.

1968   *Evolution and the genetics of populations.* Chicago: University of Chicago Press.

Wulfeck, B.

1988   Grammaticality judgments and sentence comprehension in agrammatic aphasia. *Journal of Speech and Hearing Research* 31:72–81.

Wulfeck, B., and Bates, E.

1991   Differential sensitivity to errors of agreement and word order in Broca's aphasia. *Journal of Cognitive Neuroscience* 3:258–72.

Wulfeck, B., Bates, E., and Capasso, R.

1991   A crosslinguistic study of grammaticality judgments in Broca's aphasia. *Brain and Language* 41:311–36.

Wynn, T.

1989   *The evolution of spatial competence.* Urbana: University of Illinois Press.

Yamakoshi, G., and Matsuzawa, T.

1993   A field experiment in cultural transmission between groups of wild chimpanzees at Bossou, Guinea. Paper presented at the 23d International Ethological Conference, Torremolino, Spain.

Yamamoto, N., Yamada, K., Kurotani, K., and Toyama, K.

1992   Laminar specificity of extrinsic cortical connections studied in coculture preparations. *Neuron* 9:217–88.

Yut, E.

1994   Gone fishin': Developmental changes in attention and their relation to the development of insect fishing technology in young chimpanzees in the wild. Unpublished manuscript, Department of Psychology, University of California, Los Angeles.

Yut, E., Greenfield, P. M., and Boehm, C.

1995   Gone fishin': Developmental changes in attention and their relation to the development of insect fishing technology in young chimpanzees in the wild. Poster presented at the annual meeting of the Jean Piaget Society, Berkeley, California.

Zecevic, N., Bourgeois, J.-P., and Rakic, P.

1989   Changes in synaptic density in motor cortex of rhesus monkey during fetal and postnatal life. *Developmental Brain Research* 50:11–32.

Zelditch, M. L., and Fink, W. L.

1996   Heterochrony and heterotopy: Stability and innovation in the evolution of form. *Paleobiology* 22:241–54.

Zimmerman, R. R., and Torrey, C. C.

1965   Ontogeny of learning. In A. M. Schrier, H. F. Harlow, and F. Stollnitz, eds., *Behavior of nonhuman primates,* vol. 2, 405–47. New York: Academic Press.

Zucker, E. L., and Clarke, M. R.

1992   Developmental and comparative aspects of social play of mantled howling monkeys in Costa Rica. *Behaviour* 123:144–71.

Zurif, E., and Caramazza, A.

1976   Psycholinguistic structures in aphasia: Studies in syntax and semantics. In H. Whitaker and H. A. Whitaker, eds., *Studies in neurolinguistics,* vol. 1, 260–92. New York: Academic Press.

Zurif, E., Swinney, D., Prather, P., Solomon, J., and Bushell, C.

1993   An on-line analysis of syntactic processing in Broca's and Wernicke's aphasia. *Brain and Language* 45:448–64.

Index

School of American Research Advanced Seminar Series

PUBLISHED BY SAR PRESS

PUBLISHED BY CAMBRIDGE UNIVERSITY PRESS

DREAMING: ANTHROPOLOGICAL AND
PSYCHOLOGICAL INTERPRETATIONS
Barbara Tedlock, ed.

THE ANASAZI IN A CHANGING
ENVIRONMENT
George J. Gumerman, ed.

REGIONAL PERSPECTIVES ON THE OLMEC
Robert J. Sharer & David C. Grove, eds.

THE CHEMISTRY OF PREHISTORIC
HUMAN BONE
T. Douglas Price, ed.

THE EMERGENCE OF MODERN HUMANS:
BIOCULTURAL ADAPTATIONS IN THE
LATER PLEISTOCENE
Erik Trinkaus, ed.

THE ANTHROPOLOGY OF WAR
Jonathan Haas, ed.

THE EVOLUTION OF POLITICAL SYSTEMS
Steadman Upham, ed.

CLASSIC MAYA POLITICAL HISTORY:
HIEROGLYPHIC AND ARCHAEOLOGICAL
EVIDENCE
T. Patrick Culbert, ed.

TURKO-PERSIA IN HISTORICAL
PERSPECTIVE
Robert L. Canfield, ed.

CHIEFDOMS: POWER, ECONOMY, AND
IDEOLOGY
Timothy Earle, ed.

PUBLISHED BY UNIVERSITY OF CALIFORNIA PRESS

WRITING CULTURE: THE POETICS
AND POLITICS OF ETHNOGRAPHY
*James Clifford &
George E. Marcus, eds.*

PUBLISHED BY UNIVERSITY OF NEW MEXICO PRESS

Photo by Katrina Lasko

Participants in the School of American Research advanced seminar "The Evolution of Behavioral Ontogeny," Santa Fe, New Mexico, August 1995. Front: John Gittlemen (left), Michael McKinney. Standing, from left: Terrence Deacon, Borje Ekstig, Sue Taylor Parker, Lynn Fairbanks, Elizabeth Bates, Jonas Langer, Brian Shea, Patricia Greenfield.